Semiconducting Transparent Thin Films

The authors are grateful to the following for granting permission to reproduce figures included in this book:

The authors of all figures not originated by ourselves
Akademie Verlag GmbH
American Institute of Physics
The American Physical Society
The American Vacuum Society
Chapman and Hall
The Electrochemical Society
Elsevier Science Ltd
Elsevier Science SA
Japanese Journal of Applied Physics
The Minerals, Metals and Materials Society
North-Holland Publishing Co.
Optical Society of America
Society of Photo-Optical Instrumentation Engineers

Semiconducting Transparent Thin Films

H L Hartnagel

Institut für Hochfrequenztechnik
Technische Universität Darmstadt

A L Dawar

Defence Science Centre, Delhi

A K Jain

Department of Physics, ARSD College, University of Delhi

C Jagadish

Department of Materials Engineering,
Australian National University, Canberra

Institute of Physics Publishing
Bristol and Philadelphia

British Library Cataloguing-in-Publication Data

A catalogue record for this book is available from the British Library

ISBN 0 7503 0322 0

Library of Congress Cataloging-in-Publication Data

Semiconducting transport thin films/H.L. Hartnagel . . . [et. al.].
p. cm.
Includes bibliographical references and index.
ISBN 0-7503-0322-0 (alk. paper)
1. Thin films—Materials. 2. Semiconductor films. I. Hartnagel, Hans, 1934–
QC176.83.S46 1995
621.3815′2—dc20 95-30589
CIP

Consultant Editor: S C Jain

Published by Institute of Physics Publishing, wholly owned by The Institute of Physics, London

Institute of Physics Publishing, Techno House, Redcliffe Way, Bristol BS1 6NX, UK

US Editorial Office: Institute of Physics Publishing, The Public Ledger Building, Suite 1035, 150 South Independence Mall West, Philadelphia, PA 19106, USA

Typeset by Paston Press Ltd, Loddon, Norfolk
Printed in the UK by J W Arrowsmith Ltd, Bristol

Contents

Preface

In recent years scientists have made rapid and significant advances in the field of semiconductor physics. Semiconducting materials have, of course, been the subject of great interest due to their numerous practical applications. Studies of this subject have also provided a fundamental insight into the electronic processes involved, and material processing has similarly become an increasingly important research field. Many new materials and devices which possess specific properties for special purposes have now become available, but material limitations are often the major deterrent to the achievement of new technological advances. Engineers and materials scientists are now particularly interested in developing materials which maintain their required properties even under extreme environmental conditions. The relationship between utility and fundamental materials science is now fully developed; for efficient and optimum use of materials, engineers must have an understanding of the factors that determine their properties, which are linked with their microscopic and submicroscopic structures. In general it is the aim of the engineers to find ways of improving quality and increasing productivity, whilst reducing the manufacturing costs. This can be achieved by the development of better processing techniques.

One of the most important fields of current interest in materials science is the fundamental aspects and applications of semiconducting transparent thin films. The characteristic properties of such coatings are low electrical resistivity and high transparency in the visible region. The first semi-transparent and electrically conducting CdO film was reported as early as 1907. Though early work on these films was performed out of purely scientific interest, substantial technological advances in such films were only made after 1940. The technological interest in the study of transparent semiconducting films has been generated mainly by their potential applications both in industry and research. Such films have demonstrated their utility as transparent electrical heaters for windscreens in the aircraft industry. However, during the last decade, these semiconducting transparent films have been widely used in a variety of other applications such as gas sensors, solar cells, heat reflectors, protective coatings, light transparent electrodes, laser-damage resistant coatings in high power laser technology, the photocathode in a photoelectrochemical cell, antistatic surface layers on temperature control coatings in orbiting satellites and surface layers in electro-luminescent applications. In addition, new developments in inexpensive techniques for the growth of these films on large areas appear to hold considerable promise in the area of horticulture (for glasshouses, etc) in the near future.

A large number of materials such as indium oxide, tin oxide, zinc oxide and cadmium stannate can be used for semiconducting transparent coatings. The basic properties of these films that are most important for practical applications are structure and morphology, electrical resistivity and optical transmission. Since the electrical and optical properties depend strongly on microstructure, stoichiometry and the nature of impurities present, the growth technique plays an important role. With the increasing sophistication of devices based on semiconducting transparent coatings, there has been a need for improved quality and understanding of the basic properties of these films. The mechanisms of doping and conduction are still inadequately understood. Compositions are still quoted without adequate detailed analysis, especially the oxygen concentration. Many reviews in this field which have been published up to now have concerned themselves with deposition techniques, properties and applications. There have also been a number of conferences over the years on understanding and facilitating further substantial advances in this scientifically exciting and technologically profitable area. It is felt that this is an appropriate moment in time to present a comprehensive and up-to-date status of semiconducting transparent coatings, which this book aims to do. In addition, it will provide an opportunity to identify the important areas for future experimental and theoretical studies. This book broadly gives the growth techniques and their limitations, electrical and optical properties of thin films and applications of thin films of most commonly used materials, such as tin oxide, indium oxide, indium tin oxide, cadmium stannate and zinc oxide.

The text has been divided into five chapters. The first chapter presents a brief introduction to the subject of transparent semiconducting coatings. The basic properties of the materials most commonly used for these coatings are also discussed.

Chapter 2, in general, deals with the growth techniques employed for the preparation of these coatings. These include CVD, spray pyrolysis, sputtering and resistive evaporation. Various deposition parameters which control the properties of films are discussed in detail for all these techniques. The merits and demerits of various techniques with reference to particular applications are also highlighted. This will guide the reader to select the most suitable growth technique for the device of interest.

The electrical properties of these coatings are discussed in detail in chapter 3. The role of various parameters such as substrate temperature, thickness, oxygen partial pressure, post-deposition annealing and doping in governing electrical properties is described. The reported results of various workers on the conduction processes involved in these films are incorporated into this chapter. Efforts have also been made to correlate the diverse results on the scattering mechanisms involved in these films. However, much work is still required to be done to understand fully the scattering processes involved in electrical conduction in these oxide films.

Chapter 4 begins with a brief introduction to optical constants. The optical properties of oxide films are described in detail in relation to deposition conditions. The shift in the absorption edge as a function of carrier concentration is explained on the basis of the Burstein–Moss model, electron–electron scattering and electron–impurity scattering. Optical properties such as transmission in the visible, and reflection in the IR region are compared for various oxide coatings. This will give an idea of the suitability of these coatings for device applications.

Various applications of transparent conducting oxide films are given in chapter 5. Ideally both the optical transmission and electrical conductivity of these coatings should be as large as possible in most applications. However, in practice, this is not always feasible. The important term 'figure of merit' used by different workers to compare the performance of different transparent semiconductor materials is discussed in detail. A comparison of values of 'figure of merit' for different films prepared by various techniques is made. Use of these oxide coatings for various applications such as wavelength selection, solar cells, protective layers and gas sensors is discussed in depth.

It is expected that this book will serve as an introduction to the important area of semiconductor transparent coatings, and it is hoped that it will help in the future development and implementation of growth techniques, and the exploitation of these techniques in the ever-growing number of applications.

H L Hartnagel
A L Dawar
A K Jain
C Jagadish

1. Transparent Conducting Oxides: Basic Properties

1.1 Introduction

Studies of transparent and highly conducting semiconducting oxide films have attracted the attention of many research workers due to their wide range of applications both in industry and in research. Transparent and conductive layers of some metallic oxides, such as cadmium oxide, tin oxide and indium oxide, have been known for a long time. Thin films of cadmium oxide (CdO) were first reported by Badeker [1] who prepared these films by thermal oxidation of sputtered films of cadmium. These films were reported to be transparent as well as conductive. Thin films (~100–200 Å) of metals such as Au, Ag, Cu, Fe, etc. have also been found to have similar properties, but the films, in general, are not very stable and their properties change with time. On the other hand, coatings based on semiconductor materials have a large number of applications because their stability and hardness are superior to those of thin metallic films.

Wide ranging applications of these coatings in electronic devices have generated interest in research related to the growth and characterization of these materials. The electronic devices include solar cells, solar heat collectors, gas sensors, etc. The high transparency of these materials in the solar spectrum, together with their high reflectivity in the infrared, makes them very attractive for use as transparent heat reflecting materials. Such spectrally selective films have wide applications in window insulation and thermal insulation in lamps [2–4].

Heterojunction solar cells with an integral conducting transparent layer offer the possibility of fabrication of low-cost solar cells with performance characteristics suitable for large-scale terrestrial applications. The conducting transparent film permits the transmission of solar radiation directly to the active region with little or no attenuation. In addition, the conducting transparent films can serve simultaneously as a low resistance contact to the junction and as an antireflection coating for the active region. Solar cells utilizing these types of coatings are now being widely fabricated, e.g. SnO_2/Si, In_2O_3/Si, ITO/Si and ITO/SiO_x/Si. Recent use of 1.3 μm light sources for optical communication has generated interest in near-IR photoelectric and photoemission devices. ITO can also be used as a transparent and conducting electrode for devices working in this near-IR wavelength range. Heterojunctions of this type, such as ITO/InP, ITO/GaSb or ITO/GaAlSb, may be potential candidates for such applications.

The use of these films as gas sensors is based on the fact that the conductance changes in semiconductor materials are large and are caused primarily by changes in carrier concentrations due to charge exchange with the species adsorbed from the gas phase. The electron concentrations in a semiconductor sensor can vary in the conduction band approximately linearly with pressure, over a range of up to eight decades, while variations in carrier mobility are generally small. It is this large and reversible variation in conductance with active gas pressure that has made semiconductor materials attractive for the fabrication of gas sensing electronic transducers.

A recent study [5] has shown that the application of a metallic oxide coating onto glass containers appreciably reduces the coefficient of friction of the glass surfaces, facilitating the movement of containers through high-speed fitting lines. It has now become common practice to apply these metallic oxide coatings to glass containers immediately after forging.

In addition to these main applications, transparent conducting films are now being used in a variety of other applications, such as production of heating layers for protecting vehicle windscreens from freezing and misting over [6], light transmitting electrodes in the development of optoelectronic devices [7], optical waveguide based electro-optic modulators [8], the photocathode in photoelectrochemical cells [9], antistatic surface layers on temperature control coatings in orbitting satellites [10] and surface layers in electroluminescent applications [11].

A large number of materials, e.g. In_2O_3, SnO_2, Cd_2SnO_4, $CdIn_2O_4$ and ZnO, can be used for these applications. Some of the properties of these materials are shown in table 1.1. Although most research on the fabrication of transparent coatings has been restricted to oxides of tin, indium and a combination of their oxides, some positive work has started on other materials also, such as ZnO, Cd_2SnO_4, etc. Recently, preliminary work on gallium indium oxide ($GaInO_3$) has indicated that this material can also be used as a transparent conducting material [12]. Initial studies [12] have revealed that doping of $GaInO_3$ with Ge or Sn results in properties comparable with those of ITO. However, extensive research work is required to be done to establish its utility for device applications. The work

Table 1.1 Some properties of transparent conducting oxides at room temperature.

Compound	Structure type	Cell dimensions (Å) *a*	*b*	*c*	Resistivity (Ω cm)	Band-gap (eV)	Dielectric constant	Refractive index
SnO_2	Rutile	4.7371	—	3.1861	10^{-2}–10^{-4}	3.7–4.6	12 ($E\|a$) 9.4 ($E\|c$)	1.8–2.2
In_2O_3	C-rare earth	10.117	—	—	10^{-2}–10^{-4}	3.5–3.75	8.9	2.0–2.1
ITO	C-rare	10.117–10.31	—	—	10^{-3}–10^{-4}	3.5–4.6	—	1.8–2.1
Cd_2SnO_4	Sr_2PbO_4	5.5684	9.8871	3.1933	10^{-3}–10^{-4}	2.7–3.0	—	2.05–2.1
ZnO	Wurtzite	3.2426	—	5.1948	10^{-1}–10^{-4}	3.1–3.6	8.5	1.85–1.90

on the growth and characterization of semiconducting transparent oxide films has been reviewed by many workers [2, 13–23] at various times. Holland [15] reviewed the work in this field carried out up to 1955. Vossen [17], Haacke [16] and Jarzebski and Marton [13] gave comprehensive reviews of experimental work reported up to the mid-1970s. Manifacier [21], Jarzebski [19], Chopra *et al* [22] and Dawar and Joshi [23] reported detailed surveys of the work in this area up to the early 1980s. More recently, Hamberg and Granqvist [14] reviewed the work on indium tin oxide films in detail, particularly from an application point of view. The aim of the present work is to present in detail the latest state of the art of transparent and highly conducting oxide films with special reference to SnO_2, In_2O_3, ITO, Cd_2SnO_4 and ZnO. Some of the important properties of these materials are discussed below.

1.2 Tin Oxide

SnO_2 has a tetragonal rutile structure with space group $D^{14}(P4_2/mnm)$ [24]. The unit cell contains six atoms—two tin and four oxygen—as shown in figure 1.1. Each tin atom (cation) is at the centre of six oxygen atoms (anions) placed approximately at the corners of a regular octahedron, and every oxygen atom is surrounded by three tin atoms approximately at the corners of an equilateral triangle. The lattice parameters [25] are $a = b = 4.737$ Å and $c = 3.185$ Å. The c/a ratio is 0.673. The ionic radii for O^{2-} and Sn^{4-} are 1.40 and 0.71 Å, respectively [26]. The lattice has 15 optical normal modes of vibration, five of which are Raman active (frequencies 100–800 cm^{-1}). Seven modes are IR active (frequencies 250–600 cm^{-1}) and two are inactive [27].

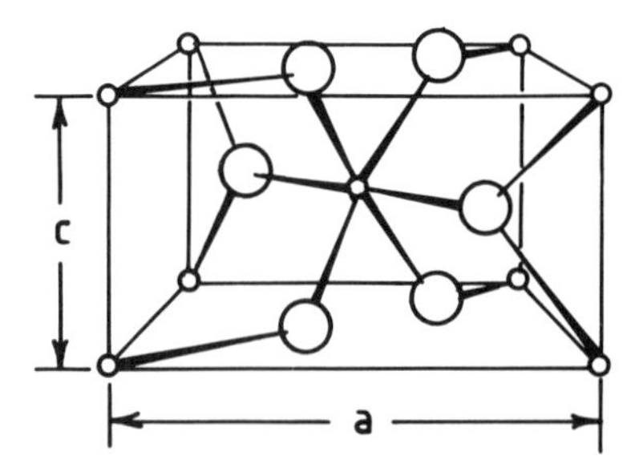

Figure 1.1 Unit cell of the crystal structure of SnO_2. Large circles indicate oxygen atoms and the small circles indicate tin atoms (from [13]).

The band structure for SnO_2 has been calculated by Robertson [28] and Munnix and Schmeits [29] using a tight bonding Hamiltonian. The band structure model of SnO_2 proposed by Robertson [28] is shown in figure 1.2. The Hamiltonian includes both Sn–O and O–O interactions. Munnix and Schmeits [29] also used parameters similar to those of Robertson [28] with slight modifications. In their case, the Hamiltonian includes all first nearest-neighbour interactions (i.e. Sn–O) as well as all second nearest-neighbour interactions (i.e. O–O and Sn–Sn). The value of the direct band-gap estimated by these authors [29] is 3.6 eV. This value of direct band-gap is in agreement with commonly accepted values obtained using optical measurements [30–32]. Using the tight bonding Hamiltonian for the bulk, Munnix and Schmeits [29] further investigated the surface electronic structure of SnO_2 (1 1 0), which is one of the natural faces of tin oxide. For these calculations, they used the scattering theoretical method as proposed by Pollman and Pantelides [33]. Figure 1.3 shows the projected bulk band structure and surface band structure of the SnO_2 (1 1 0) surface. Bound surface states and surface resonance are shown by full lines and broken lines, respectively.

If SnO_2 was completely stoichiometric, it would be an insulator or at most an ionic conductor. However, the practical material is never stoichiometric and is invariably anion deficient. This is due to the formation of oxygen vacancies in the otherwise perfect crystal. These vacancies are responsible for making electrons available for the conduction process. In the case of SnO_2, because the cation is multivalent, the creation of too many oxygen

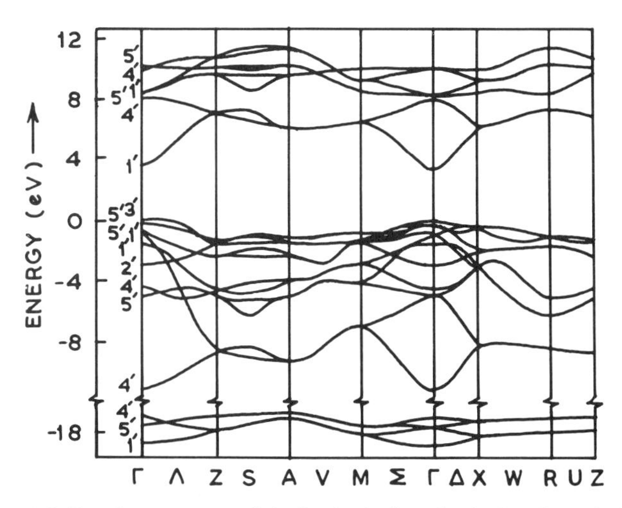

Figure 1.2 Band structure of SnO_2 including both Sn–O and O–O interactions (from [28]).

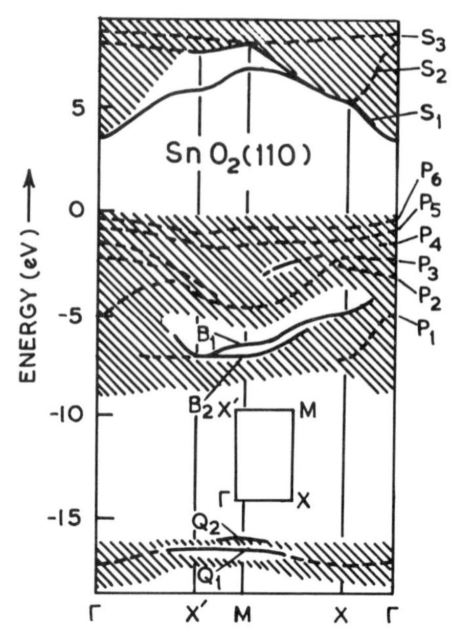

Figure 1.3 Projected bulk band structure and surface band structure of ideal SnO_2 (1 1 0) surface. Inset: irreducible part of the surface Brillouin zone (from [29]).

vacancies also results in structure change from SnO_2 to SnO. Geurts *et al* [34] studied the oxidation process of SnO to SnO_2 using Raman scattering, IR reflectivity and X-ray diffraction. It has been observed that in contrast with the rutile structure of SnO_2, the structure of SnO is layered, similar to that of PbO, with tetrahedral co-ordination D_{4h}^7. The energy gap is in the range 2.5–3 eV. SnO has four Raman-active and two IR-active vibrational modes. It has been reported [34] on the basis of the X-ray results that the oxidation of (0 0 1)-textured SnO films yields (0 1 1)-textured SnO_2 films. There is a marked similarity between the tin matrix of the SnO (0 0 1) plane and that of the SnO_2 (0 1 1) plane (figure 1.4). The inter-atomic distance d_{1-2} is 3.8 Å in SnO, whereas it is 3.7 Å in SnO_2. The distance d_{1-3} is 5.4 Å in SnO, where it is 5.7 Å in SnO_2. Therefore, very little deformation of the tin matrix is necessary to convert SnO (0 0 1) to SnO_2 (0 1 1). X-ray studies reveal that there are many intermediate oxidation states, such as Sn_2O_3 or Sn_3O_4, when SnO is converted to SnO_2. Although the intermediate states depend on the oxidation conditions, the final result is a (0 1 1)-textured SnO_2 layer.

The electrical conduction in SnO_2 results from the existence of defects in the crystal, which may act as donors or acceptors. These are generally produced because of either oxygen vacancies or interstitial tin atoms or

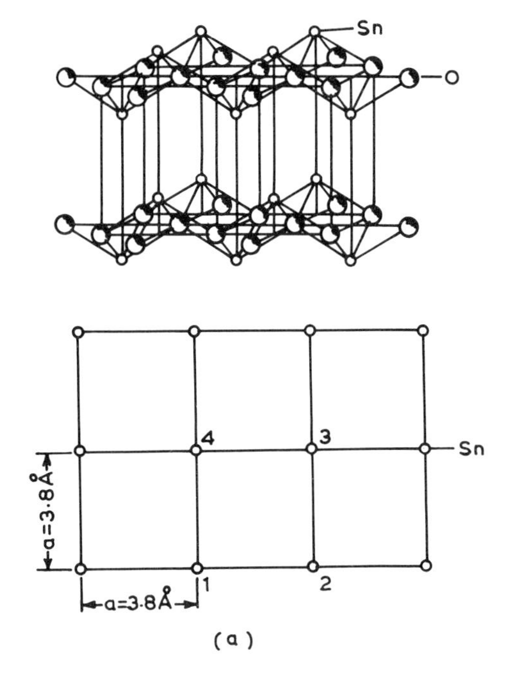

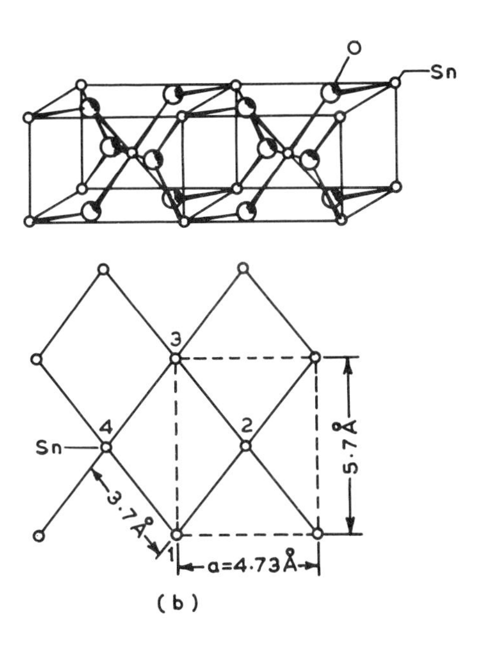

Figure 1.4 (a) Tin matrix in the (0 0 1) plane of SnO compared with (b) the tin matrix in the (0 1 1) plane of SnO_2 (from [34]).

Table 1.2 Donor/acceptor ionization energies (E_d/E_a).

Compound	Donor level (eV)	Acceptor level (eV)	Source	Reference
SnO_2	0.12	—	Cl	71
	0.034 ($N_D = 9.2 \times 10^{15}$ cm^{-3})	—	Sb	38
	0.0275 ($N_D = 9.4 \times 10^{16}$ cm^{-3})	—	Sb	38
	0.010 to 0.015 ($N_D = 2.2 \times 10^{18}$ cm^{-3})	—	Sb	38
	0.140	—	O_2	38
	0.03	—	Sn	44, 49
	0.030	—	O_2	41, 44
	0.150	—	O_2	41
	0.050	—	H_2	41
In_2O_3	0.02	—	O_2	47
	0.01	—	O_2	72
	0.1	—	—	72
	0.093	—	—	49, 50
Cd_2SnO_4	0.15	—	O_2/Cd/Sn	56
ZnO	0.02–0.04	—	H_2	73
	0.03–0.035	—	Zn	73
	0.044 ($N_D = 2.1 \times 10^{17}$ cm^{-3})	—	H_2/Zn/Li	70
	0.051 ($N_D = 5 \times 10^{16}$ cm^{-3})	—	H_2/Zn/Li	70
	0.030	—	Ga	68
	0.060	—	Ga	68
	0.150	—	Ga	68
	—	0.72	Chemisorbed O_2	68

other intentionally added impurities. Even a perfectly stoichiometric SnO_2 crystal can be made conducting by creating oxygen deficiencies by heating the sample in a slightly reducing atmosphere. Such a process is, however, reversible and the carriers so generated may be removed by heating in air or oxygen at relatively low temperatures (500–1000 °C). The other method of making SnO_2 conducting is chemical doping, for example, by fluorine or antimony. The doping effect leads to controlled valence semiconductors. The replacement of a higher valency cation by an impurity in the oxide materials increases the n-type conductivity, while the replacement of a lower valency cation produces a hole which acts as a trap in n-type semiconductors and decreases the conductivity. Similar effects can occur if the anion sites are doped with higher or lower valency impurities. This doping effect is due to a controlled valence mechanism. Sb:SnO_2 [35] or F:SnO_2 [36] can be represented in ionic notation as $Sn^{4+}_{1-x}Sb^{5+}_xO^{2-}_2e_x$ and $Sn^{4+}O^{2-}_{2-x}F^-_xe_x$, respectively.

The donor or acceptor levels observed by various workers for SnO_2 and other transparent conducting materials are listed in table 1.2. It may be observed that donor ionization energy decreases with increase in donor concentration. Such an effect is commonly observed in most semiconductors. Pearson and Bardeen [37] suggested a model for the decrease in ionization energy with increasing donor concentration by considering the electrostatic attraction between electrons and ionized donors, which increases with increase in donor concentration. According to this model,

$$E_D = E_{D0} - \beta N_D^{1/3} \tag{1.1}$$

where E_D is the donor ionization energy for a given concentration of donors N_D, E_{D0} is the donor ionization energy when the number of donors approaches zero, and β is a constant. Such a variation of E_D with N_D has been observed by Fonstad and Rediker [38] and Marley and Dockerty [39]. The results obtained by Marley and Dockerty [39] are shown in figure 1.5. Most commonly observed ionization energies in oxygen-deficient SnO_2 crystals have been ≈34 meV and 150 meV. The possible defect structure in SnO_2 has been discussed in detail by Jarzebski and Marton [13] and Mizokawa and Nakamura [40] and Marley and Dockerty [39]. Studies have revealed that this material has a double donor level; a shallow one which originates from the bulk oxygen vacancy, and a deep level which is due to the surface oxygen vacancy. Fonstad and co-workers [38, 41] also concluded that the doubly ionized oxygen vacancies predominate in SnO_2 according to the defect reaction:

$$Sn^x_{Sn} + 2O^x_O \rightleftharpoons Sn^x_{Sn} + 2V^{\bullet\bullet}_O + 4e' + O_2(g)$$

It is worth mentioning that the value of direct band-gap increases with increase in carrier concentration as a result of Burstein–Moss shift [42]. Typically, the value of band-gap increases from 3.95 eV to 4.62 eV when the carrier concentration increases from $\sim 10^{18}$ cm^{-3} to 7.9×10^{20} cm^{-3} [43].

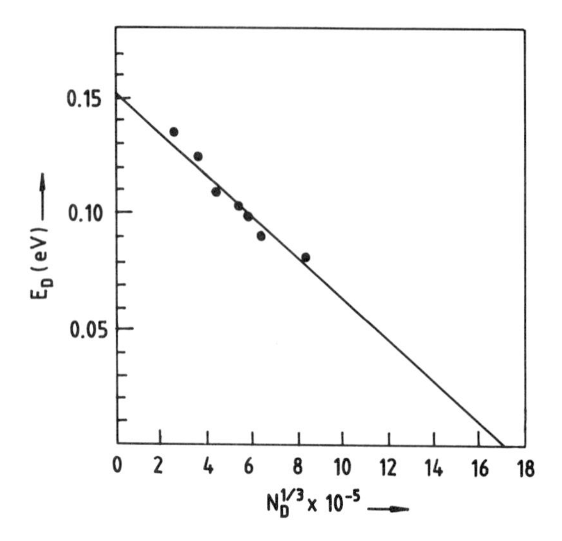

Figure 1.5 Donor ionization energy in SnO_2 as a function of donor concentration (from [39]).

1.3 Indium Oxide

In_2O_3 single crystal has the cubic bixebyte structure (also called c-type rare earth oxide structure) and belongs to the space group (T_h^7, Ia3). The lattice parameter of In_2O_3 is 10.117 Å. The co-ordination is six-fold for the In atoms and four-fold for the O atoms. One can assume that there are two crystallographically non-equivalent In sites. One of these is associated with an In–O separation of 2.18 Å, and O atoms lying nearly at the corners of a cube with two body-diagonal opposite corners unoccupied. The other is associated with non-equal In–O separations of 2.13, 2.19 and 2.23 Å, and O atoms lying nearly at the corners of a cube with two face-diagonal opposite corners unoccupied.

The unit cell contains 80 atoms and as such, the structure is highly complicated. Due to this, band structure calculations of In_2O_3 have not been made. However, Hamberg and Granqvist [14] and Fan and Goodenough [44] have proposed a simple band structure to explain the conduction mechanisms in In_2O_3. According to Hamberg and Granqvist [14], the assumed band structure (figure 1.6) has parabolic bands characterized by effective mass m_c^* for the conduction band and m_v^* for the valence band. The direct band-gap denoted by E_{g0} is 3.75 eV. The dispersions for the valence and conduction bands are given as

$$E_v^o(k) = -\hbar^2 k^2/2m_v^* \qquad (1.2)$$

$$E_c^o(k) = E_{g0} + \hbar^2 k^2/2m_c^* \qquad (1.3)$$

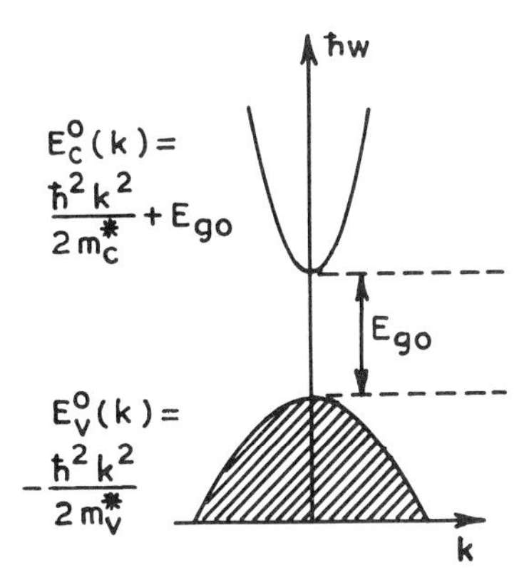

Figure 1.6 Assumed band structure of undoped In_2O_3. Shaded area denotes occupied states. Band-gap and dispersion relations are also shown (from [14]).

where k is the wave vector. Superscript o denotes unperturbed bands, i.e. for undoped In_2O_3. On the basis of the controlled valence representation of Vincent [35], Fan and Goodenough [44] suggested that the conduction band is mainly from In:5s electrons and the valence band is from O^{2-}:2p electrons. For stoichiometric In_2O_3 the Fermi energy E_F is located half-way between the energy bands. Usually In_2O_3 is somewhat reduced, and these oxygen vacancies, symbolized as V_o, give rise to shallow donor states just below the conduction band. In oxygen-deficient In_2O_3 samples, E_F lies between the donor levels and the conduction band minimum.

In_2O_3 is a non-stoichiometric compound under various conditions, with an In/O ratio larger than 2/3. This non-stoichiometry results in an n-type semiconductor or even a semimetal at high electron concentration. During crystal growth, a large number of native donors is produced because of oxygen vacancies. These donors also create an intense free-carrier absorption in the infrared reflection spectrum. Wakaki and Kanai [45] have shown that by using Mg-doped single crystals of In_2O_3, it is possible to observe the pure lattice vibrations without free-carrier dispersion. Heavy doping with Mg drastically reduces the number of free electrons due to a self-compensation effect. Far infrared transmission and reflectivity spectra studies confirmed the non-existence of the free-carrier absorption. These studies revealed the presence of 12 absorption bands. All the fundamental vibrational modes at the Γ point ($k = 0$) are decomposed into irreducible representation as

$$\Gamma_{vib} = 4A_g + 4E_g + 14T_g + 5A_u + 5E_u + 17T_u$$

where the 17 T_u modes are infrared active and the 4 A_g, 4 E_g and 14 T_g modes are Raman active. One of the 17 T_u modes is acoustic, while the remaining 16 are optical. Twelve infra-active modes were observed in their measurements and the other four modes were too weak to be observed.

The defect structure of In_2O_3 has been studied by De Wit *et al* [46, 47]. Oxygen vacancies are assumed to be the predominating point defects in In_2O_3 according to the following reaction.

$$2In_{In}^x + 3O_O^x \rightleftharpoons 2In_{In}^x + 3V_O^{\bullet\bullet} + 6e' + \tfrac{3}{2}O_2(g)$$

Frank and Kostlin [48] have suggested that in the case that some oxygen is lost from In_2O_3, the material can be represented as

$$In_2O_{3-x}(V_O^{\bullet\bullet})_x e'_{2x}$$

where x is normally less than 0.01. $V_O^{\bullet\bullet}$ denotes doubly charged oxygen vacancies and e' denotes electrons which are needed for charge neutrality on the macroscopic scale. The donor levels introduced by stoichiometric defects are generally in the range 0.008–0.03 eV, depending on the donor concentration. Weiher and co-workers [49,50] have experimentally observed the dependence of donor ionization energy on donor concentration as shown in figure 1.7. This is in good agreement with expression (1.1). It is worth mentioning that the donor level lies below the conduction band for low donor concentration. At high donor concentration an impurity band is

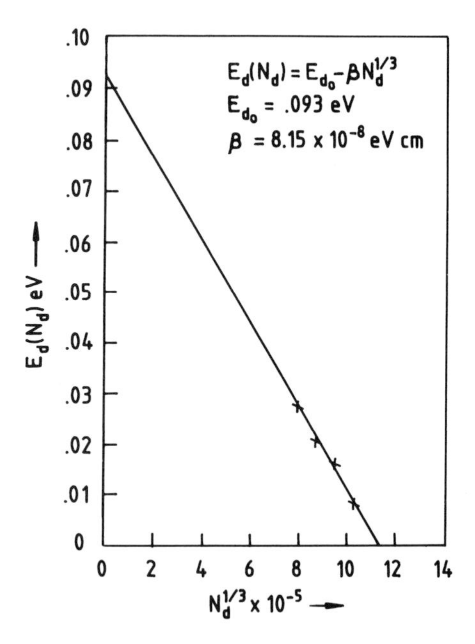

Figure 1.7 Donor ionization energy versus donor concentration for In_2O_3 (from [49]).

formed which overlaps the bottom of the conduction band, producing a degenerate semiconductor. It can be observed from figure 1.7 that the intrinsic donor level lies 93 meV below the conduction band and the degeneracy begins at a donor density of $1.48 \times 10^{18}\ \mathrm{cm}^{-3}$.

1.4 Indium Tin Oxide

Tin-doped indium oxide (ITO) films prepared by various techniques are always polycrystalline and retain a crystal structure of bulk-undoped In_2O_3. However, the lattice constant values are usually larger than those in bulk-undoped In_2O_3. The amount of increase in lattice constant is reported to depend on the deposition parameters. For example, this value strongly depends on the partial oxygen pressure $p(O_2)$ in the sputtering process [51]. For the lowest value of $p(O_2)$, $a_o = 10.15$ Å while for $p(O_2) > 5 \times 10^{-5}$ Torr, $a_o \simeq 10.23$ Å. The X-ray diffraction patterns for the films deposited at intermediate pressures consist of lines corresponding to both of these a_o values, indicating the presence of two cubic phases. For spray-deposited films, Kulaszewicz [52] observed a linear increase in lattice constant with deposition temperature in the range 823–973 K. Increases in the value of lattice constant up to 10.31 Å have been observed by many workers [52–54]. This increase in lattice constant is due to the substitutional incorporation of Sn^{4+} ions at In^{3+} sites or the incorporation of tin ions in interstitial positions. The ITO films, in general, exhibit a strong (1 1 1) or (1 0 0) preferred orientation depending on the deposition conditions.

The properties of ITO can be understood by superimposing the effect of tin doping on the host lattice of In_2O_3. The band structure of In_2O_3 is thus a basis of such a study. As already mentioned, the crystal structure of In_2O_3 is highly complicated and as such the band structure calculations of ITO have also not been made. Like In_2O_3, Hamberg and Granqvist [14] have assumed a band structure of tin-doped indium oxide (ITO). Figure 1.8 shows the assumed band structures of both In_2O_3 and ITO. In the case of ITO, a partial filling of the conduction band as well as shifts in the energy of the bands relative to their locations in In_2O_3, take place as indicated in figure 1.8. As a result of doping, there is a blocking of the lowest states in the conduction band and hence a widening of E_g, but at the same time electron–electron and electron–impurity scattering tends to decrease the gap. This is illustrated by setting $E_{g0} > W$ in figure 1.8, where W is the energy difference between the top of the valence band and bottom of the conduction band in ITO. These effects can be described by replacing the dispersions of unperturbed bands (equations (1.2) and (1.3)) by the following relations

$$E_v(k) = E_v^o(k) + \hbar\Sigma_v(k) \tag{1.4}$$

$$E_c(k) = E_c^o(k) + \hbar\Sigma_c(k) \tag{1.5}$$

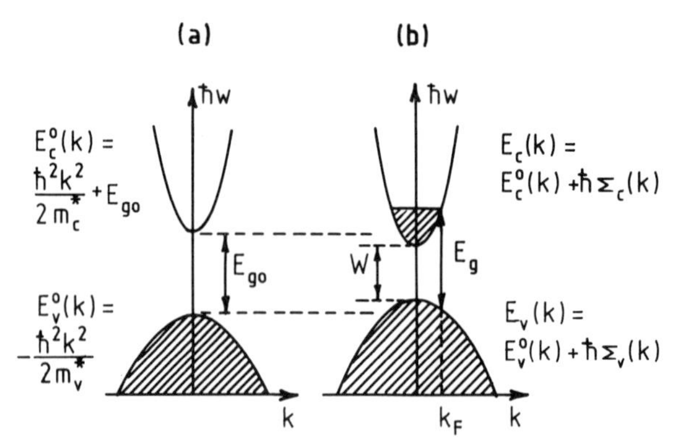

Figure 1.8 Assumed band structures of (a) undoped, and (b) Sn-doped In_2O_3. Shaded areas denote occupied states. Shift of the bands is apparent. Band-gaps, Fermi wave number (k_F) and dispersion relations are also indicated (from [14]).

where $\hbar\Sigma_v$ and $\hbar\Sigma_c$ are self-energies due to electron–electron and electron–impurity scattering.

As discussed earlier, In_2O_3 is usually oxygen deficient. These oxygen vacancies give rise to a shallow donor level just below the conduction band. In the case of Sn-doped In_2O_3 (ITO), Sn:5s level is stabilized just below E_c. Unlike oxygen vacancies V_O, a two-electron donor level, Sn:5s level is a one-electron donor level. In ITO films, V_O and Sn:5s donor levels co-exist and both contribute conduction electrons as shown in figure 1.9. When a small amount of Sn is added to In_2O_3, the Sn enters substitutionally in the cation sublattice. Hence, we have Sn^{4+} replacing In^{3+} and acting as an

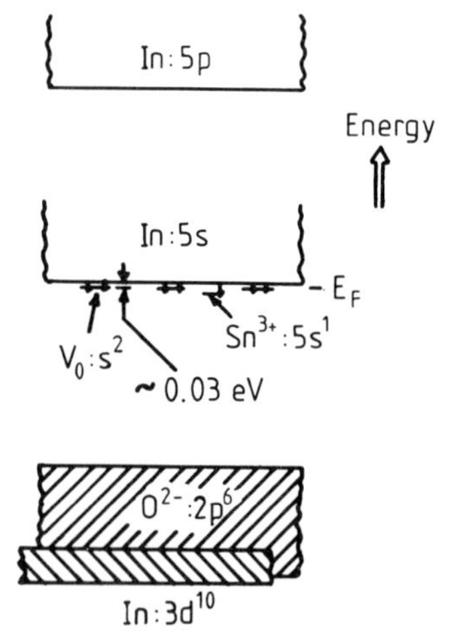

Figure 1.9 Schematic energy band model of lightly Sn-doped In_2O_3 (from [44]).

n-type donor. The new material can be represented as $In_{2-y}Sn_yO_3e_y$. The oxygen vacancies normally do not contribute when there is a large percentage of tin, particularly greater than 10 mol%. For higher values of doping (greater than 10 mol%), the compensation of part of the donor levels of non-stoichiometric origin starts, which makes the impurity band narrower and increases the activation energy for conduction. At higher Sn contents there are three types of defect; an ionizable $(\dot{Sn}_2O_i'')$ complex involving an interstitial O atom loosely bound to two Sn atoms; a non-ionizable $(Sn_2O_4)^x$ complex composed of two nearby Sn atoms which strongly bind the three closest O atoms together with an additional O atom; and a $(\dot{Sn}_2O_i'')$ $(Sn_2O_4)^x$ associate. In addition, some interstitial tin atoms also play the role of defects in In_2O_3 [55].

1.5 Cadmium Stannate

There are two known phases of cadmium stannate, namely Cd_2SnO_4 and $CdSnO_3$. Crystals of Cd_2SnO_4 are generally orthorhombic whereas $CdSnO_3$ can exist both in rhombohedral and orthorhombic structures. Rhombohedral $CdSnO_3$ has an ilmenite structure, while orthorhombic $CdSnO_3$ has a distorted perovskite structure. The orthorhombic structure of Cd_2SnO_4 is of Sr_2PbO_4 type (Pbam) with unit cell dimensions $a = 5.568$ Å, $b = 9.887$ Å and $c = 3.192$ Å [56]. Haacke *et al.* [57] and Leja and co-workers [58, 59], however, observed that sputtered Cd_2SnO_4 films, unlike crystals, have cubic spinel structure rather than orthorhombic. Leja and co-workers [58, 59] have shown that the sputtered films of this material are polycrystalline in nature and these crystallize in the inverse spinel structure of space group Fd 3m(O_h^7) with cell parameter $a = 9.172$ Å and an oxygen parameter $u = 0.384$ Å. The unit cell of this material is shown in figure 1.10. Half of the cadmium cations occupy tetrahedral sites and the remaining cations are distributed with the tin cations on the octahedral sites (Cd[SnCd]O_4). The

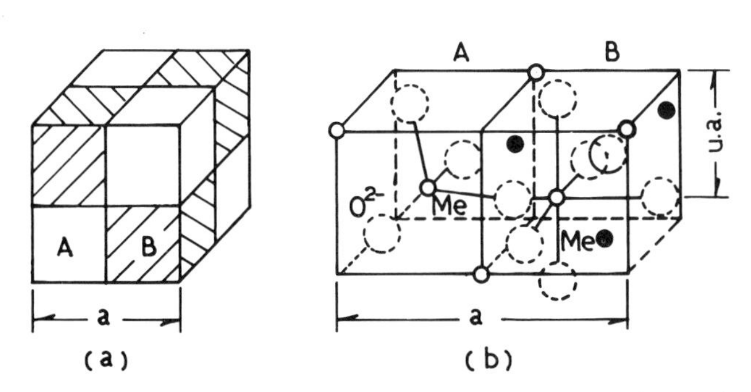

Figure 1.10 Representation of unit cell of Cd_2SnO_4: large circles, oxygen: small solid circles, tin: small hollow circles, cadmium (from [58]).

films, in general, exhibit a texture along the [1 0 0] direction perpendicular to the substrate. Mossbauer studies [58] on the material confirmed that Cd_2SnO_4 is a compound of tetravalent tin. There is no contribution from the divalent tin. To the best of our knowledge, the band structure of Cd_2SnO_4 has not been calculated to date. As-grown crystals of Cd_2SnO_4 are nearly stoichiometric and have a high value of resistivity. However, doping with Sb_2O_4, which results in $Cd_2Sn_{1-x}Sb_xO_4$, is a conducting material. The other technique to obtain a conducting phase of Cd_2SnO_4 is to create oxygen vacancies in this material. In the case of sputtered Cd_2SnO_4 films, the oxygen partial pressure can be regulated to control the number of oxygen vacancies; the higher the oxygen pressure, the lower the number of oxygen vacancies. Leja *et al* [58] have shown that the carrier concentration N decreases with increasing oxygen partial pressure in accordance with the relation $N \propto p(O_2)^{-1/\alpha}$ where $\alpha = 7.8$. The proposed reaction that permits the existence of interstitial ionized cadmium $Cd_i^{\bullet}$ and oxygen vacancies $V_O^{\bullet\bullet}$ as the point defects is

$$2Cd_{Cd}^x + Sn_{Sn}^x + 4O_O^x \rightleftharpoons 2Cd_i^{\bullet} + Sn_{Sn}^x + 4V_O^{\bullet\bullet} + 10e' + 2O_2(g)$$

where Cd_{Cd}^x, Sn_{Sn}^x, O_O^x are cadmium, tin and oxygen in lattice sites, respectively, and $O_2(g)$ is oxygen in the gas phase. The value of N is given as

$$N = 2[V_O^{\bullet\bullet}] + [Cd_i^{\bullet}]$$

where $[V_O^{\bullet\bullet}]$ and $[Cd_i^{\bullet}]$ are the concentrations of defects together with the assumption that $[V_O^{\bullet\bullet}] = [Cd_i^{\bullet}]$ gives $N \propto p(O_2)^{-1/8}$. Haacke *et al* [57] reported that the sputtered films generally contain CdO and $CdSnO_3$ phases, depending strongly on the sputtering conditions. However, the heat treatment of these films in a Ar–CdS atmosphere at 600–700 °C results in a decrease of CdO content in the films, along with an increase in the lattice parameter from 9.167 to 9.189 Å. This is possibly due to the fact that cadmium dissociates from the CdO and diffuses into the Cd_2SnO_4 lattice to form interstitial donors.

The transparent conducting oxides containing Sn or In can be prepared in different structures, all with networks of linked Sn^{4+} or In^{3+} octahedra. In SnO_2, In_2O_3 and Cd_2SnO_4, the Sn^{4+} or In^{3+} octahedra share edges, whereas in $CdSnO_3$, they share only corners. Since $CdSnO_3$ is not a transparent conductor, continuous edge sharing of Cd^{2+}, In^{3+} and Sn^{4+} [56] octahedra seems to be a necessary criterion for the formation of a transparent conducting material.

1.6 Zinc Oxide

Zinc oxide occurs in nature as the mineral zincite. Zinc oxide crystallizes in the hexagonal wurtzite (B 4-type) lattice. The zinc atoms are nearly in the position of hexagonal close packing. Every oxygen atom lies within a tetrahedral group of four zinc atoms, and all these tetrahedra point in the same direction along the hexagonal axis giving the crystal its polar symmetry. The wurtzite lattice of zinc oxide is shown in figure 1.11. The lattice constants are $a = 3.24$ Å and $c = 5.19$ Å.

The band structure of ZnO has been calculated by many workers [60–64]. A typical band structure is shown in figure 1.12. This band structure has been calculated by using a Hartree–Fock–Slater Hamiltonian employing an oxygen non-local ionic pseudopotential [60]. The calculated band structure for ZnO is shown along high symmetry lines in the hexagonal Brillouin zone. The lowest two valence bands occurring at approximately −20 eV correspond to O:2s core-like states. The next six valence bands from −6 eV to 0 eV correspond to O:2p bonding states. The first two conduction band states are strongly localized on Zn and correspond to unoccupied Zn:3s levels. The higher conduction bands are fairly free-electron-like. The fundamental band-gap calculated using band structure models [60, 63, 64] is ~3.5 eV, which is in good agreement with the experimental data. Ivanov and Pollmann [61] calculated the surface electronic structure for the non-polar (1 0 1 0) ZnO surface. The bulk electronic properties of ZnO are described by a tight bonding Hamiltonian obtained by fitting self-consistent pseudopotential bulk band structure. The surface band structures have been calculated using the scattering theoretical method [33]. The Hamiltonian includes first and second nearest-neighbour interactions, namely Zn:4s and O:2p orbitals. Figure 1.13 depicts the projected bulk band structure together with bound surface states for a non-polar (1 0 1 0) ZnO surface. Bound surface states and surface resonances are shown by full lines and broken lines, respectively. It has been reported by many workers [65–67] that thin

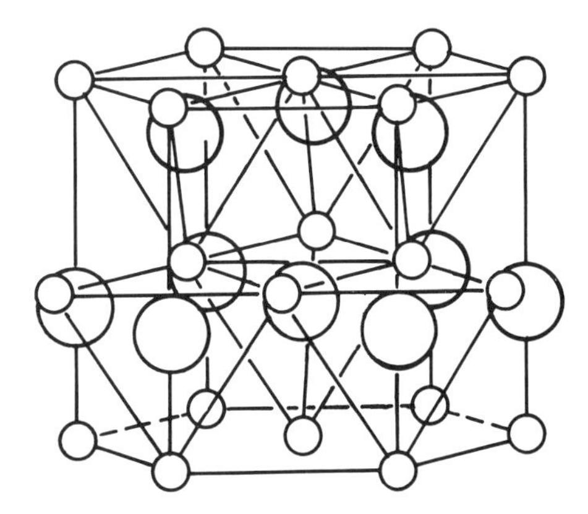

Figure 1.11 The wurtzite lattice of zinc oxide: small circles represent zinc atoms, whereas large circles depict oxygen atoms.

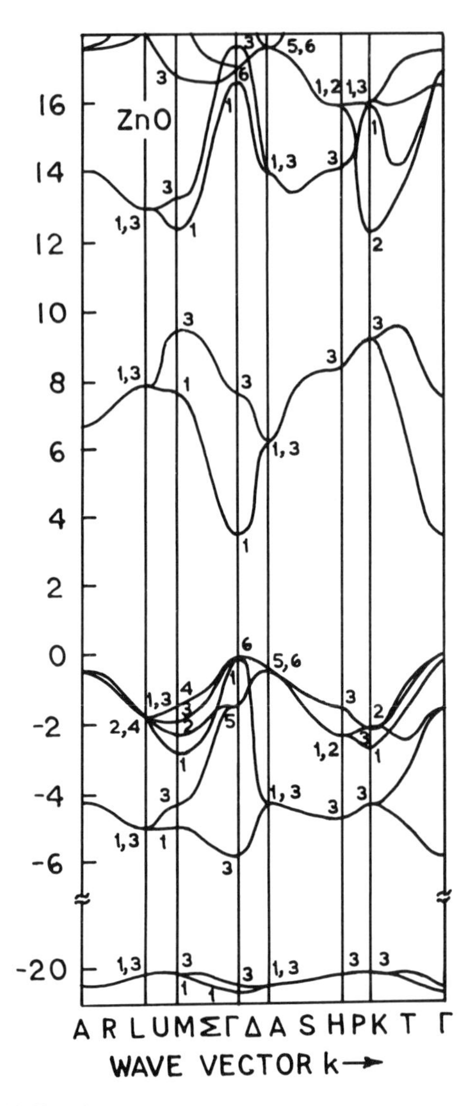

Figure 1.12 Band structure of ZnO. The energies are in electronvolts (eV) and the zero reference corresponds to valence band maximum (from [60]).

films of ZnO also retain the bulk wurtzite structure, and that grain size lies in the range 50–300 Å. The donor levels in ZnO lie in the range 0.02–0.05 eV, depending on carrier concentration. These hydrogen-like donor levels can be produced either due to oxygen vacancies (V_O) or due to incorporation of hydrogen, indium, lithium or zinc. In the case of gallium-doped ZnO films,

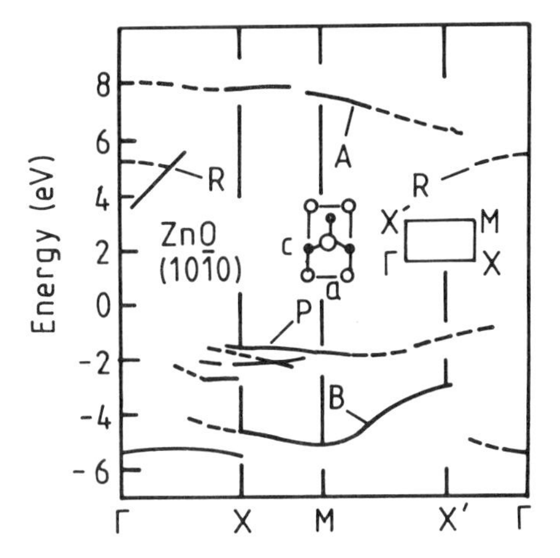

Figure 1.13 Projected bulk band structure and surface band structure for a non-polar (1 0 $\bar{1}$ 0) ZnO surface. The layer unit cell and the irreducible part of the Brillouin zone are shown as insets. Full lines show bound states and broken lines show resonances (from [61]).

Barnes *et al* [68] observed three donor levels at 30, 60 and 150 meV and an acceptor level at ~0.72 eV below the conduction band. The deep-lying acceptor level is probably due to chemisorbed oxygen. A similar type of acceptor level has also been obtained at 0.80 eV below the conduction band in undoped ZnO films [69].

1.7 Subject Texture

The need for coatings which have both low resistivities and high transparency in the visible region has led to the development of various deposition techniques. Each of these deposition processes has its own merits and demerits, e.g. sputtering techniques allow the fabrication of high quality films but the equipment cost is very high with a low production rate. On the other hand, spray techniques are very cheap but the films produced are not consistent. Also, the results reported by different workers using different techniques vary significantly. There is a wide diversity in the optimized dopant concentrations for the best quality films. This is probably due to differences in the deposition parameters and the purity of the elements. Keeping all these aspects in mind, various deposition techniques are discussed separately in chapter 2. The electrical and optical properties of these films are discussed as a function of various parameters, such as doping, temperature, oxygen partial pressure, etc. in chapters 3 and 4, respectively.

A brief description of electrical and optical parameters in semiconducting thin films is also incorporated in these chapters in order to make them self-sufficient. Chapter 5 presents various applications of these films in devices, with reference to optimization of different parameters.

References

[1] Badeker K 1907 *Ann. Phys. (Leipzig)* **22** 749
[2] Lampert CM 1981 *Sol. Energy Mater.* **6** 1
[3] Frank G, Kauer E and Kostlin H 1981 *Thin Solid Films* **77** 107
[4] Granqvist CG 1990 *Thin Solid Films* **93/94** 730
[5] Puyane R (ed) 1981 Proc. of Battele seminar on coatings on glass (Geneva, Switzerland) 18–20 September, 1980 *Thin Solid Films* **77** nos 1–3
[6] Yoshida H, Furubayashi H, Inque Y and Tonomura T 1976 *J. Vac. Soc. Japan* **19** 13
[7] Latz R, Michael K and Scherer M 1991 *Japan. J. Appl. Phys.* **30** L149
[8] Chen RT and Robinson D 1992 *Appl. Phys. Lett.* **60** 1541
[9] Tien HT and Higgins J 1980 *J. Electrochem. Soc.* **127** 1475
[10] Hass G, Heaney JB and Toft AR 1979 *Appl. Opt.* **18** 1488
[11] Miura N, Ishikawa T, Sasaki T, Oka T, Ohata H, Matsumoto H and Nakano R 1992 *Japan. J. Appl. Phys.* **31** L46
[12] Cava RJ, Hou SY, Krajewski JJ, Marshall JH, Pack WF, Rapkine DH and van Dover RB 1994 *Appl. Phys. Lett.* **65** 115
[13] Jarzebski ZM and Marton JP 1976 *J. Electrochem. Soc.* **123** 199C, 299C, 333C
[14] Hamberg I and Granqvist CG 1986 *J. Appl. Phys.* **60** R 123
[15] Holland L 1958 *Vacuum Deposition of Thin Films* (New York: Wiley) p 492
[16] Haacke G 1977 *Ann. Res. Mater. Sci.* **7** 73
[17] Vossen JL 1977 *Phys. Thin Films* **9** 1
[18] Granqvist CG 1981 *Appl. Opt.* **20** 2606
[19] Jarzebski ZM 1982 *Phys. Status Solidi* a **71** 13
[20] Kostlin H 1982 *Festkorperprobleme* **22** 229
[21] Manifacier JC 1982 *Thin Solid Films* **90** 297
[22] Chopra KL, Major S and Pandya DK 1983 *Thin Solid Films* **102** 1
[23] Dawar AL and Joshi JC 1984 *J. Mater. Sci.* **19** 1
[24] Wyckhoff RWG 1963 *Crystal Structures* vol 1, 2nd edn (New York: Wiley)
[25] Baur WH 1956 *Acta Crystallogr.* **9** 515
[26] Pauling L 1960 *The Nature of Chemical Bond* (New York: Cornell University Press)
[27] Katiyar RS, Dawson P, Hargreave MM and Wilkinson GR 1971 *J. Phys. C: Solid State Phys.* **4** 2421
[28] Robertson J 1979 *J. Phys. C: Solid State Phys.* **12** 4767
[29] Munnix S and Schmeits M 1982 *Solid State Commun.* **43** 867
[30] Dawar AL, Kumar A, Sharma S, Tripathi KN and Mathur PC 1993 *J. Mater. Sci.* **28** 639
[31] Demiryont H, Nietering KE, Surowiec R, Brown FI and Platts DR 1987 *Appl. Opt.* **26** 3803

[32] Agekyan VD 1977 *Phys. Status Solidi* a **43** 11
[33] Pollmann J and Pantelides ST 1978 *Phys. Rev.* B **18** 5524
[34] Geurts J, Rau S, Richter W and Schmitte FJ 1984 *Thin Solid Films*, **121** 217
[35] Vincent CA 1972 *J. Electrochem. Soc.* **119** 515, 518
[36] Manifacier JC, Szepessy L, Bresse JF, Perotin M and Stuck R 1979 *Mater. Res. Bull.* **14** 163
[37] Pearson GL and Bardeen J 1949 *Phys. Rev.* **75** 865
[38] Fonstad CG and Rediker RH 1971 *J. Appl Phys.* **42** 2911
[39] Marley JA and Dockerty RC 1965 *Phys. Rev.* A **140** 304
[40] Mizokawa Y and Nakamura S 1975 *Japan. J. Appl. Phys.* **14** 779
[41] Samson S and Fonstad CG 1973 *J. Appl. Phys.* **44** 4618
[42] Burstein E 1954 *Phys. Rev.* **93** 632
Moss TS 1964 *Proc. Phys. Soc.* B **67** 775
[43] Shanthi E, Dutta V, Banerjee A and Chopra KL 1980 *J. Appl. Phys.* **51** 6243
[44] Fan JCC and Goodenough JB 1977 *J. Appl. Phys.* **48** 3524
[45] Wakaki M and Kanai Y 1986 *Japan. J. Appl. Phys.* **25** 502
[46] De Wit JHW 1977 *J. Solid State Chem.* **20** 143
[47] De Wit JHW, Van Unen G and Lahey M 1977 *J. Phys. Chem. Solids* **38** 819
[48] Frank G and Kostlin H 1982 *Appl. Phys.* A **27** 197
[49] Weiher RL 1962 *J. Appl. Phys.* **33** 2834
[50] Weiher RL and Ley RP 1966 *J. Appl. Phys.* **37** 299
[51] Fan JCC, Bachner FJ and Foley GH 1977 *Appl. Phys. Lett.* **31** 773
[52] Kulaszewicz S 1981 *Thin Solid Films* **76** 89
[53] Lehmann HW and Widmer R 1975 *Thin Solid Films* **27** 359
[54] Vossen JL 1971 *RCA Rev.* **32** 269
[55] Ovadyahu Z, Ovryn B and Kraner HW 1983 *J. Electrochem. Soc.* **130** 917
[56] Shannon RD, Gillson JL and Bouchard RJ 1977 *J. Phys. Chem. Solids* **38** 877
[57] Haacke G, Mealmaker WE and Siegel LA 1978 *Thin Solid Films* **55** 67
[58] Leja E, Stapinski T and Marzalek K 1985 *Thin Solid Films* **125** 119
[59] Stapinski T, Leja E and Pisarkiewicz T 1984 *J. Phys. D: Appl. Phys.* **17** 407
[60] Chelikowsky JR 1977 *Solid State Commun.* **22** 351
[61] Ivanov I and Pollmann J 1980 *Solid State Commun.* **36** 361
[62] Jaffe JE, Pandey R and Kunz AD 1991 *J. Phys. Chem. Solids* **52** 755
[63] Rossler U 1969 *Phys. Rev.* **184** 733
[64] Bloom S and Ortenburger I 1973 *Phys. Status Solidi* b **58** 561
[65] Roth AD and Williams DF 1981 *J. Appl Phys.* **52** 6685
[66] Aranovich J, Ortiz A and Bube RH 1979 *J. Vac. Sci. Technol.* **16** 994
[67] Call RL, Jaber NK, Seshan K and Whyte JR Jr 1980 *Sol. Energy Mater.* **2** 373
[68] Barnes JO, Leary DJ and Jordan AG 1980 *J. Electrochem. Soc.* **127** 1636
[69] Leary DJ, Barnes JO and Jordan AG 1982 *J. Electrochem. Soc.* **129** 1382
[70] Hutson R 1957 *Phys. Rev.* **108** 222
[71] Miloslavskii VK 1959 *Opt. Spectrosc.* **7** 154
[72] Muller HK 1968 *Phys. Status Solidi* **27** 723
[73] Heiland G, Mollwo E and Stockmann F 1959 *Solid State Phys.* **8** 191

2. Growth Techniques

2.1 Introduction

Transparent oxide films with a large energy band-gap exhibit high electrical conductivity, high optical transmittance in the visible region and high reflectance in the IR region. These unique properties of the transparent conducting oxide films make them suitable for a variety of applications. On this account, various techniques for the growth of these films have been intensively investigated during the recent past. The growth technique plays a significant role in governing the properties of these films, because the same material deposited by two different techniques usually has different physical properties. This is due to the fact that the electrical and optical properties of these films strongly depend on the structure, morphology and the nature of the impurities present. Moreover, the films grown by any particular technique may often have different properties due to the involvement of various deposition parameters. The properties, however, can be tailored by controlling the deposition parameters. It is, therefore, essential to make a detailed investigation of the relationship between the properties of these films and the method of deposition.

There are various methods to grow semiconducting thin films and these have been described in detail in standard text books [1–4]. Some of these techniques have been employed for the growth of transparent conducting thin films. These studies have been reviewed from time to time by many workers [5–10]. The specific techniques that have been used to grow the transparent conducting oxide films include chemical vapour deposition, hydrolysis or spray, evaporation of oxide materials, reactive evaporation of metals in the presence of oxygen, plasma-assisted reactive evaporation, flash evaporation, sputtering of oxide targets, reactive sputtering of metals, ion-beam sputtering, ion plating, etc. Each of these processes has its own merits and demerits. For example, sputtering techniques allow the fabrication of high quality films but they have high equipment cost and relatively low production rates. On the other hand, spray techniques are very cheap but the films produced are inconsistent.

The purpose of this chapter is to give a brief account of the commonly used techniques, particularly suitable for the growth of these films. The growth parameters for various techniques for the fabrication of these films are discussed in detail.

2.2 Various Deposition Techniques: A Brief Description

In this section we shall discuss numerous deposition techniques that can be employed to grow transparent conducting oxide films.

2.2.1 Chemical vapour deposition

Chemical vapour deposition (CVD) is one of the most important techniques for producing thin films of semiconductor materials. This technique involves a reaction of one or more gaseous reacting species on a solid surface (substrate). In this process, metallic oxides are generally grown by the vaporization of a suitable organometallic compound. A vapour containing the condensate material is transported to a substrate surface, where it is decomposed, usually by a heterogeneous process. The nature of the decomposition process varies according to the composition of the volatile transporting species. The decomposition condition should be such that the reaction occurs only at or near the substrate surface and not in the gaseous phase, to avoid formation of powdery deposits which may result in haziness in the films.

A typical CVD system [11] is shown in figure 2.1. The vapours of a volatile compound are carried by a carrier gas, e.g. O_2, N_2, Ar, from a hot bubbler through a heated line to a reaction chamber to which oxygen or water vapour is introduced. In this growth chamber the vapours decompose and the homogeneous oxide films form at the preheated substrate surface. The quality of the films depends on various parameters, such as substrate temperature, gas flow rate and system geometry. In order to obtain the best quality films all these parameters should be optimized and controlled. The substrate temperature and gas flow rate largely determine the deposition rate. If the substrate temperature is low, carbon occlusions are found in the films as a result of incomplete oxidation of the organic material. If the substrate temperature is too high, excessive diffusion of surface impurities during film growth can occur. In addition, radiations from the substrate heater may preheat the gas in the reaction chamber. This will result in decomposition of organometallic compounds in the gas phase rather than at

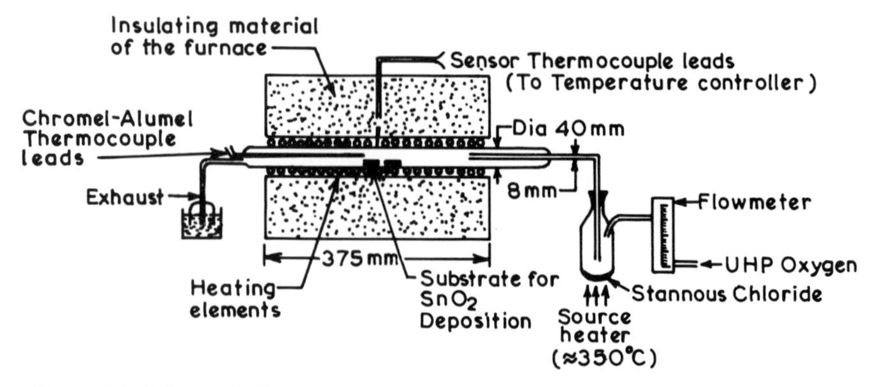

Figure 2.1 Schematic diagram of the experimental CVD setup for the deposition of SnO_2 films by the reaction of stannous chloride vapour and oxygen (from [11]).

the substrate surface, thereby producing a powder-like deposit instead of a smooth film. The gas flow and system geometry determine the uniformity of the films deposited over large areas. Thus, it is important to introduce the gas into the reaction chamber in a controlled manner both in terms of gas inlet location and flow rate. Most reaction chambers contain baffles and/or planetary rotating substrate holders to ensure that the gas flow in the vicinity of the substrate is as uniform as possible. Fiest *et al* [12] discussed a number of reactor designs for CVD for the growth of homogeneous and uniform films.

A CVD setup suitable for continuous coating of glass ribbons is shown in figure 2.2. Substrates as large as 30 cm in width are carried on a conveyor belt through a furnace with one or more CVD coating stages in the hot zone. The flat configuration has been adopted for atmosphere control by appropriate inert gas purging through narrow access slots. The reactants are injected through a narrow slot running across the width of the conveyor belt. The coating thickness in this process depends on the belt speed and reactant concentration. A coating thickness uniformity of ±3% over a width of 10 cm has been reported [13].

The main advantages of using the CVD process are simplicity, reproducibility and the ease with which it can be adopted for large-scale production without requiring vacuum as an essential requirement for deposition. Moreover, due to the low cost of equipment used, the cost of production of thin films by this technique is reasonably low. Using this technique, films of high purity, stoichiometry, and structural perfection can be obtained. However, the morphology of the films deposited by CVD is strongly influenced by the nature of the chemical reaction and the activation mechanism.

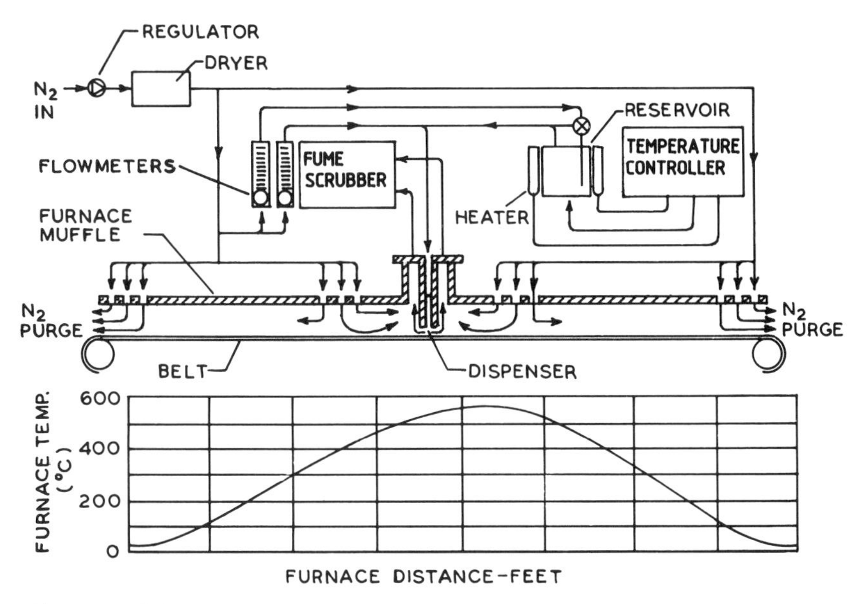

Figure 2.2 Schematic diagram of an in-line CVD furnace for continuous deposition of tin oxide and similar coatings (from [13]).

2.2.2 Spray pyrolysis

Spray pyrolysis is one of the relatively simple and cheap methods which can easily be adopted for mass production of large-area coatings for industrial applications. The method has, for many years, been widely used for preparation of transparent conducting oxide films. Spray pyrolysis is based on the pyrolytic decomposition of a metallic compound dissolved in a liquid mixture when it is sprayed onto a preheated substrate. The method depends on the surface hydrolysis of metal chloride on a heated substrate surface in accordance with the reaction.

$$MCl_x + yH_2O \rightarrow MO_y + xHCl$$

in which M is any host metal such as Sn, In, Zn, etc. of the oxide films. The conventional spray pyrolysis [10] technique is shown in figure 2.3. The atomization of the chemical solution into a spray of fine droplets is accomplished by the spray nozzle with the help of a filtered carrier gas. The carrier gas and solution are fed into the spray nozzle at predetermined pressures and flow rates. The substrate temperature is maintained constant using a feedback circuit which controls a primary and an auxiliary heater power supply. Large-area uniform coverage of the substrate is achieved by scanning either the spray head and/or the substrate, employing an

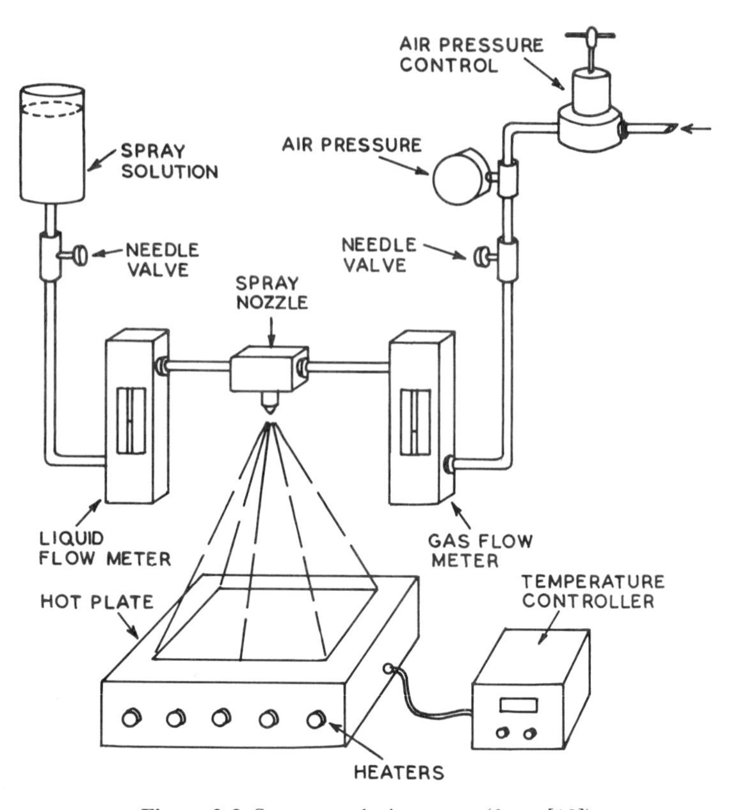

Figure 2.3 Spray pyrolysis system (from [10]).

electromechanical arrangement. The geometry of the gas and the liquid nozzles strongly determine the spray pattern, size distribution of droplets and spray rate, which in turn determine the growth kinetics and hence the quality of the films obtained. The other important control parameters that affect the quality of the films are the nature and temperature of the substrate, the solution composition, the gas and solution flow rates, the deposition time and the nozzle-to-substrate distance.

In general, the films grown at substrate temperatures less than 300 °C are amorphous in nature, whereas at higher substrate temperatures, polycrystalline films are formed. In order to obtain films with good conductivity, it is normally essential that complete oxidation of the metal be avoided. This is generally achieved by adding an appropriate reducing agent such as propanol, ethyl alcohol or pyrogallol. The decomposition products of these organic materials lead to reduction of the oxide film, thereby creating anion vacancies. High substrate temperatures increase the deposition rate and favour chemical reduction (anion vacancies). The type and quantity of the organic material mixed with the starting metallic chlorides determines the amount of reduction.

(a) *Modified spray techniques*
Attempts have been made to modify the above-mentioned simple spray system in order to produce homogeneous and reproducible films. One such system is based on the pyrolysis of aerosol produced by ultrasonic spraying. Such an ultrasonic sprayer is shown in figure 2.4. The system consists of an ultrasonic atomizer spray nozzle, a pyrolysis furnace and an exhaust system. The ultrasonic atomizer comprises a glass vessel containing the liquid to be sprayed, and fitted at the bottom with a ceramic transducer excited by a high frequency (~1 MHz) generator. A power supply of 100–200 W is normally required to spray one litre per hour of liquid at a frequency of 800 to 1000 kHz. The diameter of the droplets d generated by the ultrasonic vibrations depends on the frequency of vibration 'f' in accordance with the relation [15]

$$d = K(8\pi\sigma/\rho f^2)^{1/3}$$

where K is a constant that depends on the system, σ is the surface tension and ρ is the density of the fluid. The variation in diameter of the droplet with frequency is shown in figure 2.5. Since the size of the droplet in the pyrosol process is a critical parameter in deciding the homogeneity of the films, the use of ultrasonic vibrations to generate the aerosol has advantages; the size of the droplet can be controlled and it is highly reproducible. Moreover, the distribution in size is quite narrow in comparison with the distribution usually encountered in pneumatic spraying. Blandenet *et al* [15] used this system to coat SnO_2 on flat surfaces as well as tubes and shaped glasses.

A modified spray technique employed to coat large-area substrate [16, 17] is shown in figure 2.6. It consists of a hot plate, the temperature of which can be stabilized between 200 and 650 °C, and an X–Y device which allows the spray to cover the whole working surface by successive travels of the nozzle.

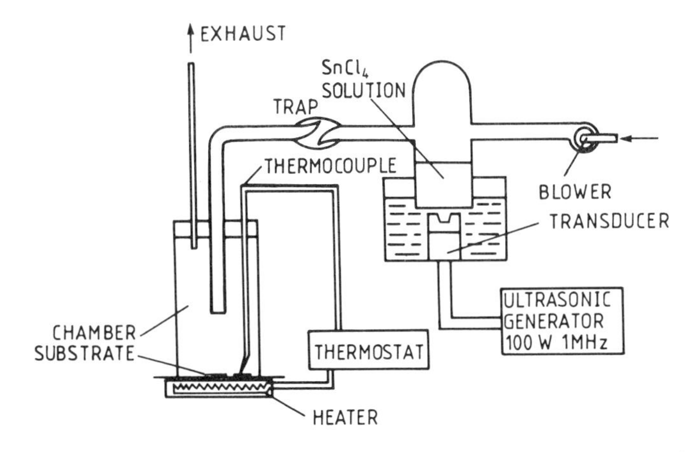

Figure 2.4 Experimental setup of spray pyrolysis using an ultrasonic sprayer (from [14]).

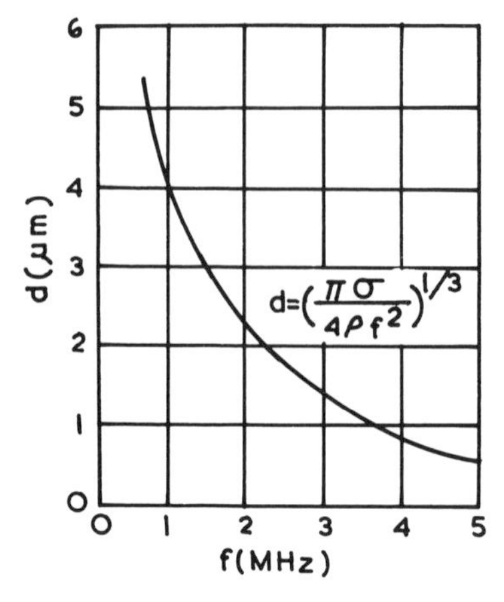

Figure 2.5 Mean diameter of droplets versus ultrasonic frequency (from [15]).

The nozzle is moved at such a rate that the temperature remains constant (within 5 °C) over the substrate and allows large-area (10 cm × 10 cm) homogeneous deposits. The liquid and gas flow rates are controlled by flow meters. For coating large areas, the low deposition efficiency of spray pyrolysis has an adverse effect on the cost of the material. Only a small fraction of material supplied is deposited on the substrate. 'Corona spray pyrolysis', which uses electrophoretic transportation, enhances the efficiency to more than 80% rendering it possible to use spray pyrolysis economically. The schematic diagram of this process is shown in figure 2.7.

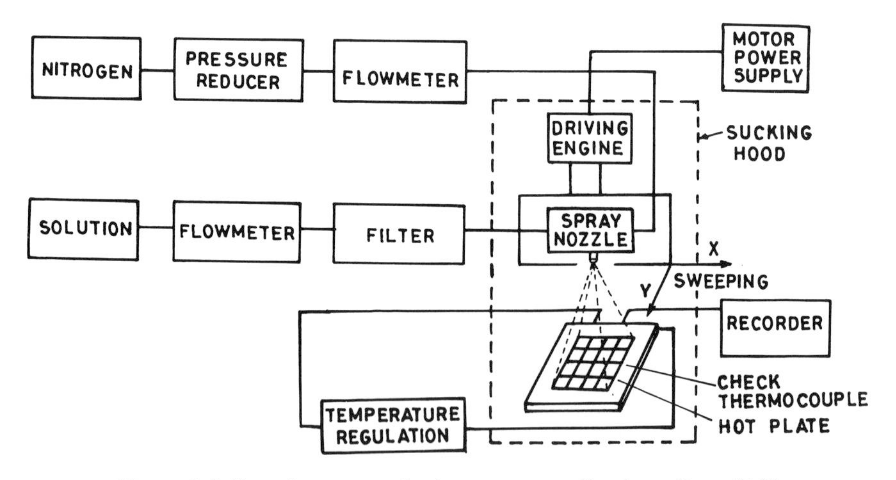

Figure 2.6 Spraying system for large-area applications (from [16]).

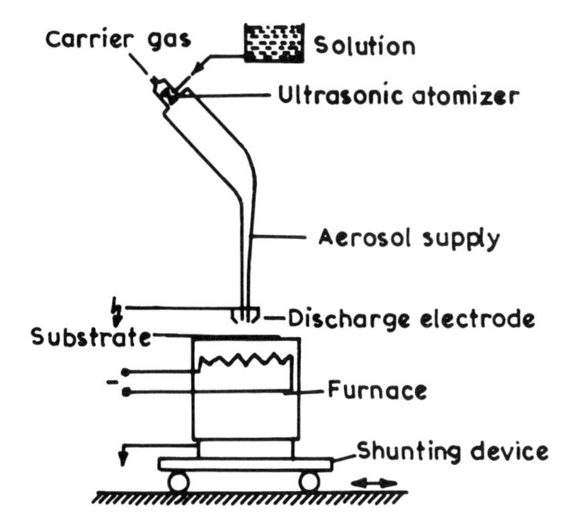

Figure 2.7 Schematic diagram of corona spray pyrolysis (from [18]).

The atomization of the solution is carried out with a piezoelectric transducer, using compressed air as a carrier gas. The droplets are charged in the electric field between the discharge electrodes and the substrate and are transported to the substrate by electric forces [18, 19].

2.2.3 Vacuum evaporation

Vacuum evaporation, although one of the most widely used techniques for the growth of semiconducting thin films, has not been extensively used for the growth of transparent conductors. The basic setup is shown in figure 2.8. A resistively heated tungsten or tantalum source, or an electron beam heated source, can be used to evaporate the charge. The substrate heater is placed above the substrate to heat it to the required temperature. Oxygen or an argon–oxygen mixture can be admitted through the calibrated leak valve. The important control parameters are the substrate temperature, evaporation rate, source-to-substrate distance and oxygen partial pressure. The transport conducting oxide films can be evaporated in three ways: (i) by directly evaporating metal oxides, e.g. SnO_2, In_2O_3, Cd_2SnO_4; (ii) by reactive evaporation of the metal in the presence of oxygen, or (iii) post-oxidation of metal films. When oxide materials are evaporated there is always some deficiency of oxygen in the films. Either the films must be evaporated in the partial pressure of oxygen or some post-deposition heat treatment in air is essential to achieve good quality films. In reactive evaporation, the corresponding metal or alloy is evaporated at rates of 100–300 Å min^{-1} in oxygen atmosphere onto substrates heated to about

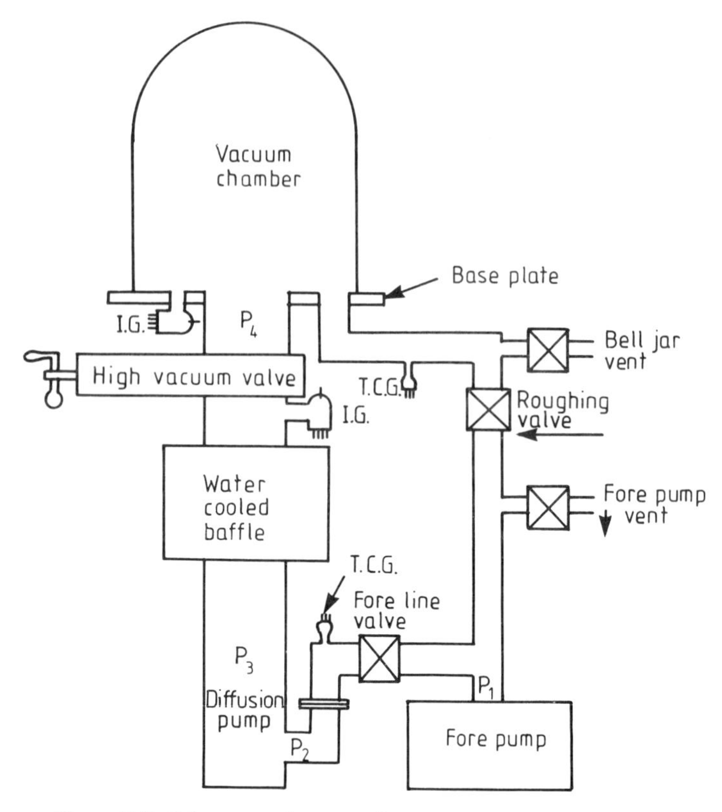

Figure 2.8 Schematic diagram of a vacuum evaporation system.

400 °C. The partial pressure of the oxygen is achieved by first creating a vacuum $\sim 10^{-6}$ Torr and then introducing oxygen such that the pressure increases to 10^{-4} Torr. Recently activated reactive evaporation has been employed to grow better quality transparent conducting films at a higher growth rate (500 Å min^{-1}) [20]. In this method, the reaction between the evaporated species and the gas is activated by employing a thermionically assisted plasma in the reaction zone. A dense plasma is generated by using a thoriated tungsten emitter and a low voltage anode assembly. In the case of post-oxidation of metal films, the conductivity and transparency of the films are controlled mainly by the oxidation temperature, which is usually in the range 350–500 °C.

2.2.4 Sputtering

Sputtering is one of the most versatile techniques used for the deposition of transparent conductors when device-quality films are required. Compared with other deposition techniques, the sputtering process produces films with higher purity and better-controlled composition, provides films with greater adhesive strength and homogeneity and permits better control of film thickness. The sputtering process involves the creation of a gas plasma (usually an inert gas such as argon) by applying voltage between a cathode and anode. The cathode is used as a target holder and the anode is used as a substrate holder. Source material is subjected to intense bombardment by ions. By momentum transfer, particles are ejected from the surface of the cathode and they diffuse away from it, depositing a thin film onto a substrate. Sputtering is normally performed at a pressure of 10^{-2}–10^{-3} Torr.

Normally there are two modes of powering the sputtering system. In a DC sputtering system, a direct voltage is applied between the cathode and the anode. This process is restricted to conducting targets, such as tin or indium. In RF sputtering, which is suitable for both conducting and insulating targets, a high frequency generator (13.56 MHz) is connected between the electrodes. Currently, however, use of magnetically enhanced gas discharge (magnetron sputtering) is on the rise. Magnetron sputtering is particularly useful where high deposition rates and low substrate temperatures are required.

Both reactive and non-reactive forms of DC, RF and magnetron sputtering have been employed. Sputtering from oxide targets to form conducting transparent oxide films is significantly different from reactive sputtering of metal targets. The control of film stoichiometry has been found to be much easier with oxide targets, which avoids the need for high temperature and post-deposition annealing. In DC sputtering of mixed oxide targets, the control of stoichiometry is virtually automatic, whereas preferential sputtering of deposited film material in RF sputtering usually requires that the target composition be suitably adjusted to compensate for the material lost at the substrate. In the case of reactive sputtering of metallic targets, the deposition rate and structural properties of transparent conducting films strongly depend on the oxygen content in the sputtering chamber. Figure 2.9 shows a typical plot of sputtering rate against the partial pressure of the

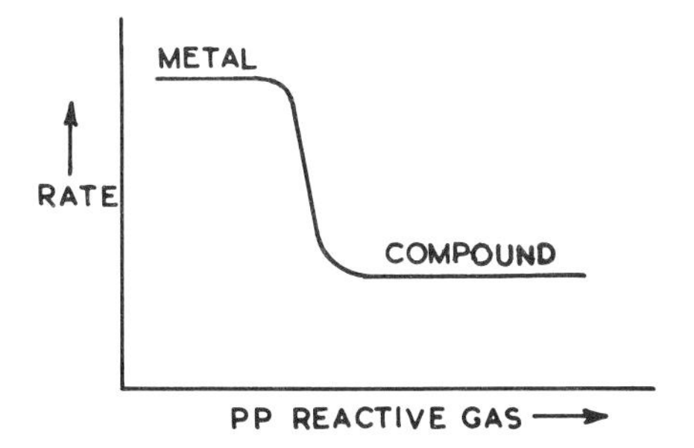

Figure 2.9 Plot of sputtering rate against partial pressure of reactive gas (from [21]).

reactive gas, i.e. oxygen, for a constant power level. At a low partial pressure, metal is sputtered from the target and the oxidation reaction takes place only on the substrate. At higher partial pressures, oxidation of the target face occurs and the sputtering rate drops rapidly since oxides sputter generally much more slowly than pure metals. Figure 2.10 shows the relation between the partial pressure of oxygen and the oxygen flow rate for a fixed power level. The straight line indicates the position before plasma ignition. The hysteresis effect is quite evident in the sputtering curve. As O_2 flow is initially increased, most of the gas is consumed by the sputtered film. When point B is reached, an oxide forms on the target surface and thus a rapid transition occurs. If the oxygen flow rate is again reduced to point C, the metallic face of the target is exposed and pumping by the metallic film again predominates. Operation on the lower curve (A to B) results in a metallic-rich film which requires annealing in the presence of oxygen to become transparent. Operation on the upper curve (D to C) produces an oxidized film which becomes conducting on annealing in a reducing atmosphere. To ensure reproducibility, it is important to avoid operation near the transition points.

Usually films sputtered at room temperature are amorphous, but an increase in substrate temperature or post-deposition heat treatment improves crystallinity and grain size. The quality of the films depends on various deposition parameters such as sputtering rate, substrate temperature, sputtering gas mixture, sputtering pressure, etc. The application of a negative bias to the substrate can increase the purity of the film by resputtering poorly bonded surface atoms.

Sputtering may be carried out in a variety of systems, which may differ in sputtering configuration, geometry, target type, etc. Experimental sputtering systems usually have small targets and low production rates, whereas commercial production systems have large targets and rapid substrate

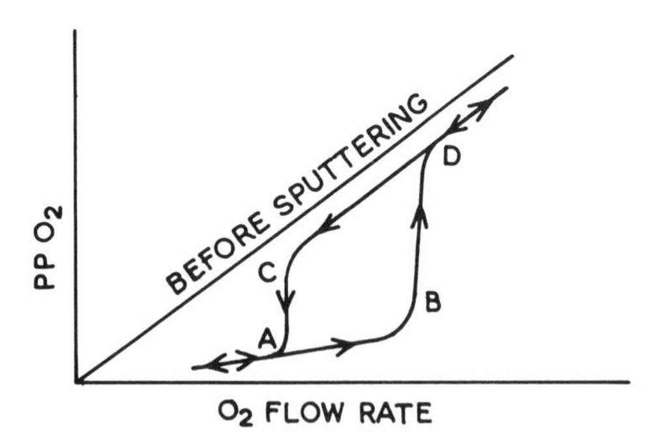

Figure 2.10 General form of the response of the partial pressure of oxygen to the oxygen flow rate prior to and during ITO sputtering (from [21]).

transport to maximize production rates. The basic sputtering process, however, remains the same irrespective of the sputtering system used. A basic sputtering system is shown in Figure 2.11. Typical conditions for DC sputtering of these materials are: cathode voltage, 1–5 kV; cathode current density, 1–10 mA cm^{-2}; cathode substrate distance, 2–4 cm; working gas pressure, 10^{-2} Torr. For RF sputtering, power 200–1000 W is delivered to the target from an RF generator of 13.56 MHz via an impedance matching network. These conventional sputtering processes, while offering simply controlled and reproducible film growth rates, suffer from two main disadvantages: (i) low deposition rates; and (ii) substantial heating of the substrate due to its bombardment by secondary electrons from the target. By employing a magnetic field to confine these electrons to a region close to the target surface, the heating effect is substantially reduced and the plasma is intensified, leading to greatly increased deposition rates. The source target is usually a water-cooled magnetron cathode, and a sufficiently uniform magnetic field of 0.02–0.1 tesla is required to confine the electrons near the target surface. The magnetic field is parallel to the target surface and orthogonal to the electric field. Discs of various target materials may be fixed to the cathode surface. In conventional magnetron sputtering systems, there is always resputtering of the film due to bombardment of high energy O^- ions This problem is particularly serious in the case of oxide films, in

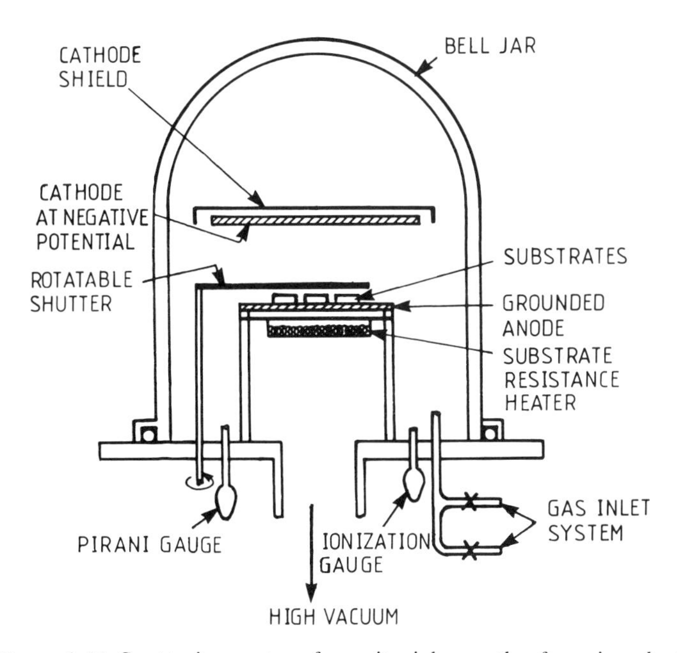

Figure 2.11 Sputtering system for epitaxial growth of semiconductor films.

which the film's composition and optical properties are greatly influenced by resputtering [22]. Matsuoka *et al* [22] used a double-ended magnetron system to overcome this problem in ZnO. The system used by them is shown in figure 2.12. Typical values of magnetic field and other parameters used by Matsuoka *et al* [22] are; B = 0.1 tesla, $V_{DC} = 500$ V, $V_{AC} = 100$ V_{rms} (50 Hz), current density = 10 mA cm^{-2}. In a double-ended magnetron system, the magnetic field is perpendicular to the target. If the substrates are placed outside the region between the targets, the ions ejected from one target will sputter the other target. The substrate and detectors are fixed to the underface of an earthed plate 6 cm above the cathode. A three-position shutter serves either to shield all active surfaces, or to expose just the quartz crystal or all the surfaces. Its position is indicated to the microprocessor by means of light emitting diode detectors.

A specially designed magnetron sputter gun (figure 2.13) offers distinct advantages over the conventional planar magnetron system. It provides electrical insulation between the source and the substrate, allowing the use of high power density without subjecting the substrate to electron bombardment. It thus avoids contamination of the films. Another important feature

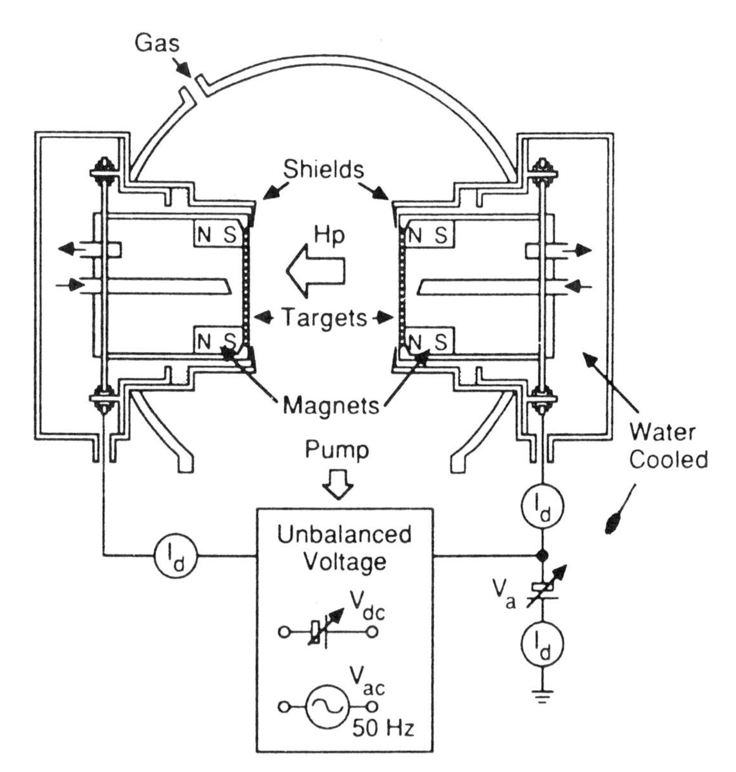

Figure 2.12 Schematic diagram of a double-ended magnetron system (from [22]).

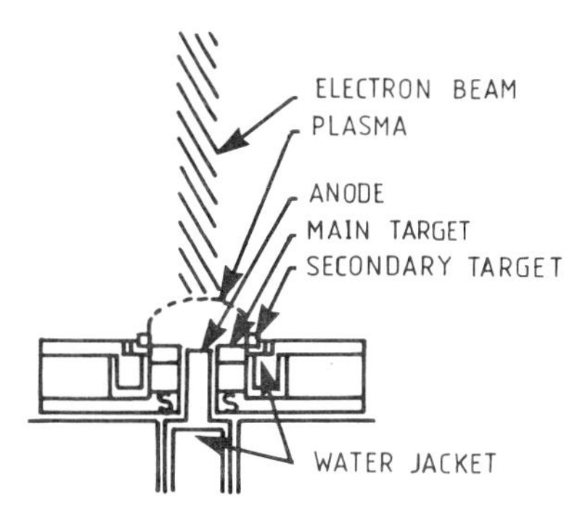

Figure 2.13 A magnetron sputter gun with specially designed cathode assembly. Electron flow from the high density plasma in the sputter gun is also shown (from [23]).

is that the anode is cooled by central water cooling, while the cathode, on thermal expansion, is cooled by a confining water jacket. This cooling further prevents high substrate temperatures through radiant heating and also eliminates the need for backing plates and their associated problems. Since the sputtered cathode is only a slip fit into the water jacket, targets can be changed easily without breaking water connections. The system has been used [23] to deposit pure *c*-axis oriented ZnO transparent and smooth films up to 8 μm thick. Although these films have been prepared with the specific aim of using them in surface acoustic wave devices, the system can be used to produce films for transparent conductor applications, either by varying the plasma composition or by post-deposition annealing in a reducing atmosphere.

2.2.5 Ion-assisted deposition techniques

Structural, electrical and optical properties of thin films depend mainly on the energies of the deposition species. In conventional thermal evaporation processes, atoms of the source material have energies less than 1 eV. Although sputtering enhances these energies to ~10 eV, the deposition energy is still insufficient to produce coatings that can tolerate the extreme conditions encountered in many wear-resistant applications. The properties and the structure of the deposited films can be improved by increasing the substrate temperature during deposition or by applying a bias to the substrate. However, the nature of the substrate limits the temperature that can be used. For such applications, where deposition of highly adherent coatings is required at low temperatures, ion-assisted deposition techniques are promising alternatives [24–30]. The deposition energies in these processes are of the order of a few hundred electron volts. The ion-assisted deposition techniques can be broadly divided into two categories: (a) ion-beam deposition; and (b) ion plating.

(a) *Ion-beam deposition*

Ion-beam deposition can be used in two different configurations, namely sputtering and evaporation. In the case of ion-beam sputtering, the ions of the non-reactive gas are used to bombard the target, whereas in the case of ion-beam evaporation, the ions of a suitable reactive gas are allowed to impinge on the substrate.

Unlike conventional sputtering systems, ion-beam sputtering involves minimal intrinsic heating and electron bombardment. Moreover, in this process the arrival rate of depositing species can be controlled precisely. Figure 2.14 shows a typical ion-beam sputtering system used for deposition of ITO films. An argon ion beam from a Kaufman ion source is directed onto the hot pressed ITO target to supply the depositing species. The argon ion beam source is usually capable of giving beams within maximum ion

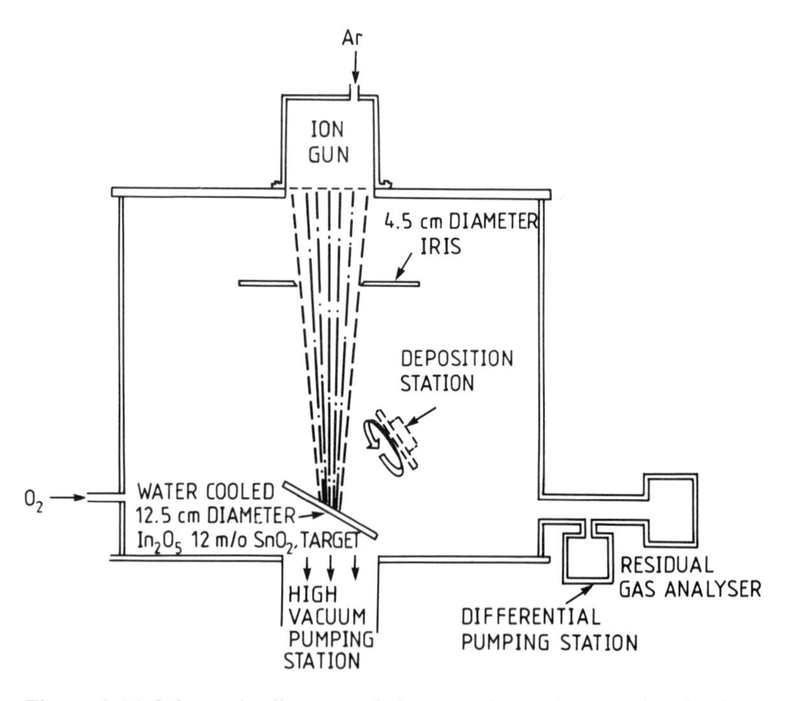

Figure 2.14 Schematic diagram of the experimental setup for the ion-beam deposition of ITO films (from [29]).

energies up to 1000–1500 eV and beam currents up to 30–50 mA. In the system shown in figure 2.14, a base pressure of 5×10^{-7} Torr is achieved and the sputtering is carried out in an oxygen partial pressure of $1–7 \times 10^{-5}$ Torr. The ions are usually incident on the target at approximately 45° to maximize the sputter yield.

Recently, a dual ion-beam sputtering technique has been used for the deposition of a variety of films. Two different ion sources are used; one for sputtering the metallic species and the other for providing the reactive gas ions. Figure 2.15 shows a typical dual ion-beam sputtering system. Two Kaufman ion sources are used to supply metal ions and reactive ions to the substrate. The metal atom flux is sputtered by the target ion beam and the reactive ion flux is supplied to the growing film by the substrate ion beam. This technique provides comprehensive control of deposition parameters during growth. Quantitative deposition parameters, such as arrival rate ratio of metal atom and reactive species, and incident ion energy, are system independent.

In the case of ion-beam evaporation, a thermal evaporator, i.e. electron beam or a tungsten boat, is used for the evaporation of metallic atoms. An ion-beam gun of reactive gas (O_2) with energies up to 500 eV is used. The oxygen ions are directed towards the substrate. The arrival rate of metal atoms and oxygen ions decide the composition and stoichiometry of the film, which can be controlled. A typical system is shown in figure 2.16. Although this system has been mainly used for the growth of copper oxide films [30], it appears to be a viable method for the deposition of transparent conducting oxide films.

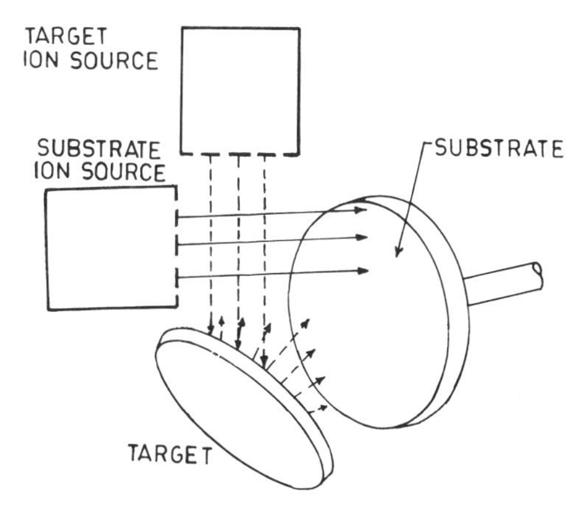

Figure 2.15 Dual ion-beam sputtering system.

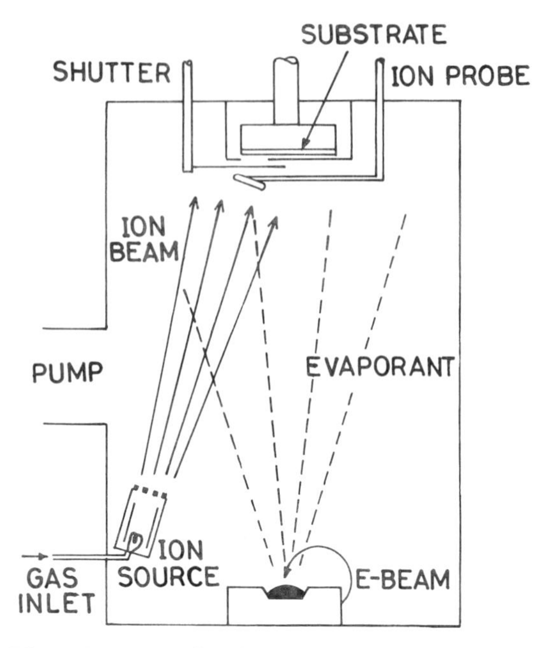

Figure 2.16 A typical ion-beam evaporation system.

(b) *Ion-plating*

In the ion-plating method, deposition energies are significantly enhanced by continuously bombarding the substrate and growing film with energetic particles. Ion plating can be used in an evaporative or sputter mode. In both cases the deposition species are ionized after leaving the source. A high negative bias voltage is applied to the substrate, which creates an ionizing plasma discharge and accelerates the resulting ions to the substrate. Figures 2.17 and 2.18 depict typical ion-plating systems for evaporative and sputter modes, respectively. The system used in the evaporative mode comprises a vacuum chamber containing an evaporative source and a high voltage cathode. The deposition chamber is initially evacuated to $\sim 10^{-6}$ Torr and subsequently filled with argon, raising the pressure to $\sim 10^{-2}$–10^{-3} Torr. A DC voltage of -500 to -3000 V is applied to the cathode in order to create plasma. Plasma generation is often enhanced by the incorporation of a third electrode, i.e. a hot filament, which introduces additional electrons into the plasma region. For evaporation of materials, an electron beam is usually employed. As electron beams operate most efficiently at pressures $<10^{-4}$ Torr, the electron beam system is housed in a differentially pumped chamber adjacent to the main chamber. The electron beam is deflected into the main chamber through a narrow opening by a magnetic field so that it is made to impinge on the target.

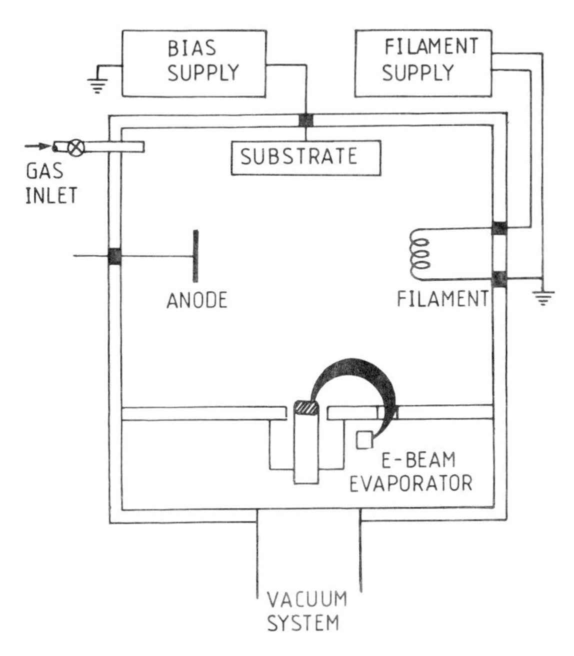

Figure 2.17 Schematic diagram of ion plating using evaporation mode.

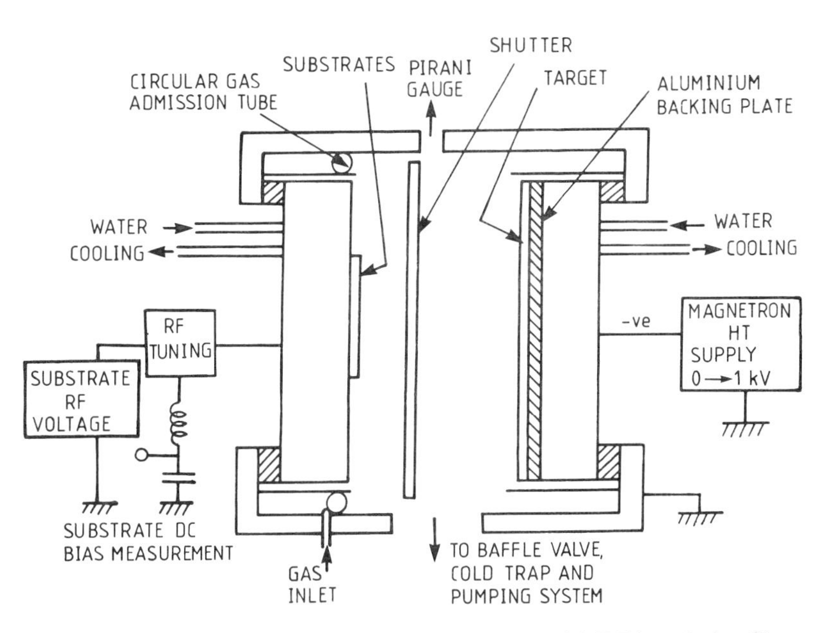

Figure 2.18 Reactive magnetron sputtering system with RF ion plating (from [28]).

Figure 2.18 shows a reactive magnetron sputtering type system with ion-plating facility. In this case the reactive gas is activated by ionization and driven towards the substrate by coupling an RF field to the substrate. The power of the applied ion-plating RF field is normally in the range of 100–500 W. In general, the partial pressure of oxygen in the system is $\sim 1.5 \times 10^{-3}$ Torr and that of argon 3.5×10^{-3} Torr for reactive ion plating of transparent conducting oxide films. Ridge *et al* [27] extended this process to continuous coating of transparent conducting materials on flexible plastic sheets. The system used is shown in figure 2.19. Coating on large areas of plastic sheets is usually done by pulling the sheet substrates tightly over a cooled drum and transferring the plastic from an initial roll to a final roll over the rotating drum.

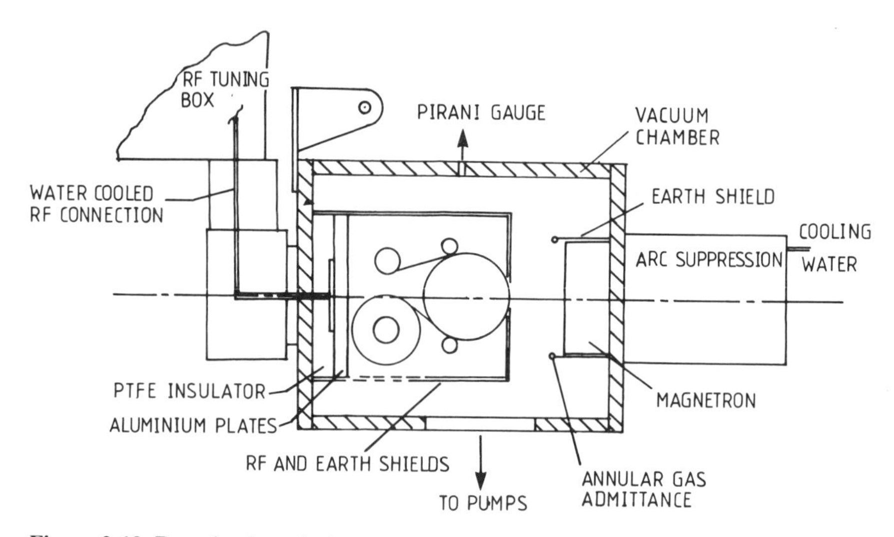

Figure 2.19 Reactive ion-plating system for continuous coatings (from [27]).

2.2.6 Other methods

In addition to the conventional growth techniques discussed, electroless deposition, sol–gel, anodization and laser-assisted deposition have also been used for the growth of semiconducting transparent thin films.

(a) *Electroless chemical growth technique*

Although this technique is mainly suitable for the growth of chalcogenide films, it has also been explored for the deposition of various oxide films. In this process, the substrate is immersed in an aqueous solution of metal chloride. Solid phases of metal hydroxide or metal hydrous oxide are formed, which on heating yield the metal oxide. The important parameters which control the deposition process are the composition of the initial solution and its pH value. Figure 2.20 shows the dependence of film thickness on deposition time with pH as a parameter. It can be observed that the pH value of the solution controls the ultimate thickness achieved in the films; the lower the pH value, the higher is the film thickness.

(b) *Sol–gel technique*

The sol–gel dip coating technique has certain important features: it is simpler and cost-effective and it allows coating of complex shapes and large surfaces. The substrates are inserted into a solution containing hydrolysable metal compounds (organometallic) and then pulled out at a constant speed into an atmosphere containing water vapour. In this atmosphere, hydrolysis and condensation processes take place. Water and carbon groups are removed by baking at temperatures ~500 °C and transparent metal oxide layers are obtained.

The reaction stages leading to oxide formation are

$$M(OR)_n + nH_2O \rightarrow M(OH)_n + nROH$$

$$2M(OH)_n \xrightarrow{\text{heat}} M_2O_n + nH_2O$$

Where M represents the metal and R an alkyl group.

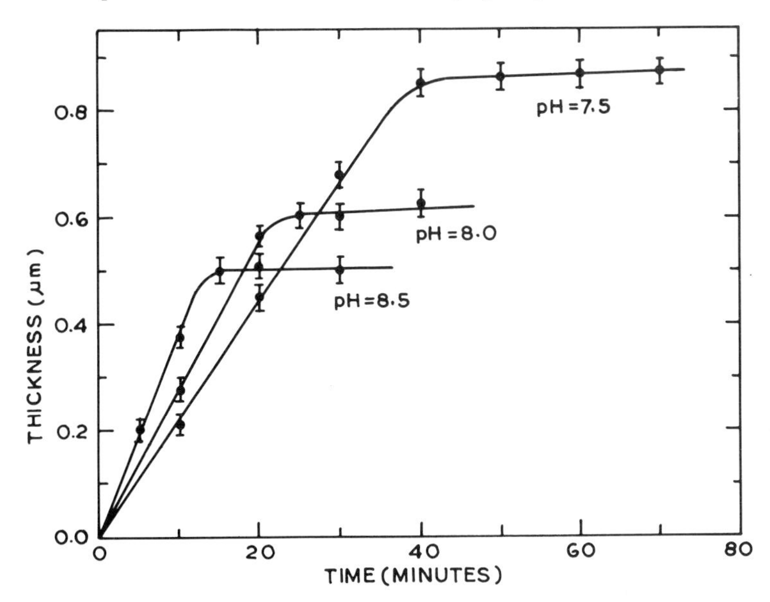

Figure 2.20 Variation of film thickness with time at different pH values for SnO_2 films grown using an electroless deposition technique (from [31]).

The conductivity in undoped oxide films grown by this technique is generally low (1–10 Ω^{-1} cm^{-1}). However, the films are useful for applications which do not depend on high conductivity, but do require a cost-effective technique, for example, fluorescent lamp envelope coatings and transparent grounding films required in cathode ray tubes.

(c) *Laser-assisted deposition techniques*
The laser evaporation deposition technique has been successfully used to grow high T_c superconductors [32, 33] and semiconductor films [33–35] of good quality. The main advantage of this method is that it can be used to grow highly oriented films at low substrate temperatures. This is due to the fact that the particles ejected using lasers have sufficient kinetic energy to arrange the structure of the films on the substrate. Recently Dai *et al* [36] applied this technique to grow well-oriented SnO_2 films. Lasers most often used for these applications are excimer and Nd:glass. Typical energy densities are in the range 20–200 J cm^{-2}. Excimer lasers of low energy density (10–50 mJ) have also been used for photochemical reactions leading to the formation of conducting transparent films [37, 38].

(d) *Anodization*
Anodization has been used as a simple and efficient method for converting metals into their oxides. However, attempts to anodically oxidize tin or indium have been only partially successful. The metal to be oxidized is employed as an anode and is dipped in an electrolyte from which it attracts oxygen ions. The oxygen ions combine with the metallic atoms to form oxide molecules. The rate of film growth depends on the temperature and the kind of electrolyte used. Anodization can be carried out either at constant current or at constant voltage conditions. It should be mentioned that it is not possible to grow films of large thickness using this technique.

2.3 Growth of Tin Oxide Films

A number of thin film deposition techniques employed to deposit tin oxide films, doped as well as undoped, are discussed in this section. Emphasis is placed on interpretation of the film properties that have been reported by various workers as related to film deposition parameters.

2.3.1 SnO_2: CVD

In this technique tin oxide films are usually deposited by vaporization of suitable organometallic compounds and their *in situ* oxidation with O_2, H_2O

and H_2O_2. These organometallic compounds should be volatile, thermally stable at a temperature sufficiently high to produce an adequate vapour pressure and should dissociate at higher temperatures. Stannous and stannic chloride [11, 39–42], tetramethyl tin [39, 43] dimethyl tin dichloride [44] and dibutyl tin diacetate [45, 46] are the most commonly used organometallic compounds for the growth of tin oxide films. The heterogeneous reactions leading to the formation of SnO_2 are

$$SnCl_4(g) + O_2(g) \rightarrow SnO_2(s) + 2Cl_2(g)$$

$$SnCl_4(g) + 2H_2O(g) \rightarrow SnO_2(s) + 4HCl(g)$$

$$SnCl_2(g) + O_2(g) \rightarrow SnO_2(s) + Cl_2(g)$$

$$(CH_3)_4Sn(g) + 8O_2(g) \rightarrow SnO_2(s) + 6H_2O(g) + 4CO_2(g)$$

$$(CH_3)SnCl_2 + O_2 \rightarrow SnO_2(s) + 2CH_3Cl$$

$$(C_4H_9)_2Sn(OOCCH_3)_2 + O_2 \rightarrow SnO_2(s) + 2CH_3COOC_4H_9$$

Chemical vapour deposition of SnO_2 films using these reactions have been extensively studied by many workers [11, 13, 39–62]. The growth kinetics of SnO_2 films involving the reactions of $SnCl_4$ and $(CH_3)_4Sn$ have been studied in detail by Ghoshtagore [39]. It has been observed [39] that in this process the surface reactions of both tin chloride and tetramethyl tin take place with an adsorbed oxygen atom or water vapour. For the surface reaction of tin chloride, however, Ghoshtagore [39] assumed that four adjacent symmetrically adsorbed water molecules are needed for every colliding gaseous tin chloride species to react at the surface. In his experiments no evidence of homogeneous gas-phase volume reaction was found in the temperature range 670 to 1080 K and tin chloride partial pressures ranging from 2.6×10^{-3} to 4.2×10^{-2} Pa (1.9×10^{-5} to 3.15×10^{-5} Torr). Instead, such undesirable volume reactions were found for tin chloride and tetramethyl tin oxidation above 1150 and 760 K, respectively. In his kinetic investigations of heterogeneous reactions involved in the growth of SnO_2 films, it was observed that the deposition rate is strongly dependent on oxygen partial pressure, substrate temperature and stannic chloride partial pressure. Typical results are shown in figures 2.21 and 2.22. It has also been observed that the rate of reaction is independent of substrate temperature T_s in the temperature range 400–800 °C. However, Muranoi and Furukoshi [50] and Advani *et al* [49] reported that the deposition rate is a strong function of reaction temperature. Figure 2.23 shows the variation of deposition rate as a function of substrate temperature for $SnCl_4$–H_2O, $SnCl_4$–H_2O_2 and $SnCl_4$–O_2 reaction systems. It can be observed that the systems based on H_2O and H_2O_2 are much faster than the O_2-based systems. Though H_2O-based systems have a lower deposition rate than H_2O_2-based systems, the electrical properties of the films produced by the former system are

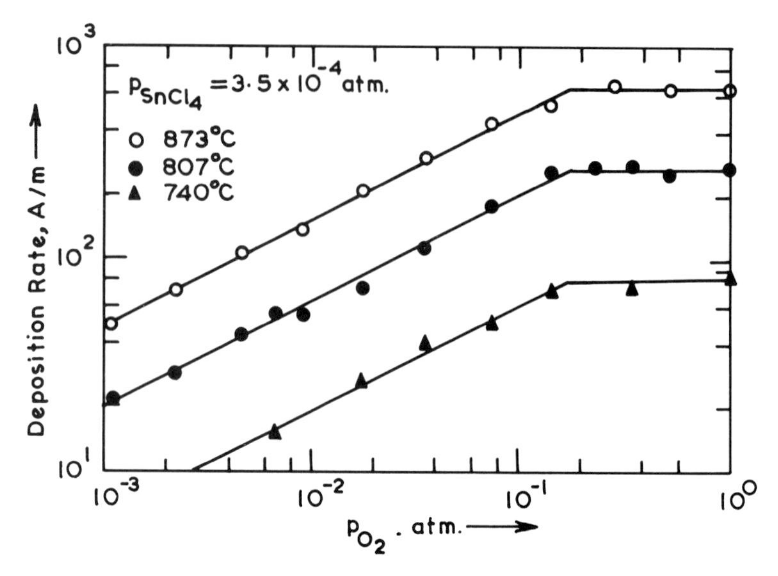

Figure 2.21 Oxygen partial pressure dependence of SnO_2 deposition at different temperatures for the $SnCl_4 + O_2$ reaction (from [39]).

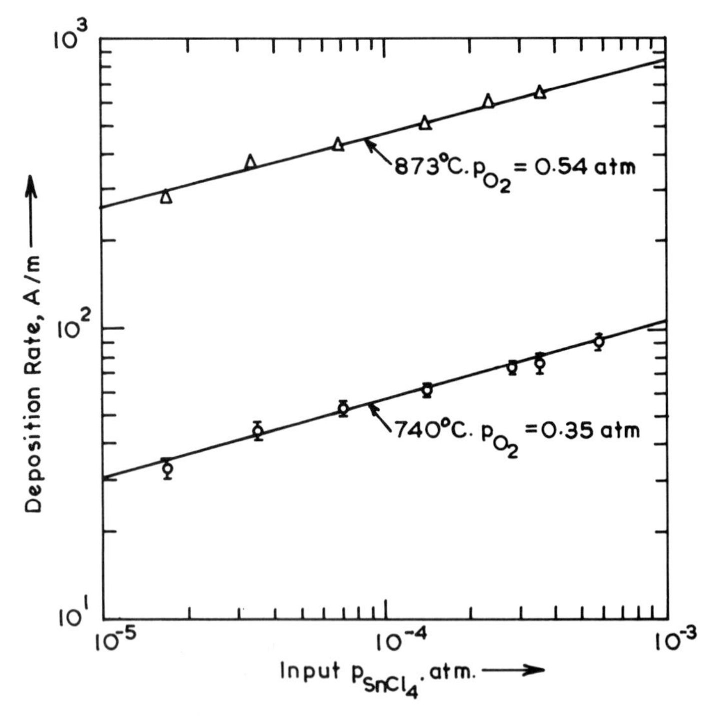

Figure 2.22 SnO_2 deposition rate as a function of stannic chloride partial pressure (from [39]).

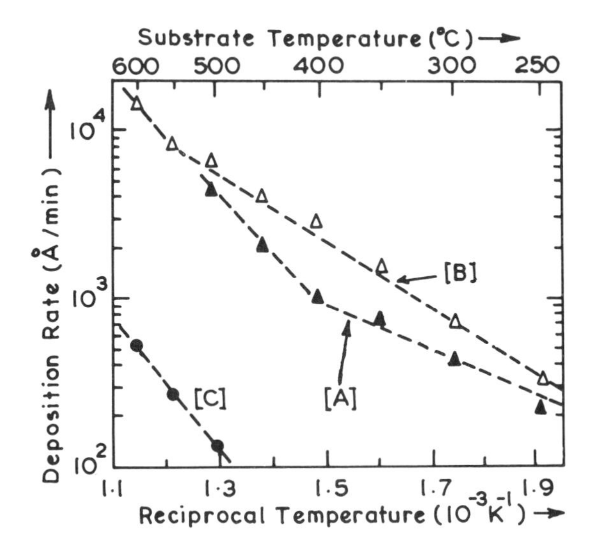

Figure 2.23 The deposition rate of stannic oxide films versus reciprocal substrate temperature: (A) $SnCl_4$–H_2O system; (B) $SnCl_4$–H_2O_2 system; (C) $SnCl_4$–O_2 system (from [50]).

reported [50] to be superior. Typically the resistivity in the films are 10^{-3} Ω cm and 10^{-2} Ω cm for H_2O and H_2O_2-based systems, respectively. The higher value of resistivity of the H_2O_2-based system is due to absorption of H_2O_2 into the film during growth. The transmission of films grown by both these systems are in the range 80–95%. It can be further observed from figure 2.23 that in the case of H_2O and H_2O_2-based systems, tin oxide can be grown at temperatures as low as 250 °C, which is not possible with O_2-based systems. Advani *et al* [49], on the basis of thermodynamic analysis, predicted the possibility of growing SnO_2 films at low temperatures. Their results are depicted in figure 2.24. There is a good agreement between theoretical and experimental results, which allows the prediction of growth rate as a function of temperature and input reactant ratios. The deposition rate is a sensitive function of temperature and the molar ratio B of H_2O and $SnCl_4$. At a given temperature, the growth rate increases with increase in the value of molar ratio B.

Srinivasamurthy and Jawalekar [47] used oxygen as reactant as well as carrier gas for the growth of SnO_2 films using $SnCl_2$ as a source material. The substrate temperature was varied between 400 and 500 °C and the gas flow rate (f) was varied from 1.35 to 2.50 l min^{-1}. They studied the morphology of the films using scanning electron microscopy. It has been reported that uniformity of the films increases with increasing T_s at constant gas flow rates. When T_s is decreased below 400 °C, the quality of the films deteriorates. On

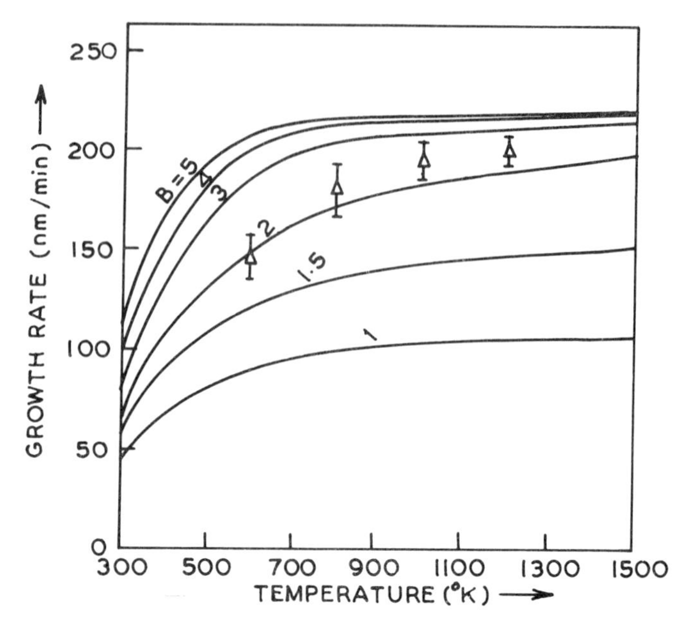

Figure 2.24 Equilibrium deposition rate (nanometer per minute) of thin films of SnO_2 as a function of temperature, and with B (n H_2O/n $SnCl_4$) as a parameter: ———, theoretical; △, experimental, ($B = 3$). The bars around each experimental point represent the variation in thickness of films prepared at these temperatures (from [49]).

the other hand, the uniformity of the films decreases with increase of gas flow rate at constant substrate temperature. Figure 2.25 shows the effect of gas flow rate on mean grain size of films deposited at different substrate temperatures. This figure shows that at a given T_s, the mean grain size decreases as gas flow rate increases, which is more pronounced at lower T_s. On the other hand, an increase in T_s at fixed gas flow rates results in an increase in grain size. It has been observed [47] that the grain size of the films varies from 0.083 to 0.303 μm. Lou *et al* [48], however, observed that the flow rate of oxygen has little effect on the structural and electrical properties of CVD-grown SnO_2 films.

Kane *et al* [45,46] used O_2 as reactant and N_2 as the carrier gas for the growth of undoped SnO_2 films using dibutyl tin diacetate as organometallic compound, and it was found that the addition of water vapour improved the film properties considerably. Optimum film properties were produced when the Sn source was at 98 °C and the substrate was kept at 420 °C. A large N_2 flow was needed to minimize gas phase nucleation of particles. Though the flow rates of O_2 and H_2O were not critical, the optimum conditions were found to be dependent on time. That is, for a given set of temperature–flow

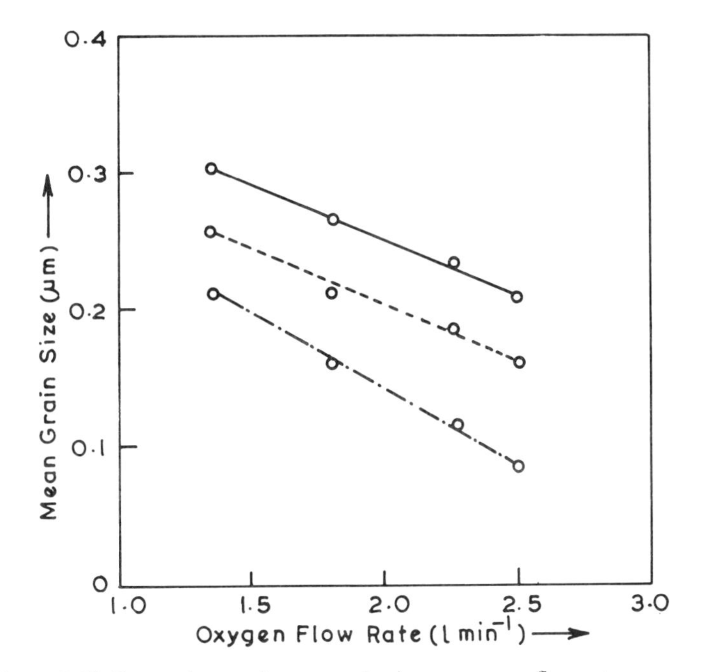

Figure 2.25 Dependence of mean grain size on oxygen flow rate: ——, deposition temperature 500 °C; – – – –, deposition temperature 450 °C; –·–·–, deposition temperature 400 °C (from [47]).

conditions, the sheet resistivity of the films decreases with deposition time, reaches a minimum and then increases as more material is deposited. This behaviour is due to competition between increasing film thickness and the combined effects of film oxidation and diffusion of acceptors from the substrate.

In films grown on soda lime glass substrates, the diffusion of alkali metal cations is a dominant process. These mobile alkali cations act as a p-type doping agent in the n-type oxide films thereby degrading the properties of the films. The effect of the alkali metal contamination is so strong that the mere presence of soda lime glass in the reaction chamber during tin oxide deposition significantly affects the properties of the films coated on other substrates. Special surface treatment can, however, be used to avoid contamination of the substrate by alkali metal ions [14]. Two different treatments, namely surface ion depletion and acid leaching, can be adopted for such treatment. Surface ion depletion of soda lime glass is accomplished by the application of an electric field to the glass, resulting in sufficient alkali ion depletion.

Instead of oxygen, an RF plasma activated oxygen can be used to react with tetramethyl tin (TMT) to produce high quality SnO_2 films [54]. The main advantage of the process is that the films can be grown at ambient temperatures and growth rates as high as 160 Å min^{-1} have been achieved. The deposition of the films is performed in a vertical reactor, 10.5 cm in diameter with RF excitation at 3.75 MHz to create oxygen plasma in the system. Electronic grade TMT is transported into the reactor by direct evaporation from a liquid source held at room temperature. At this temperature, TMT has a vapour pressure of 120 Torr and its vapour is controlled by means of a needle valve placed between the source and the reaction chamber. In actual operation, the system is pumped down to a base pressure of under 10^{-3} Torr. TMT is introduced into the chamber until the partial pressure is 0.1 Torr. Oxygen is introduced to bring the system pressure to 0.5 Torr and then RF power (2000 V, 0.2 A) is applied.

The CVD technique has also been used to deposit tin oxide films continuously onto glass ribbons [51, 52]. Kalbskopf [51] observed that the deposition rate is proportional to the concentration of reactants in the carrier gas and the molar ratio of H_2O to $SnCl_4$ in the vapour phase. Deposition rates as high as 1 $\mu m\ s^{-1}$ have been obtained with this technique. However, a white haze appears in and on the layers when a $SnCl_4$–H_2O mixture with N_2 is used and the extent of haziness is proportional to the deposition rate. In order to suppress haze, 2% HF is used instead of pure H_2O, the $H_2O/SnCl_4$ molar ratio is kept at unity and 10 mol h^{-1} of each component is injected per meter nozzle length. The electrical resistivity of the films is observed to be 10^{-2} Ω cm. This value can be reduced to 10^{-4} Ω cm by using a 40% H_2–60% N_2 mixture as the carrier gas.

Most of the work on CVD-grown tin oxide films indicates that the films are, in general, polycrystalline in nature. It is, however, possible to grow single-crystal thin films of SnO_2 on single-crystal substrates such as rutile titania (TiO_2). Vetrone and Chung [61] produced such films on (1 1 0) oriented TiO_2 using tetramethyl tin and oxygen as reactants. The substrate temperature was kept between 450 and 700 °C. Since film growth is sensitive to substrate cleanliness, smooth and polished titania substrates were cleaned in NaOH, distilled water and boiling isopropyl alcohol. XRD studies (figure 2.26) revealed that only (1 1 0) type peaks from both films and substrates were present. It should be mentioned that titania orientation influences the structure of the grown tin oxide films. The films grown on other orientations of titania substrates, (1 0 0), (0 0 1) and (1 1 1), are reported to be highly oriented polycrystalline.

Tin oxide films grown on fused quartz substrate using CVD techniques are always polycrystalline in nature [61, 63]. However, the preferred orientation is a strong function of growth temperature and deposition rate. Figures 2.27(a), (b) show that the degree of preferred orientation decreases with increasing growth rate, and more rapidly at higher growth temperatures. Typically the films grown at a rate less than 0.3 $\mu m\ h^{-1}$ and at temperatures

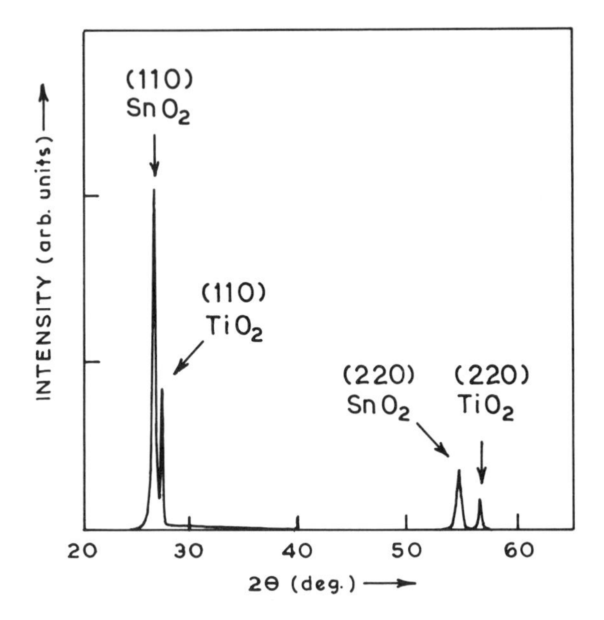

Figure 2.26 Typical X-ray diffraction spectrum for SnO_2 films deposited on TiO_2 (110) (from [61]).

450 °C or less have only (1 1 0) type peaks, whereas films grown faster than 0.8 μm h^{-1} or at temperatures above 650 °C are randomly oriented.

(b) *Doped SnO_2: CVD*
In order to improve the properties of SnO_2 films, doping with various elements such as antimony, phosphorus, arsenic, etc. has been tried [43, 45, 59, 64–66]. Antimony has been widely used for doping SnO_2 due to its substitutional occupancy without affecting the lattice parameters. Kane *et al* [45] grew antimony-doped SnO_2 films on various substrates, such as fused quartz and borosilicate, using a CVD technique. Antimony was added in the form of antimony chloride ($SbCl_3$, $SbCl_5$), which was kept in a separate container, and its vapour transported into the reaction chamber using N_2/Ar as carrier gas. The ratio of Sb to Sn was controlled by adjusting the flow rates of the carrier gas. The optimum dopant concentration was found to be within the range 0.6–2.7 mol% Sb in SnO_2. It should be mentioned that although organometallic compounds of antimony are superior to antimony chlorides due to the elimination of the possibility of halide contamination of SnO_2, due to their high thermal stability, these organometallic compounds cannot be used for doping purposes.

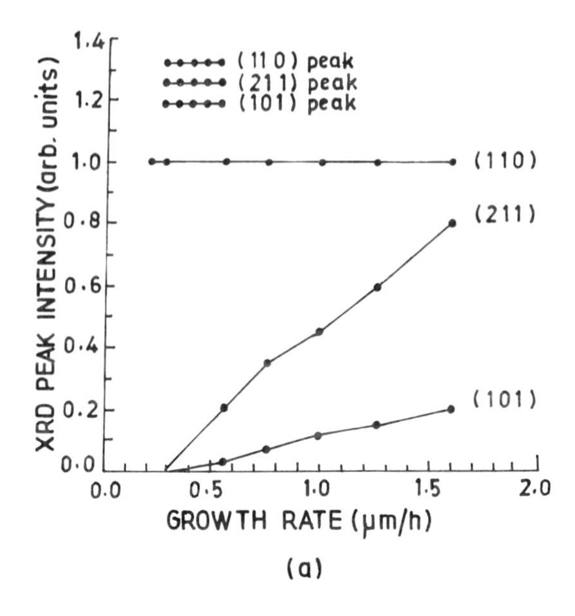

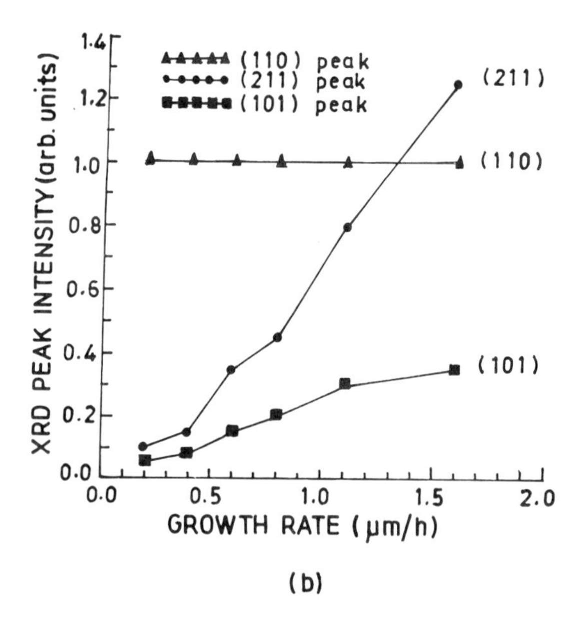

Figure 2.27 Growth rate dependence of integrated intensity of major X-ray diffraction peaks (normalized to (1 1 0) peak) for SnO_2 films deposited on quartz at (a) 450 °C, and (b) 650 °C (from [61]).

Phosphorus-doped SnO_2 films have been grown employing a CVD technique using tetramethyl tin as the base organometallic compound, argon as the carrier gas and oxygen as the reactant gas [43]. P_2O_5 has been used for phosphorous doping. Reproducible films of phosphorous-doped SnO_2 can be fabricated under the following typical conditions: deposition temperature = 500 °C, argon flow rate in the TMT bubbler = 15 ml min^{-1}, oxygen flow rate = 48 ml min^{-1}. The optimum properties are achieved for PH_3/TMT mole ratio of 0.01. Annealing of these films in forming gas at 300 °C for 5 min further increases the conductivity two-fold. The CVD technique has also been used to fabricate arsenic-doped SnO_2 films by introducing arsine gas in argon [59]. The flow rates of arsine/argon and TMT/argon are adjusted in order to vary the arsenic concentration in the films. A typical graph showing the effect of (arsine/TMT) ratio on the growth rate is shown in figure 2.28.

2.3.2 SnO_2: spray pyrolysis

(a) *Undoped SnO_2: spray*

Transparent conducting films of tin oxide have been grown by many workers [9, 14–19, 55, 64, 67–105] using this technique. The base solution may be stannic/stannous chloride, stannic bromide, tetrabutyl tin, diacetyl tin chloride or di-ammonium tin chloride dissolved in one of the alcohols, such as ethanol, methanol or propanol. These solvents have high volatilities and low surface tensions. The addition of HCl or acetic acid to the base solution

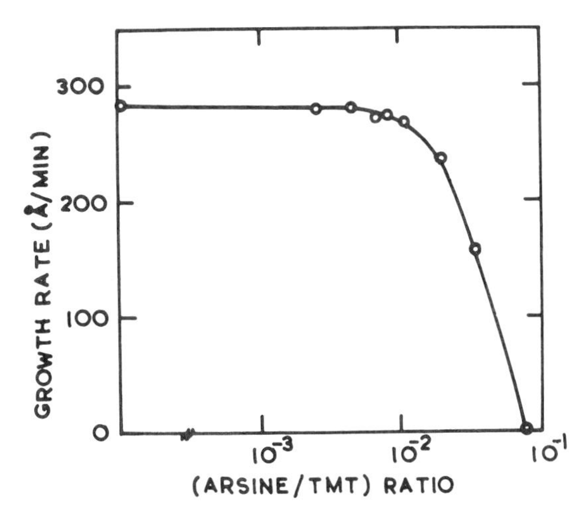

Figure 2.28 Growth rate as a function of AsH_3/TMT partial pressure ratio (from [59]).

has been found to improve the quality of the films. The carrier gases used most commonly are O_2, N_2, argon and air.

In general, the system employed by different workers is similar to the one shown in figure 2.3. This system does not have any preheating of spray solution. Some workers have, however, used a furnace placed between the substrate and the spraying nozzle to preheat the spray solution. Although preheating does not affect the quality of the films, it prevents vapours from condensing before the reaction takes place. The typical conditions for the growth of SnO_2 films are: spray mixture 0.7 M $SnCl_4$, 10 M propanol, 0.2 M HCl; substrate temperature 340–540 °C, spray rate 10 cm^3 min^{-1}, carrier gas flow rate ~6 l/min, distance between substrate and nozzle ~20–30 cm.

Substrate temperature is an important parameter in spray deposited tin oxide films. When the substrate temperature is higher than 350 °C, there is one-to-one correspondence between layer thickness and the amount of solution sprayed. However, the growth of the layers exhibit some peculiarities when grown at substrate temperatures less than 350 °C. For instance, [67] the thickness of the films never exceeds 0.05 μm, irrespective of the amount of solution sprayed. The possible reason for this peculiarity in this temperature range is that the upper part of the substrate is heated to a temperature very close to the limiting temperature at which the reaction occurs. If a small amount of solution is sprayed, the substrate can provide the necessary heat for the reaction to take place. On the other hand, if a large quantity of solution is sprayed, the substrate cools due to vaporization of the solution. Due to the poor thermal conductivity of the substrate, the heat does not reach the top surface of the substrate quickly enough and the reaction stops. The remaining solution is simply vaporized. It is, however, possible to grow thick oxide layers, even at low temperatures, if the process is carried out sequentially by depositing layers, typically 100 Å thick. The films are generally polycrystalline when grown at substrate temperatures higher than 350 °C and the average grain size is ~250 Å. The transmission of the films also improves with increase in deposition temperature. Typically transmission increases from 77 to 84% on increasing deposition temperature from 340 to 540 °C [68]. The maximum observable conductivity in such films is of the order of 10^2 Ω^{-1} cm^{-1}.

Although most workers have used propanol/ethanol with acetic acid/HCl as solvent, ethyl acetate [69, 70] can also be used to produce highly transparent SnO_2 thin films. Badawy *et al* [70] used a solution of 0.7 mol $SnCl_4$ in ethyl acetate to grow haze-free films. The spray system used by them is slightly different from the conventional system and is shown in figure 2.29. In order to obtain homogeneous films, the spray is carried upward in the reactor so that only very small droplets are carried by the gas flow, thus preventing the larger droplets from reaching the substrates. The deposition is carried out in a closed system rather than in ambient atmosphere. The driving gas (N_2/H_2 – 95:5) flows at a rate of 5 l min^{-1} carrying 1.4 g min^{-1} of

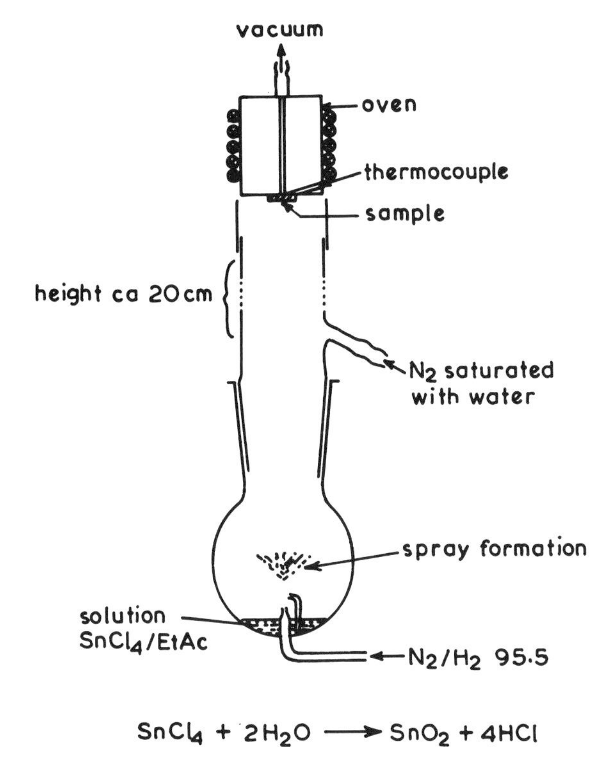

Figure 2.29 Schematic representation of the spray apparatus used to prepare SnO_2 films (from [70]).

the solution. Nitrogen saturated with water is added at a rate of $1\,l\,min^{-1}$ thus making the spray contents $1.7 \times 10^{-4}\,mol\,l^{-1}$ $SnCl_4$ and H_2O, i.e. it contains only 50% of the stoichiometric amount of water in accordance with the relation

$$SnCl_4 + 2H_2O \rightarrow SnO_2 + 4HCl.$$

Figure 2.30 shows the typical growth rates of SnO_2 films at specific temperatures. These rates are independent of the film thickness, indicating that the film growth process on the SnO_2 surface progresses unchanged with time, thus confirming the homogeneous nature of the films during growth. SnO_2 films with conductivity $160\,\Omega^{-1}\,cm^{-1}$ have been grown at 350 °C using this solution mixture.

Another spray system [71] designed so that an aerosol with only a very small drop size reaches the substrate is shown in figure 2.31. Filtered compressed air is used as the carrier gas and the flow is controlled by the use of a flow meter. Substrate temperature affects growth rate and the overall

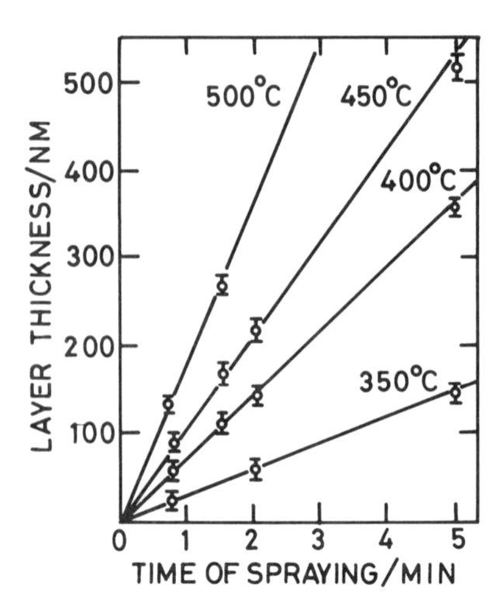

Figure 2.30 Growth of SnO_2 films with time of spraying at the indicated substrate temperatures: each 1.7×10^{-4} mol $SnCl_4$ and H_2O per dm^3 spray. Flow rate 6 dm^3 min^{-1} (from [70]).

thickness of the films significantly. Figure 2.32 shows the effect of substrate temperature on the optical thickness of tin oxide films for identical aerosol flow rates and spray times. The film thickness apparently increases with temperature, which is due to the fact that the surface reaction is incomplete at lower temperatures resulting in evaporation of the constituents from the surface.

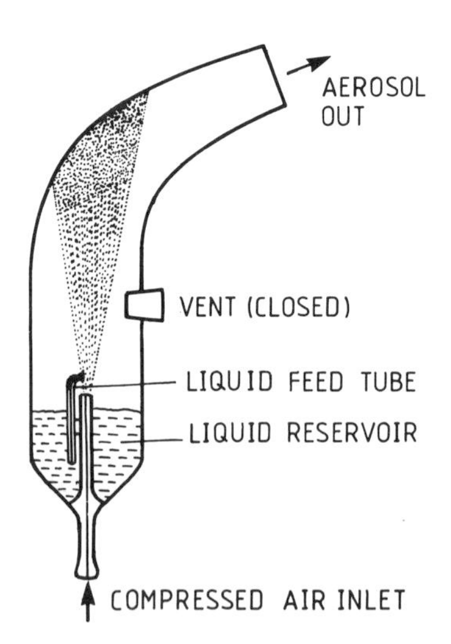

Figure 2.31 Schematic view of the aerosol generator for pyrolytic spraying (from [71]).

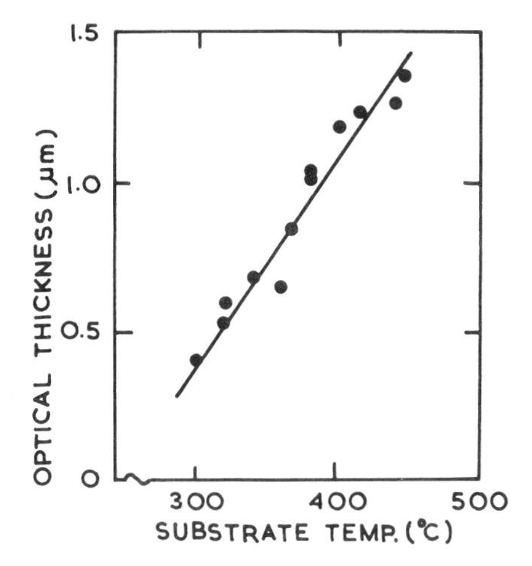

Figure 2.32 The relation between substrate temperature and optical thickness of undoped SnO_2 films under otherwise identical conditions (from [71]).

The nature of the glass substrates significantly affects the properties due to the diffusion of p-type alkali ions into SnO_2 films in the case of soda lime glass [9, 45, 72]. Moreover, owing to the low softening temperature of these glass substrates, some warpage occurs if the processing temperature exceeds 500 °C: this further enhances the diffusion of alkali ions. Either the use of better substrates, such as fused quartz, or special chemical treatment of the surface of the soda lime glass to deplete the surface of alkali ions improves the properties of the films. Specific methods for alkali ion depletion have been discussed in section 2.3.1.

Post-deposition heat treatment in vacuum or in other ambients can be used to improve the properties of these films. Improvements are for the following reasons:

(i) recrystallization of the amorphous part of the film;
(ii) change of highly resistive SnO phase to low resistive SnO_2 phase;
(iii) reorientation of existing crystallites;
(iv) change of frozen-in stresses;
(v) oxygen chemisorption/desorption mechanisms at grain boundaries.

Efforts have been made to improve the properties of SnO_2 films by annealing [43, 59, 73–75] them in different atmospheres, such as air, vacuum, oxygen, forming gas (80% H_2 and 20% N_2), hydrogen, etc. Irradiation of these films with low energy laser sources has also been used to improve the structural and electrical properties. Dawar *et al* [76] observed an improvement in crystallinity and an increase in grain size in laser-irradiated SnO_2 films. Nd:YAG laser pulses (1.06 μm) of various energy densities (2–50 mJ cm^{-2}) with pulse width 20 ns have been employed for irradiation. Figure 2.33 shows the XRD patterns for as-deposited and laser-irradiated films. The films irradiated with laser pulses show better orientation, with (1 1 0) as the dominating plane. The peak height corresponding to this plane increases significantly with laser irradiation, whereas the peak width decreases. This shows an improvement in the laser-irradiated films. Typically the grain size increases from 220 Å for as-deposited film to ~850 Å for films irradiated with a laser of energy density 30 mJ cm^{-2}. Islam and Hakim [73] annealed SnO_2 films at 250 °C, both in air and in vacuum. They observed no change in optical properties, whereas there is a significant improvement in electrical properties. This is due to the oxygen chemisorption/desorption mechanisms at the grain boundaries. Shanthi *et al* [74] reported the annealing of SnO_2 films in vacuum, air and oxygen. It has been observed that vacuum annealing at 400 °C and oxygen annealing at 240 °C are the most effective in improving the film properties. On the other hand, annealing in air and oxygen at 400 °C has been found to deteriorate the film properties. This may be due to filling of oxygen vacancies, thereby decreasing conductivity. Siefert [18] reduced SnO_2 films in CO–CO_2 atmosphere at 440 °C in order to create more oxygen vacancies, thereby enhancing the conductivity of the films.

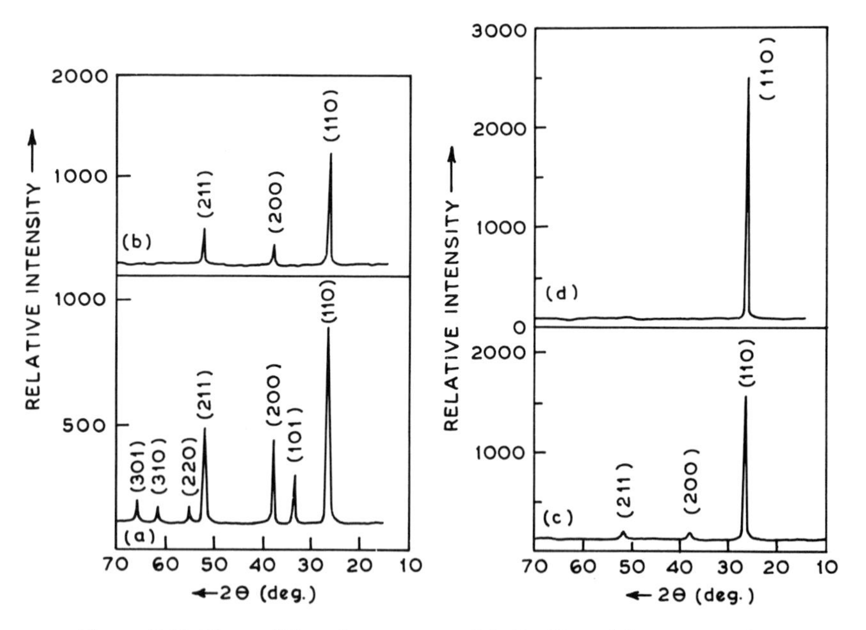

Figure 2.33 X-ray diffraction spectra of SnO_2 films: (a) as-grown film; (b) 10 mJ cm^{-2}, 30 pulses; (c) 20 mJ cm^{-2}, 30 pulses; (d) 30 mJ cm^{-2}, 30 pulses (from [76]).

(b) *Doped SnO_2: Spray*
As mentioned earlier in section 2.3.1, doping of SnO_2 films has been found to change the properties of the films. In addition to the intentional doping, chlorine also produces unintentional doping in SnO_2 films prepared from $SnCl_2$ or $SnCl_4$. Most studies reveal that the Cl content in the films is a strong function of deposition temperature [45, 77, 78]. Typical chlorine concentration profiles along the film thickness for spray-deposited films grown at different temperatures (studied using secondary ion mass spectroscopy (SIMS)) are illustrated in figure 2.34. It can be observed from this figure that the chlorine content strongly decreases as the deposition temperature increases. Figure 2.35 shows the variation of chlorine concentration versus inverse deposition temperature. For comparison, data of other workers for SnO_2 films grown by different techniques are also shown. Carlson [78] grew SnO_2 films using a DC glow discharge technique, whereas Kane *et al* [45] used a CVD technique. It can be seen that the chlorine concentration is a function of growth temperature irrespective of the growth technique used. However, Aboaf *et al* [55] did not notice such a temperature dependence of chlorine contents in spray-deposited SnO_2 films.

The spray pyrolysis technique has also been effectively used to grow doped SnO_2 films. The most widely used dopants are Sb, In, and F. Antimony can be introduced into SnO_2 by mixing $SnCl_4$ with $SbCl_3$ or $SbCl_5$. Fluorine can be incorporated in the films by adding HF, NH_4F or fluoroacetic acid to $SnCl_4$ solution. Many workers [68, 71, 79–83] have used this spray

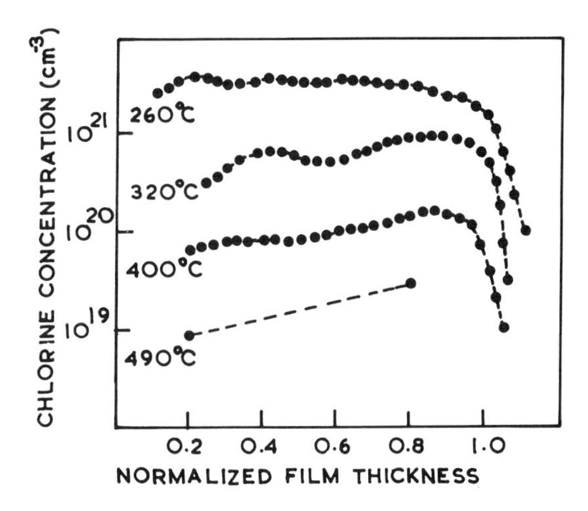

Figure 2.34 Chlorine concentration versus normalized film thickness for SnO_2 sample grown at different temperatures as shown. Primary ion energy $E_p = 2.1$ keV and current density $I_p = 2.5 \times 10^{-5}$ A cm^{-2} (from [77]).

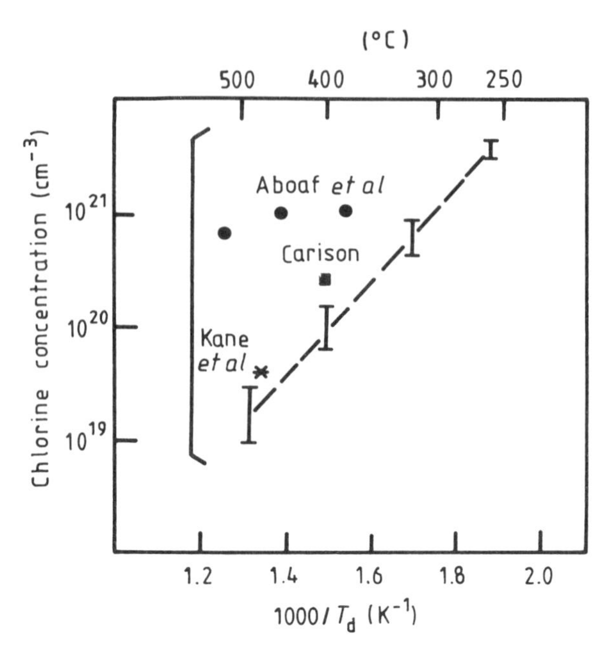

Figure 2.35 Chlorine concentration versus inverse deposition temperatures for samples grown at four different temperatures. Measurements by Aboaf *et al* [55], Carlson [78] and Kane *et al* [45] are also indicated in the figure (from [77]).

technique to grow Sb-doped SnO_2 films on various substrates, such as silicon, fused quartz, borosilicate glass, soda lime glass, etc. A typical spray solution has the composition 40 g $SnCl_4$: 5 H_2O, 0.4 g $SbCl_3$, 2 cm^3 HCl, 20 cm^3 H_2O and ethyl alcohol. HCl is usually added to obtain complete dissolution of $SnCl_4$. A small quantity of reducing agent such as pyrogallol improves the film quality. For example, the addition of 2.5 cm^3 of 2% pyrogallol added to 18 cm^3 of mixed $SnCl_2$ and $SbCl_3$ solution results in films with sheet resistivity as low as 65 $\Omega\,\square^{-1}$ [79]. Generally, the films grown on substrates at temperatures below 623 K are amorphous and films grown at higher substrate temperatures are polycrystalline. In order to obtain homogeneous thick films, repetitive spraying of solution with a short duration (1 s) is carried out. Kaneko and Miyake [80] prepared Sb-doped SnO_2 films on different substrates, namely fused quartz, borosilicate glass and soda lime glass, at 600 °C using an aqueous solution of a mixture of $SnCl_4$ and $SbCl_3$. It was observed that the films grown on fused quartz and borosilicate glass were of comparable quality, whereas the films grown on soda lime glass were of relatively poor quality. SEM studies revealed that the films grown on quartz and borosilicate glass had better crystallinity and a smooth and structureless surface. The surface of the films on soda lime glass substrates, on the other hand, differed, and many sharp cornered pits appeared on the surface.

It is important to note that the concentration of antimony in the films is always less than that in the spray solution. Since it is not possible to have exactly the same thermodynamic parameters for different pyrolytic reactions, the composition of the films is not simply related to the composition of the spray solution. In general, the correlation can be established empirically for each system. Shanthi *et al* [81] correlated the dopant concentration of antimony in spray-deposited SnO_2 films onto glass substrates with that in the solution using Auger analysis. Figure 2.36 shows Sb concentration in the films as a function of antimony concentration in the solution. It can be seen from this figure that the antimony concentration in the films is 5–50% lower than that in the solution. It should be mentioned that the nature of the substrate may also significantly change the thermodynamic parameters for the pyrolytic reaction, thereby changing the ratio of Sb to Sn content in the films.

The effect of the addition of antimony in SnO_2 films is to improve both the grain size and the surface topography of the films. Typically the average grain size in undoped SnO_2 films is ~250 Å, whereas the addition of up to

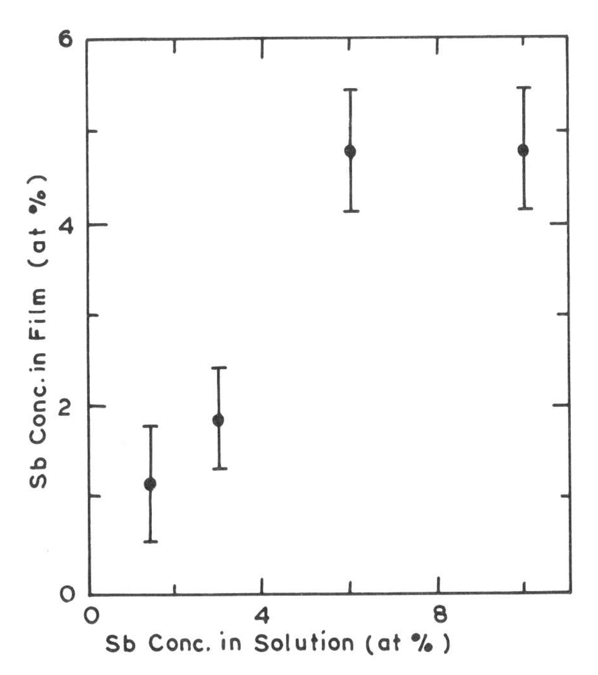

Figure 2.36 Variation of the antimony concentration in films versus antimony concentration in the solutions (from [81]).

1.4 mol% antimony increases the grain size to 600 Å [68]. Further increase in antimony concentration, however, does not appreciably change the grain size. The addition of up to 3 mol% antimony results in smoother surfaces. However, at higher antimony concentrations the surface becomes coarse and the crystalline nature deteriorates. The optimum antimony concentration ranges between 0.4 and 3 mol%.

Fluorine is a most effective dopant due to its comparable ionic radius with that of oxygen (F^-: 1.17 Å, O^{2-}: 1.22 Å) thereby substituting the anionic sites and acting as a donor. Fluorine doping is achieved by adding NH_4F, trifluoroacetic acid or HF to the growth solution [14, 16, 71, 72, 84–90]. Typical ratios [72] of various contents in the solution for the fabrication of good quality fluorine-doped SnO_2 films are: $SnCl_4$:$5H_2O$ – 0.329, H_2O – 0.329, CH_3CH_2OH – 0.329 and NH_4F – 0.013. The optimum substrate temperature and gas flow rate are 500 °C and 3 l min^{-1}, respectively. It should be mentioned that the actual atomic ratio F/Sn present in the films is lower by a factor of 10 than that of the solution, which is due to the volatile nature of fluorine compounds. The results of SIMS measurements made on these films are shown in figure 2.37. It can be observed that in addition to

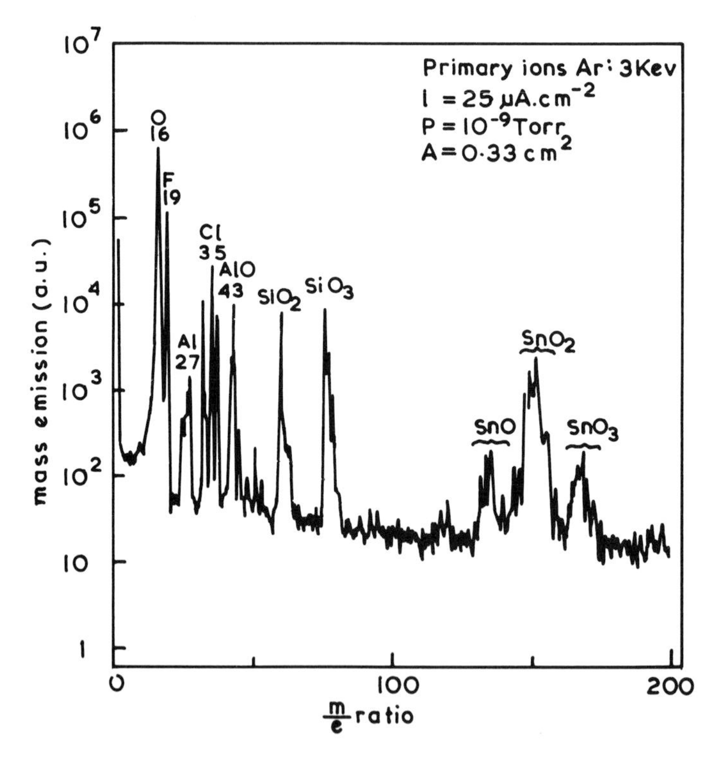

Figure 2.37 SIMS measurements of an SnO_2:F layer on a borosilicate glass substrate (from [72]).

SnO_2 there are other materials present, such as SnO, SnO_3, SiO_2, SiO_3, Na, K, etc. These impurities are introduced through the pyrex glass substrates as the substrate temperature is very high, and not through the spraying solution used. This result has been substantiated from SIMS results on films grown on silicon [84], as well as films grown on glass at lower substrate temperature using a pyrosol technique [14].

X-ray diffraction studies [85–87] reveal that no additional peak corresponding to fluorine is present in the films if the concentration of fluorine is less than 10%. In general, the crystallinity of SnO_2 films is significantly improved as a result of fluorine doping. Typically the crystalline size increases by 70–80% as a result of 10% fluorine doping. An increase in substrate temperature further improves the grain size. The preferred orientation planes in these films are usually (1 2 1), (1 1 2), (1 1 0), (2 0 0) and (3 0 1) depending upon the substrate temperature, film thickness and/or dopant concentration. For example, films containing low dopant concentration show preferred orientation of the (1 2 1) and (1 1 2) planes, whereas films with higher concentration exhibit only the (1 2 1) plane [86]. Afify *et al* [85] and Agashe *et al* [87] reported that the (2 0 0) plane is very sensitive to substrate temperature and film thickness. The effect of film thickness in 10%-fluorine-doped SnO_2 films on the crystallinity for the (2 0 0) plane is shown in figure 2.38. It can be observed that the crystallinity improves appreciably with increase in film thickness.

Very little work has been done on indium-doped SnO_2 films as compared to Sb and F-doped films. The possible reason may be that the doping of In in SnO_2 enhances resistivity, thereby making these films less attractive as transparent conductors. Carroll and Slack [91] deposited indium-doped thin films of SnO_2 using a spray pyrolysis technique. Indium doping is accomplished using $InCl_3{:}4H_2O$ in a finely dispersed solution of $SnCl_4{:}5H_2O$ in ethanol. The addition of even a small quantity of indium (up to 1 mol%) is found to cause an increase in resistivity for all films grown at both high and low substrate temperatures. The rise in resistivity in these films with the addition of indium can be explained on the basis of the controlled valency model suggested by Vincent [92].

Numerous other dopants have also been tried in tin oxide films using this technique. These include cadmium, bismuth, molybdenum, boron, tellurium, etc. These dopants have, however, been found to be relatively ineffective in enhancing conductivity in spray-deposited films.

2.3.3 SnO_2: vacuum evaporation

A wide range of vacuum evaporation techniques has been used to deposit tin oxide thin films. These films have been prepared by direct evaporation of tin oxide sources using electron beam [106–108] and flash evaporation [109]

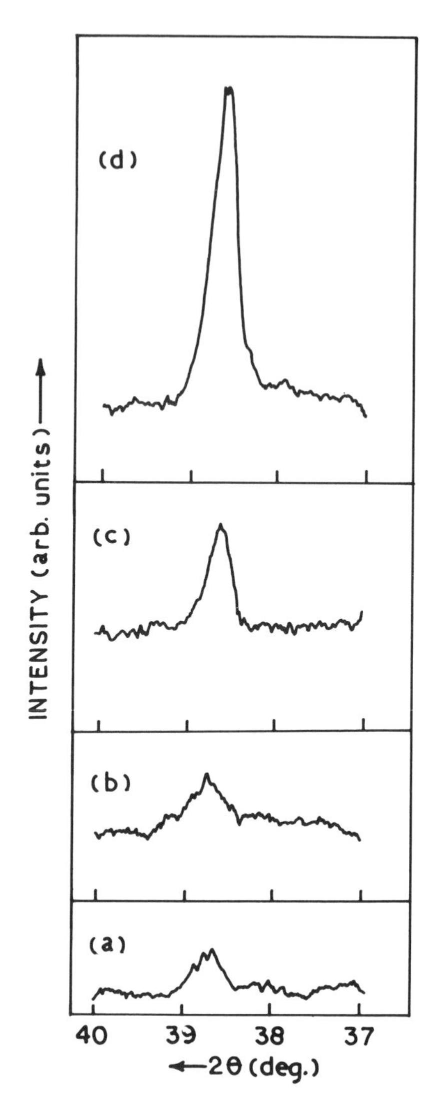

Figure 2.38 Effect of film thickness on crystallinity of SnO_2 films: (a) undoped SnO_2, $t = 85$ nm; (b–d) 10% fluorine-doped SnO_2 with $t =$ (b) 85 nm, (c) 172 nm and (d) 350 nm (from [85]).

techniques or from a tin source by reactive evaporation [110–112] or by annealing vacuum-deposited tin films in oxygen atmosphere [113]. Post-oxidation is generally necessary in order to obtain conducting transparent tin oxide thin films. It has been observed that the as-evaporated films mainly have SnO phase, which is produced by decomposition of SnO_2 molecules during evaporation. These SnO films are polycrystalline with a grain size of

500–1000 Å and the crystallites are preferentially oriented with their *c*-axis perpendicular to the substrate. The oxidation of SnO to SnO_2 is carried out in an oxygen atmosphere at temperatures in the range 200–650 °C. Geurts *et al* [106] observed that the total conversion of SnO to SnO_2 takes place at 650 °C. Figure 2.39 shows X-ray diffraction spectra depicting the various stages of oxidation. It can be observed that in the as-grown SnO films, the strongest *c*-axis peak (0 0 2) is observed at $2\theta = 37°$. The SnO (0 0 2) reflections decrease with increase in temperature while new peaks appear at 31, 33 and 33.8°. However, the peaks at 31° and 33° disappear at higher oxidation temperatures, whereas the peak at 33.8°, corresponding to SnO_2, continues to grow.

The angle of incidence of SnO_2 vapour onto the substrate significantly affects the growth behaviour of these films [107,108]. Following heat treatment between 300 and 350 °C, there is a marked difference in the properties of the films grown at different incidence angles. Films grown at incidence angles less than 60° are highly resistive and yellowish brown in colour and are rich in SnO phase. Films grown at larger angles become transparent and show low resistivity because of the presence of SnO_2 phase.

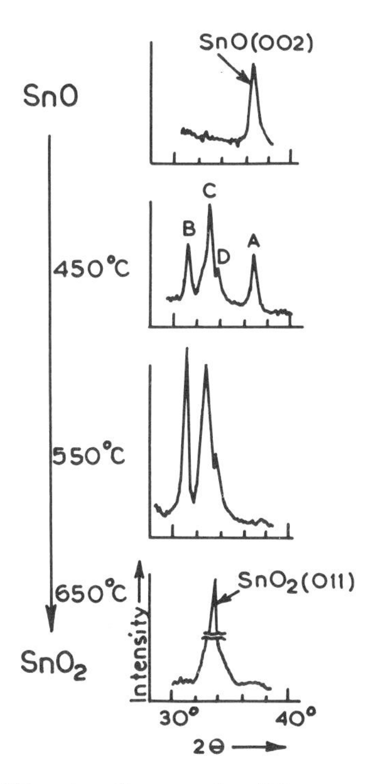

Figure 2.39 X-ray diffraction diagrams for different stages of oxidation, here denoted by temperature (from [106]).

2.3.4 SnO_2: sputtering

(a) *Undoped*

(*i*) *Sputtering of tin target*. Reactive sputtering of tin has been extensively used to grow good quality tin oxide films. The main advantage of this process is that the material can be supplied to the substrate in the desired proportion and with sufficient energy to ensure reaction of the elements and the formation of a dense structure. Usually tin is sputtered in the presence of an oxygen–argon mixture. The rate of sputtering and the ratio of oxygen to the total gas mixture decide the extent of tin oxidation. Films grown at lower substrate temperatures (<200 °C) are generally amorphous in nature and require post-deposition annealing, either in air or in vacuum. Annealing of these films improves their crystallinity and conductivity markedly. Even the films grown at higher temperatures (>200 °C), show an improvement in crystallinity on being annealed. Various workers [115–120] have employed this process to prepare SnO_2 thin films. Typical parameters used in the process are: applied DC voltage, 1–2 kV; discharge current: 10–40 mA; distance between target and substrate, 3–5 cm; gas pressure, 10^{-2} Torr; and oxygen concentration in the gas mixture, 20 to 50%. Giani and Kelly [115] grew SnO_2 films on a number of substrates, e.g. Al, Ta, KCl, air annealed SnO_2. The gas pressure was of the order of 10^{-2} Torr and the gas used was a 1:1 mixture of oxygen and argon. Applied DC voltage ranged from 1000 to 3000 V and the discharge current was typically 10 mA. The results indicate that the films prepared onto KCl or Ta at temperatures up to 200 °C are amorphous whereas films deposited onto SnO_2 sintered pellets at 200 °C are finely polycrystalline. The films, however, improve their crystalline nature on being annealed in vacuum at a temperature of 400–450 °C. Annealing also results in an increase in grain size to about 200–1500 Å. The oxide films prepared by Sinclair *et al* [116] using this technique are reported to be amorphous and electrically insulating. Annealing the films in air, however, makes the films conducting and crystalline with an average crystallite size 400 Å.

Argon–oxygen reactive gas mixture ratio plays an important role in the formation of SnO_2 phase. Figures 2.40 and 2.41 show the phase contents of SnO_2 and the stoichiometric factor x in SnO_x versus oxygen concentration, respectively. Both the α-SnO and β-Sn phases are also present in addition to SnO_2 phase, for oxygen concentrations up to 10%. SnO_2 phase, however, becomes more prominent only at oxygen concentrations exceeding 10%. It can be observed from figure 2.41 that the stoichiometry factor increases monotonically with increasing oxygen concentration in the reactive mixture. However, a value of 2 for x is probably not reached even by sputtering in a pure oxygen atmosphere.

Conventional diode sputtering is too slow and the operating pressure is too high for precise control of the properties of the films due to the presence

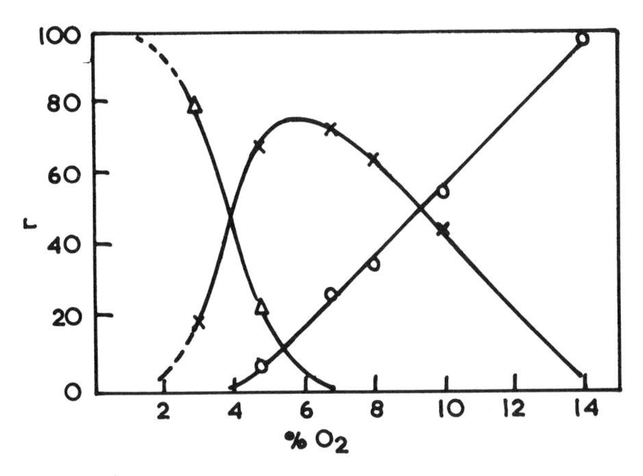

Figure 2.40 The phase content of SnO_x films versus oxygen concentration in the Ar–O_2 reactive mixture: ○, SnO_2; ×, SnO; △, Sn (from [117]).

of contaminants. Magnetron sputtering has been used [118, 119] to overcome this problem in growing SnO_2 films. Stedile *et al* [118] used this technique to grow SnO_2 films by reactive sputtering in an Ar–O_2 plasma. The sputtering conditions were: $V_{DC} = 5000$ V, $i_{DC} = 40$ mA, $P(Ar) = 2 \times 10^{-2}$ Pa (1.5×10^{-4} Torr), $P(O_2) = 1 \times 10^{-2}$ Pa (7.5×10^{-5} Torr), distance between target and substrate = 5 cm, $T_{substrate} < 100$ °C, rate of growth ~70 Å min^{-1}. In a reactive sputtering process the voltage–current diagram is characterized by a transition from the oxide sputter mode to the metallic layers sputtering mode. Figure 2.42 shows typical discharge characteristics of magnetron sputtered SnO_2 layers. For a given reactive gas

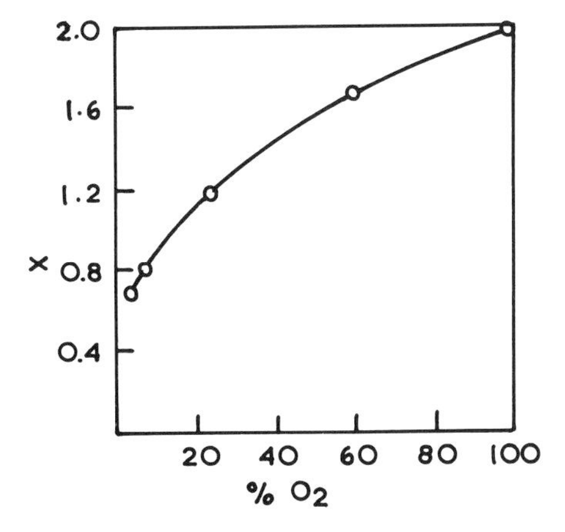

Figure 2.41 The stoichiometry factor 'x' of SnO_x films versus oxygen concentration in the Ar–O_2 reactive mixture (from [117]).

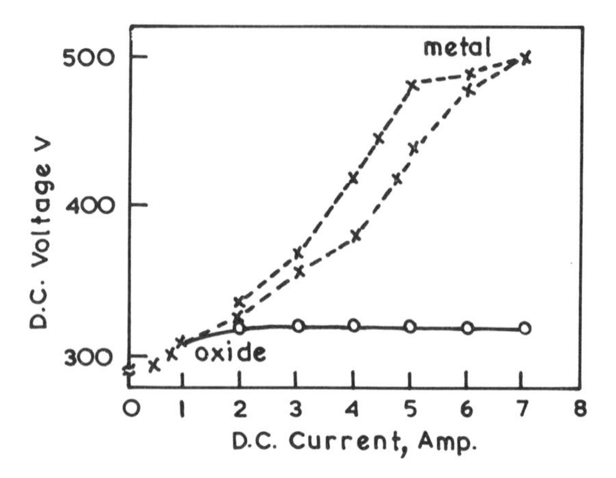

Figure 2.42 Voltage–current characteristic for high rate sputtering of Sn:SnO_2 (– – – –, $p(O_2) = 2.4 \times 10^{-3}$ mbar (1.8×10^{-3} Torr); ———, $p(O_2) = 4.8 \times 10^{-3}$ mbar (3.6×10^{-3} Torr)) (from [119]).

flow, the transition from oxide layers to metallic layers is indicated by a change in the discharge voltage. Munz *et al* [119] further observed that the deposition rate of oxide layers depends on the oxygen partial pressure. For low oxygen pressure, i.e. up to 4×10^{-3} to 5×10^{-3} mbar (3–3.73×10^{-3} Torr), the transition process from metal to metal oxide is prominent and the experimental conditions are not very stable. Above an oxygen partial pressure of 5×10^{-3} mbar (3.75×10^{-3} Torr), the SnO_2 films deposited are fully oxidized and transparent. The deposition rate, however, drastically reduces from 120 Å s^{-1} for metallic layers to 25 Å s^{-1} for oxide layers.

(*ii*) *Sputtering of tin oxide targets*. Tin oxide films have also been prepared by sputtering a tin oxide target rather than a metallic tin target. However, the presence of oxygen is essential in order to have predominantly SnO_2 phase instead of SnO in these films [121–124]. The incorporation of oxygen at lattice sites helps in the formation of crystalline SnO_2 phases and improves crystallinity. De and Ray [121] observed that the SnO_x films grown by RF magnetron sputtering, at lower substrate temperatures and in a pure argon plasma, always have a dominant SnO phase. Introduction of oxygen and an increase of substrate temperature enhances the growth of SnO_2 phase. Typically for an oxygen partial pressure 1.25×10^{-3} Torr and a substrate temperature of 300 °C, tin oxide films have a predominant SnO_2 phase. The grain size of the deposited films at 300 °C is 80–100 Å, which increases to 140–200 Å for films grown at 450 °C. Muranaka *et al* [112], however, observed that SnO_2 films are produced only at high substrate temperatures

(420 °C). Below 420 °C the films are either amorphous or contain α-SnO and β-Sn phases.

The rate of deposition is a strong function of RF power and the total pressure (p) of argon and oxygen during sputtering. Croitoru *et al* [122] studied such effects for SnO_2 films prepared by an RF sputtering technique using a SnO_2 target. The argon pressure during sputtering was varied between 0.67 and 6.70 Pa (5.025 and 50.25×10^{-3} Torr). The RF power was varied between 300 and 800 W. A variation in the amount of oxygen in the SnO_2 films was achieved through reactive sputtering employing oxygen partial pressures $p(O_2)$ of 0.067 and 0.134 Pa (5.025×10^{-4} and 1.005×10^{-3} Torr). The variation of deposition rate as a function of RF power is shown in figure 2.43. For total pressure $p = 1.34$ Pa (1.005×10^{-2} Torr), the rate of deposition varies linearly with RF power, irrespective of oxygen partial pressure. For $p = 6.7$ Pa (5.025×10^{-2} Torr) and $p(O_2) = 0.134$ Pa (1.005×10^{-3} Torr), the dependence of the rate of deposition on RF power is, however, not linear. Auger studies show that x in SnO_x is almost unity and is independent of RF power for films deposited in a pure argon atmosphere. The value of x, however, changes with the RF power for both $p(O_2) = 0.067$ Pa (5.025×10^{-4} Torr) and $p(O_2) = 0.134$ Pa (1.005×10^{-3} Torr) and tends to approach a value of 2 for the latter case for RF power of about 0.8 kW. These results are shown in figure 2.44. Croitoru

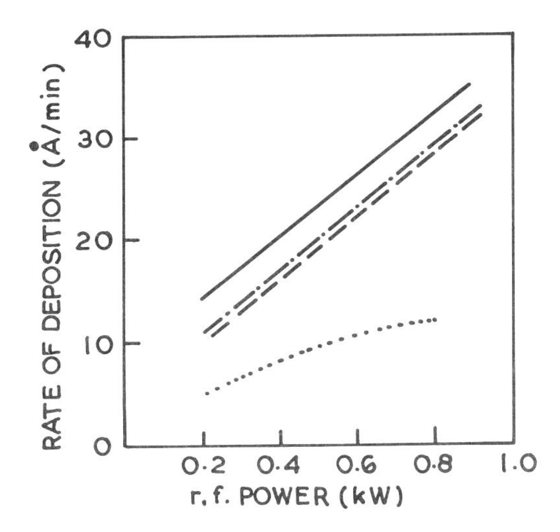

Figure 2.43 Rate of deposition versus RF power for $p = 1.34$ Pa (1.005×10^{-2} Torr) (———, –·–·–, – – – –); for $p = 6.70$ Pa (5.025×10^{-2} Torr) (· · · · ·): ———, $p(O_2) = 0$ Pa; –·–·–, $p(O_2) = 0.067$ Pa (5.025×10^{-4} Torr); – – – –, · · · · ·, $p(O_2) = 0.134$ Pa (1.005×10^{-3} Torr) (from [122]).

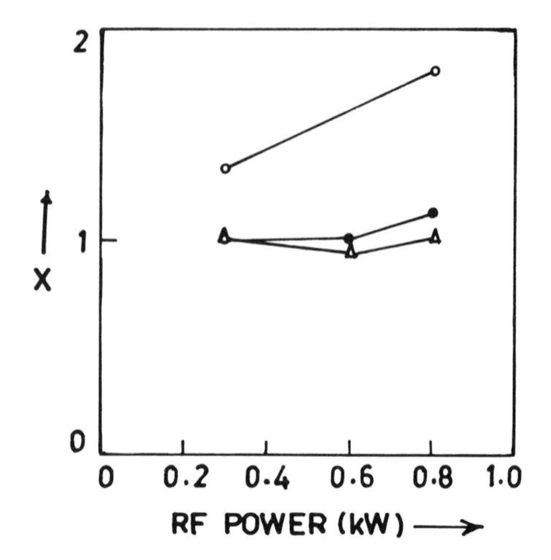

Figure 2.44 Variation of x in SnO_x as a function of RF power for total pressure $p = 1.34$ Pa (1.005×10^{-2} Torr); △, $p(O_2) = 0$ Pa; ●, $p(O_2) = 0.067$ Pa (5.025×10^{-4} Torr); ○, $p(O_2) = 0.134$ Pa (1.005×10^{-3} Torr) (from [122]).

et al [123] also observed that the density of RF sputtered films is a strong function of oxygen content in the sputtering system. For low values of oxygen partial pressure, the density of the films is as low as 4.2 g cm^{-3} as compared to 7.0 g cm^{-3} for bulk SnO_2. Increase of oxygen partial pressure, however, increases the density of the films, but density never approaches a value equal to that observed for bulk samples.

(b) *Doped*

Antimony-doped SnO_2 films have been prepared by reactive sputtering using either an Sn–Sb target or a SnO_2–Sb_2O_3 target [116,131–139]. In general the quality of the films can be improved by post-deposition annealing of as-sputtered films in the temperature range 400–600 °C for a time ranging from 15–60 min. Long-term annealing, however, makes the films insulating. Vaynshteyn [131] observed that the best quality Sb-doped SnO_2 films can be sputtered using a Sn cathode with 10 at% Sb. The other sputtering parameters are: voltage, 3.6 kV; current, 0.4–0.5 mA cm^{-2}; total pressure, 40–80 mTorr with 60% O_2 in Ar; and substrate temperature of 300 °C. Lehmann and Widmer [132] also observed that 10 mol% of Sb_2O_3 in SnO_2 produces optimum films using RF sputtering. As mentioned earlier, in RF sputtering, a substantial amount of resputtering can occur, including preferential resputtering of growing film [133]. Since the sputtering yield of Sb is about 2.5 times higher than that of Sn, the target composition should ordinarily contain excess Sb to compensate for resputtering at the substrate.

Sabnis and Moldoran [134] also observed that the dopant concentration in the films is considerably less than that in the target.

Gas composition influences the structural properties of Sb-doped SnO_2 films. Suzuki and Mizuhashi [135] observed that the films prepared in pure argon at a high rate of 1000 Å min^{-1} using RF magnetron sputtering are yellowish brown, which is the characteristic colour of SnO. The presence of oxygen in the deposition process, however, produces glass-like transparent films. X-ray studies indicate that the films grown in pure argon have a polycrystalline structure with the (1 1 1) plane parallel to the substrate. This (1 1 1) line diminishes rapidly with increasing oxygen gas concentration and disappears almost completely at 10% oxygen. With increasing oxygen concentration, the (1 0 1) line starts appearing and becomes very strong at 10% oxygen level. At 20% oxygen, this (1 0 1) line disappears and instead (1 1 0) and (2 1 1) diffraction lines start appearing. The presence of the (1 0 1) diffraction line in $Ar:O_2$ (90:10) atmosphere has not been observed by other workers. Vossen and Poliniak [136] reported the (0 0 2) orientation while Hecq and Portier [137], Lehmann and Widmer [132] and Tohda *et al* [138] reported the (1 1 0) orientation. It seems that the preferred (1 0 1) orientation may be a characteristic structure of magnetron sputtering. It may also be mentioned that no antimony compound is found in these Sb-doped SnO_2 films, indicating that complete substitution of Sb in the SnO_2 lattice is achieved during deposition [135, 136].

2.3.5 SnO_2: other techniques

An electroless deposition technique has been used to produce thin layers of undoped and Sb-doped SnO_2 [6, 31]. For the growth of SnO_2 films [31], a solution of 1 mol $SnCl_4$: 20 cm^3; 6 mol NH_4F: 25 cm^3; and 0.1 mol $AgNO_3$: 0.01 cm^3 in 100 cm^3 water is prepared and the substrates are immersed in it. An appropriate amount of 1 mol NaOH is added to adjust the pH value between 7.5 and 8.5. After 20–40 min the substrates are taken out, washed, dried and heated in vacuum at 170 °C for 2–3 h. For antimony-doped films 1 mol $SbCl_3$ (0.2–2.0 cm^3) is added in the solution. Figure 2.45 shows the variation of antimony concentration in the films with antimony concentration in the solution. It is clearly evident that all of the antimony does not go into the films because some of the antimony ions, available in the bulk of the solution, react with anions (OH^- ions) and precipitate as antimony hydroxide. X-ray diffraction studies reveal that the films are polycrystalline and the average grain sizes for SnO_2 and $Sb:SnO_2$ are 300 and 650 Å, respectively. Raviendra and Sharma [31] observed a resistivity $\sim 10^{-2}$ Ω cm and transmission ~80% in their films.

The sol–gel technique has been applied by a few workers [140–144] to grow thin films of SnO_2. Stannous chloride is dissolved in ethyl or isopropyl

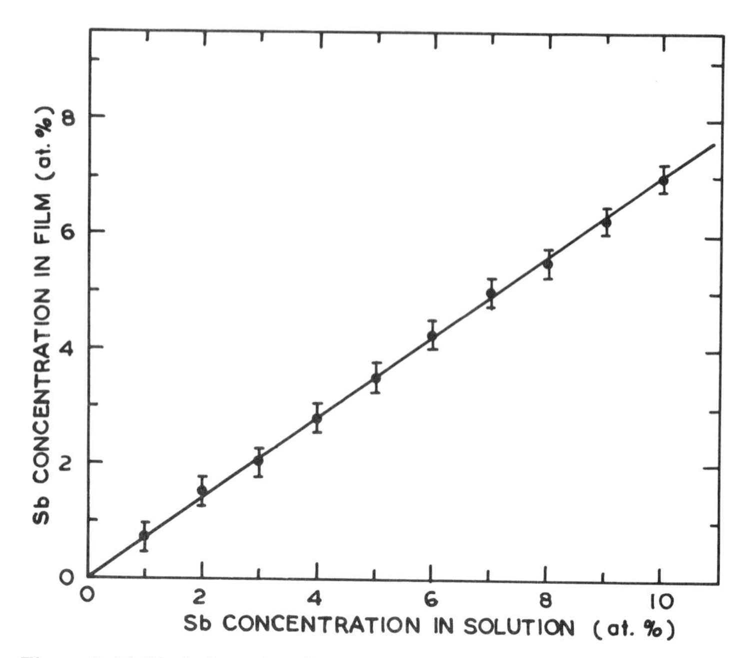

Figure 2.45 Variation of antimony concentration in film with antimony concentration in solution for SnO_2 films grown using an electroless deposition technique (from [31]).

alcohol. The solution is well stirred, refluxed for one hour and aged for another two hours. After hydrolysis, the substrates to be coated are dipped in the solution and pulled out at constant speed. Firing at 500 °C gives transparent conducting coatings.

The laser evaporation deposition technique has been used to prepare conducting and transparent tin oxide films at room temperature using a SnO_2 powder pressed target [36]. A Nd:glass laser (1.06 μm) operating at 1–10 pulses per second and having a pulse width of 300 μs is focused onto the target with an energy density 20–150 J cm^{-2}. The base pressure in the system is 5×10^{-7} Torr. X-ray diffraction studies reveal that the films have preferred orientations for the (2 0 0) plane. Typically the resistivity and transmittance of these films are $3 \times 10^{-3}\ \Omega$ cm and 75%, respectively. The average grain size in the films is 2000 Å. Kunz *et al* [38] used an excimer laser (Ar–F) to grow SnO_2 films by photochemical reaction of $SnCl_4$ and N_2O vapour. The reactions involved are

$$SnCl_4 + h\nu(6.42\ \text{eV}) \rightarrow SnCl_3 + Cl$$

$$N_2O + h\nu(6.42\ \text{eV}) \rightarrow N_2 + O.$$

The laser energy density determines whether SnO_2 or $SnOCl_2$ is formed as a result of reaction between $SnCl_3$ and O. SnO_2 is formed only when the energy density exceeds 20 mJ cm^{-2}. Highly transparent films with resistivity of 4×10^{-2} Ω cm have been produced.

The anodization technique has been used to grow SnO_2 films [115, 147]. The anode is either a Sn sheet or a vacuum-deposited Sn film, while the cathode is either aluminium or a nickel plate. Giani and Kelly [115] used 330 g l^{-1} of ammonium pentaborate in ethylene glycol and water as an electrolyte solution. The films were formed at a voltage of 50 V and current of 10 mA cm^{-2}. The effect of adding water to the anodizing solution has been systematically studied. Electrolyte without any water yields opaque white films. The increase of water in the electrolyte slows down the process but the quality of the films improves. The films formed with water contents of 300 ml l^{-1} show interference colours and are the best films. Further addition of water again results in opaque films. The typical decrease in anodizing current with increase of anodizing time is shown in figure 2.46. X-ray studies show that the films are continuous and polycrystalline in nature. Annealing of these films at 200–300 °C improves the grain size and crystallinity. The bath used by Dhar *et al* [147] for tin anodization contained 1.5 g of Na salt of EDTA in 100 ml of CH_3OH and 100 ml of distilled water. Anodization was carried out at 30 °C. An average current density of the order of 3 mA cm^{-2} was found to be satisfactory for the conversion of Sn to SnO_2, yielding a completely transparent layer. However, at higher current densities, instead of being transparent, the tin layer was converted to a brownish mass, probably of SnO. Typical values of resistivity and transmission in these films were 10^{-2} Ω cm and 90% respectively.

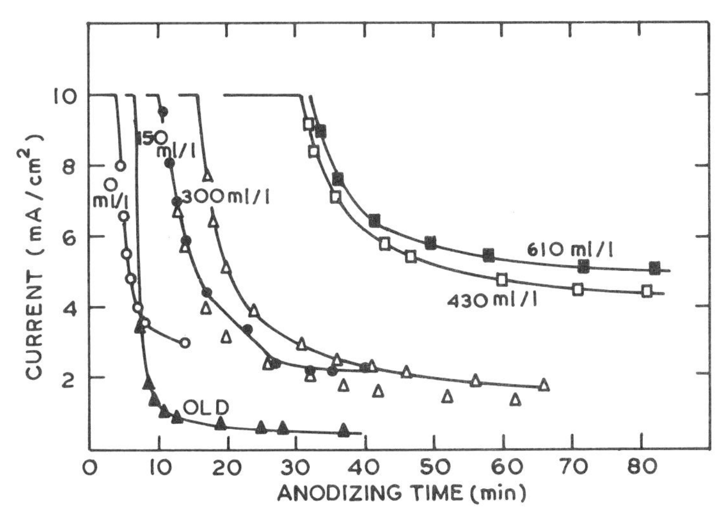

Figure 2.46 Current–time curves for anodizing Sn when both current and voltage are present (from [115]).

2.3.6 SnO_2: technique comparison

In this section, a comparison is made between the various growth techniques in relation to different growth parameters and electrical and optical properties. Table 2.1 lists the results for undoped and doped SnO_2 films prepared by different techniques. It can be seen that the CVD technique requires deposition temperatures of 400–600 °C, whereas a relatively lower temperature is sufficient in the case of the spray process. On the other hand, it is possible to grow films at room temperature using a sputtering and laser-assisted deposition technique. Figure 2.47 gives a graphical representation of the transparent conducting properties of undoped and doped SnO_2 films. Results on films with resistivity greater than 10^{-2} Ω cm and/or transmission less than 70% have not been included in this graph. It is quite evident that, compared to undoped SnO_2 films, doped SnO_2 films, particularly fluorine doped, prepared by a spray pyrolysis technique have better electrical and optical properties. It should be mentioned that arsenic-doped SnO_2 films seem to have comparable properties, but due to the limited available data, it is difficult to assess their suitability for transparent conducting applications.

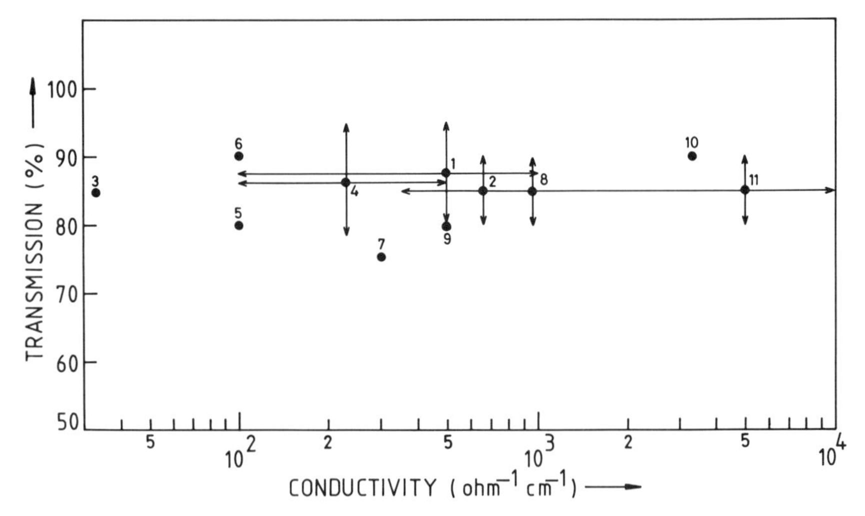

Figure 2.47 Graphical representation of transparent conducting properties of doped and undoped SnO_2 films grown using different techniques. (a) Undoped: (1) CVD, (2) spray, (3) evaporation, (4) sputtering, (5) electroless, (6) anodization, (7) laser. (b) Doped: (8) Sb–SnO_2: spray, (9) Sb–SnO_2: sputtering, (10) F–SnO_2: CVD, (11) F–SnO_2: spray. Bar shows the variation in the compared data.

2.4 Growth of Indium Oxide Films

Compared with SnO_2 films, very little work has been reported on the growth of In_2O_3 films. The techniques that have been employed to deposit indium oxide films include CVD, spray, sputtering, etc. [16, 20, 21, 28, 37, 114, 149–184]. The effects of various deposition parameters on the properties of these films are discussed in this section.

2.4.1 In_2O_3: CVD

Since the cost of indium halides and organometallic compounds of indium is very high, the CVD process has not been employed very often. Earlier work [151, 152] on the growth of In_2O_3 films by a CVD technique using indium acetyl acetonate $In(C_5H_7O_2)$ indicates that the films produced by this process are highly resistive ($\sim 10^{12}\ \Omega$ cm). However, recently Maruyama and Fukui [153, 154] have grown low resistance ($\sim 10^{-2}$–$10^{-3}\ \Omega$ cm) indium oxide films using a CVD technique. Films prepared using indium acetylacetonate and oxygen at deposition temperatures from 350–500 °C are transparent and conducting. In the absence of oxygen, transparent insulating films are obtained. On the other hand, films grown by thermal decomposition of 2-ethylhexanoate in an inert gas atmosphere (N_2) at a deposition temperature between 350 and 500 °C are conducting and transparent. In this case, there is no need to supply oxygen to obtain transparent conducting films. All the films grown are polycrystalline in nature. Figure 2.48 shows a typical XRD pattern of In_2O_3 films grown using indium acetylacetonate on borosilicate glass. A preferred (2 2 2) plane orientation is quite evident in this figure. When films are grown using 2-ethylhexanoate $In(C_7H_{15}\text{–}COO)_3$ additional peaks corresponding to (2 1 1) and (6 2 2) are also observed in

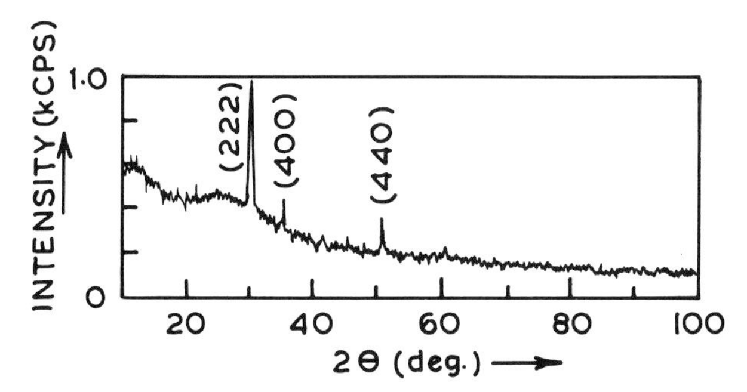

Figure 2.48 XRD diffraction pattern of In_2O_3 film on borosilicate glass grown by a CVD technique (from [153]).

Table 2.1 Properties of SnO_2 films grown using different techniques.

Process	Substrate temperature (°C)	Rate (Å min^{-1})	Resistivity (Ω cm)	Transmission (%)	Remarks	Reference
SnO_2						
CVD	250–400	—	$10–10^{-3}$	80–95	$SnCl_4:H_2O$	50
			$10^2–10^{-2}$		$SnCl_4:H_2O_2$	
CVD	600	—	10^{-2}	80	Process suitable for glass ribbons	51
CVD	480–680	—	$4 \times 10^{-2}–10^{-3}$	~90	—	17
CVD	450–600	—	0.6	84	Low pressure CVD compatible with silicon technology	58
CVD	450	300	3×10^{-2}	95	—	43
CVD	420	—	5×10^{-3}	>90	$O_2 + H_2O$ vapour used in CVD	46
Spray	410	—	10^{-2}	80–85	—	77
Spray	400	—	10^{-3}	90	—	95
Spray	390	—	2.6×10^{-3}	80	—	71
Reactive evaporation	480	60–120	0.03–40	>85	$O_2 = 3–5 \times 10^{-3}$ Torr	111
Sputtering of tin target	<100	66	7.2×10^{-2}	—	Sputtered in Ar/O_2 Ga^+ ion implantation and air annealed	118
Sputtering of tin target	(Room)	1800	3×10^{-3}	75	Ar/O_2 mixture	126
Sputtering of tin target	450	70	6.1×10^{-3}	95	$O_2 = 1.25 \times 10^{-3}$ Torr	121
Sputtering magnetron	—	—	10^{-2}	~80	$Ar–O_2$	127
Anodization	—	—	10^{-2}	~90	—	147
Laser evaporation	(Room)	24 Å pulse	3×10^{-3}	75	Nd:Glass laser 300 μs	36

Electroless	—	—	10^{-2}	80	—	31
SnO_2:Sb						
Spray	346–446	250–450	8×10^{-4}	80	—	64
Spray	600–680	—	$65\ \Omega\ \square^{-1}$ [a]	~90	—	79
Spray	500–700	—	7.5×10^{-3}	90	—	83
Spray	600	—	$\sim 10^{-3}$	85	—	80
Spray	540	—	2×10^{-3}	80	1.4 mol% Sb	68
Spray	440	—	10^{-3}	80	1 mol% $SbCl_3$	71
Sputtering magnetron	400	—	2×10^{-3}	80	10% O_2–Ar	135
SnO_2:F						
CVD	570	3600	3.3×10^{-4}	90	—	66
Spray	400	—	4.6×10^{-4}	85	F/Sn = 0.5 at%	16
Spray	400	—	$\sim 5 \times 10^{-4}$	~85	—	81
Spray	400	—	5.4×10^{-4}	90	—	88
Spray	400–500	—	9×10^{-3}	91	—	89
Spray	400	—	1.1×10^{-3}	88	3 mol% NH_4F	71
Spray	350–500	—	10^{-4}	85	—	
Spray	450	—	4.5×10^{-4}	90	10% F	85
Spray	450	200	4.3×10^{-4}	80	Air as carrier gas	148
Spray	450	250	9.0×10^{-4}	90	N_2 as carrier gas	148
Spray (ultrasonic)	300–450	—	5×10^{-4}	~80	F/Sn = 10 wt%	14
SnO_2:P						
CVD	450	300	5×10^{-3}	95	—	43
CVD	400	—	7.5×10^{-4}	83	—	52
SnO_2:Mo						
Reactive evaporation	480	60–120	3×10^{-3} to 2×10^{-2}	>85	$O_2 = 3–5 \times 10^{-3}$ Torr	111

[a]Sheet resistance.

XRD studies [154]. The transmittance observed in these CVD-grown In_2O_3 films is ~85%.

Recently, transparent conducting films of fluorine-doped and sulphur-doped indium oxide have been prepared using a CVD technique [155, 156]. For fluorine-doped In_2O_3 films, indium 2-ethylhexanoate and indium fluoride have been used as source materials [155]. The incorporation of fluorine into the films increases the growth rate, conductivity and transmission. Typically the resistivity and transmission increase from 10^{-1} Ω cm and 82% to 2.89×10^{-4} Ω cm and 85%, respectively on fluorine doping. Figure 2.49 shows the variation of doping rate with reaction temperature. The deposition rate of In_2O_3:F is about three times that of In_2O_3 films. Nomura *et al* [156] prepared sulphur-doped In_2O_3 films by vapour phase decomposition of indium thiolate. The indium thiolate readily decomposes into indium sulphide at 200 °C and this indium sulphide oxidizes at 300 °C to indium oxide. Resistivity and transmission in such films are 5.2×10^4 and 90%, respectively.

2.4.2 In_2O_3: spray pyrolysis

Various systems employed for the growth of transparent conducting films by spray pyrolysis as described earlier in section 2.2.2 can be used for the

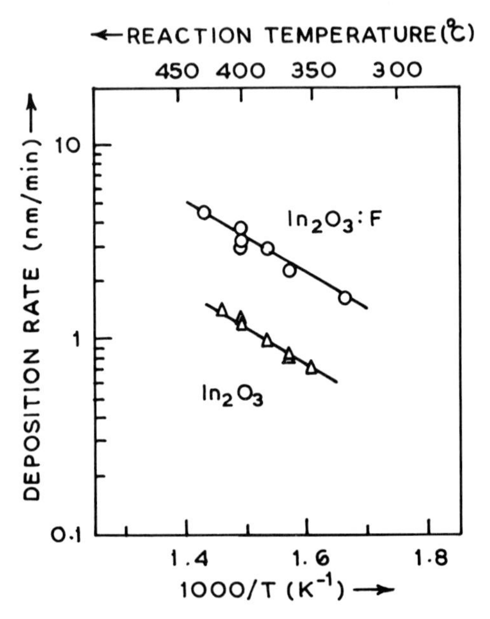

Figure 2.49 Deposition rate versus reaction temperature for In_2O_3 and fluorine-doped In_2O_3 films (from [155]).

growth of In_2O_3 films. Transparent conducting films of indium oxide can be deposited by spraying a solution prepared by dissolving either $InCl_3$, $InCl_2$ or indium acetyl acetonate in a solvent such as butyl acetate, butanol, acetylacetone or methanol-water. The films are generally prepared at substrate temperatures in the range 400–500 °C. Groth [157] prepared In_2O_3 films using solutions of $InCl_3$ in butyl acetate and in butanol. It has been observed that the films are cloudy if the pyrolytic reaction takes place in the hot air layer above the substrate. To avoid cloudiness in the films, the reaction should occur at the substrate. Raza *et al* [158] prepared thin films of indium oxide using an aerosol stream containing an atomic solution of $InCl_3$ in butyl acetate (293.3 g of $InCl_3$:$4H_2O$ dissolved in 1 l of butyl acetate). The solution was sprayed onto glass substrates preheated to a temperature of about 500 °C. The carrier gas used was air and the flow rate of the spray 30 cm^3 min^{-1}. Pommier *et al* [16] used $InCl_3$ solution dissolved in methanol-water to grow In_2O_3 films. Even though $InCl_3$ dissolves completely in methanol, water is added in order to complete the hydrolysis process. The optimized composition of the solution to produce best quality films was 4 g of $InCl_3$ dissolved in 100 cm^3 of methanol.

A fogging system [159] as shown in figure 2.50 has also been used to grow

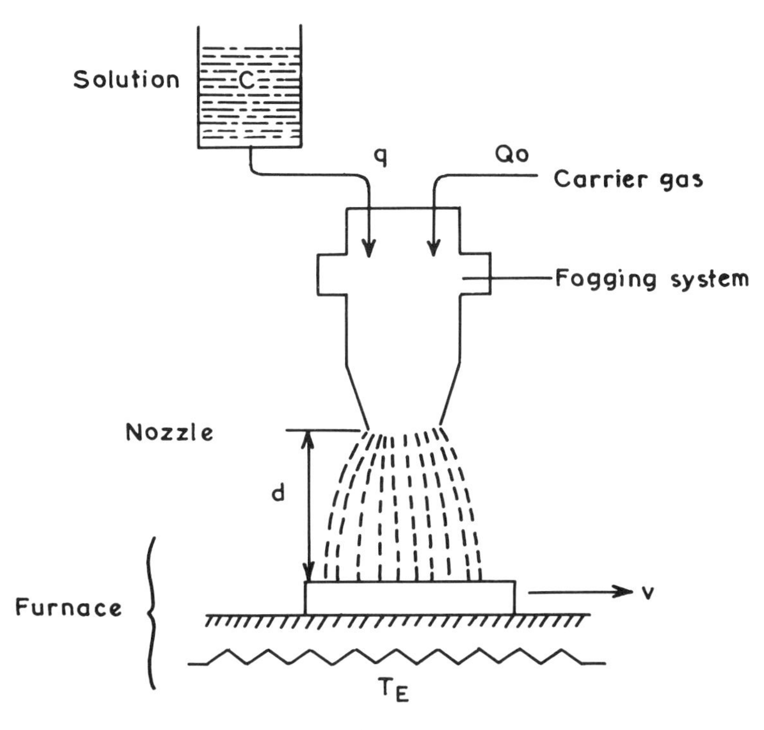

Figure 2.50 Schematic diagram of equipment for chemical spray deposition (from [159]).

indium oxide films. The main advantage of this system is that films can be grown at relatively lower temperatures. Fogging systems, as shown in the figure, can be of different types: a pneumatic atomizer or a piezoelectric transducer. Of the two, the one involving a piezoelectric transducer is considered better for the following reasons: (i) the carrier gas flow can be set at very low values, thus minimizing the cooling effect on the substrate; and (ii) at a frequency of 1 MHz, the droplet diameter is very small, typically between 2 and 8 μm. This results in good quality, homogeneous films.

2.4.3 In_2O_3: vacuum evaporation

Vacuum evaporation is the most commonly used technique for the growth of indium oxide films [161–169]. Films can be grown from an indium source by reactive evaporation or by the annealing in oxygen of vacuum-deposited indium films. Pan and Ma [161, 162] produced transparent conducting films of In_2O_3 using thermal evaporation of 90% In_2O_3–10% In placed in an alumina crucible. It was observed that incorporation of metallic indium in the evaporation source not only significantly enhanced the rate of evaporation, but also significantly improved the electrical and optical properties of the films. Initially the vacuum chamber was evacuated to a base pressure of 1×10^{-6} Torr. Pure O_2 was then admitted to the chamber to a pressure of 2×10^{-4} Torr. The substrates were maintained at a temperature of 320–350 °C. X-ray diffraction data suggest that the films are polycrystalline. The important effect observed in their experiment is that when films are grown from a crucible and a Ta or W heater, the conductivity of the films is an order less, which may be due to the possible chemical reaction between the indium oxide vapour and the Ta or W heater. It is suggested that either a proper combination of crucible and heater element (single unit) or a crucible taller then the top edge of the heater should be used to avoid this problem. The resistivity and transmission of the films grown under optimum conditions were 1.8×10^{-4} Ω cm and >90%, respectively.

A novel activated reactive evaporation technique has been employed in order to enhance the reactivity of indium vapours with oxygen [20,163]. Such an increase in reactivity is achieved by generating a dense plasma using a thoriated tungsten emitter and a low voltage assembly. Nath and Bunshah [20] used this technique in which a resistively heated tungsten/tantalum source was used for evaporating indium. The substrates were maintained at a temperature of 350 °C. A mixture of Ar–15% O_2 was used as a reactive gas and the pressure during deposition was 1×10^{-4} Torr. X-ray diffractometer measurements on In_2O_3 showed a preferred orientation of (1 0 0) direction. Typical values of resistivity and transmission observed in these films were 3.2×10^{-4} Ω cm and >96%, respectively. Lau and Fonash [163], however, observed that the substrate temperature should be less than 200 °C, higher temperature results in high resistivity films. The indium oxide films prepared by them were, however, stable at room temperatures but degraded after vacuum or hydrogen annealing at high temperatures.

Growth of indium oxide films by annealing the vacuum-evaporated indium films has not been very successful. Sundaram and Bhagavat [164] used this method to grow In_2O_3 films by evaporating indium and subsequently annealing the films in oxygen atmosphere at 300 °C and 400 °C for a period of 0.5 to 4 h, depending on the thickness of the films. Complete oxidation of thick films (5000 Å) has never been achieved even after heating at 400 °C for a period of 6 h. This may be due to the decrease in diffusion coefficient of oxygen through the oxide layers at that temperature. X-ray diffraction studies reveal that films oxidized at 300 °C are polycrystalline and the crystallinity and grain size improve with oxidation temperature.

2.4.4 In_2O_3: sputtering

Initial experiments on the growth of pure indium oxide films using reactive sputtering resulted in nearly stoichiometric films, which were highly resistive and as such, were not useful as transparent conductors. However, later efforts using this technique [21, 28, 114, 170–180] resulted in the growth of conducting indium oxide films.

Indium oxide films can be deposited by sputtering either a metallic indium target or an indium oxide target in the presence of oxygen. The stoichiometry and rate of sputtering are strongly dependent on the oxygen content in the system. In reactive sputtering of a metallic target, the surface of a metal target is strongly oxidized, but further sputtering causes dissociation and extraction of the oxygen and metal layer by layer. The oxygen atoms extracted from the surface may be substituted by others from the environment. For this reason the sputtering rate of the target decreases significantly. The variation of rate of sputtering as a function of oxygen concentration for In_2O_3 films is shown in figure 2.51. This figure also shows the deviation in the film stoichiometry η as a function of oxygen concentration. The deviation of film stoichiometry has been determined by comparing with a standard film made with an oxygen concentration sufficient for complete oxidation. The lowest oxygen concentration for which deviation from film stoichiometry does not occur is 8%. These workers also calculated the concentration of oxygen in the gas mixture as a function of the discharge power for various working pressures and the results are shown in figure 2.52. It is obvious from this figure that, for a given oxygen concentration in the gas mixture, if the sputtering power is greater than a certain value, it is impossible to completely oxidize the sputtered indium.

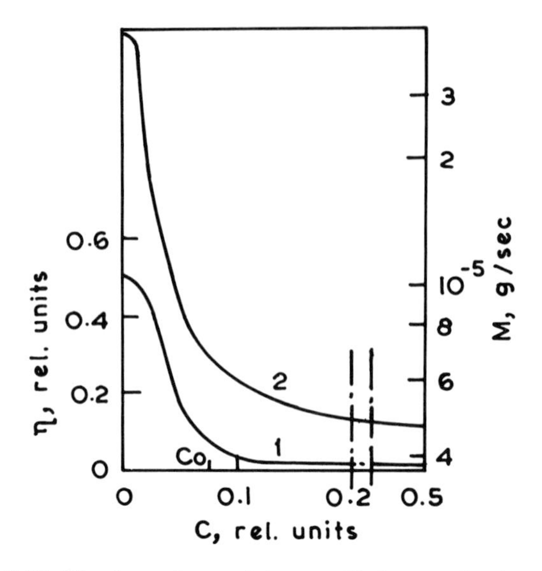

Figure 2.51 The dependence of the rate *M* of sputtering from an indium cathode under constant discharge power (curve 1) and the dependence of the relative normalized deviation of the indium oxide film composition from stoichiometry (curve 2) on the relative oxygen volume *C* in the gas mixture (from [170]).

In_2O_3 films have also been grown using In_2O_3 targets [171, 172]. The effect of substrate bias on the deposition rate and structural properties of RF sputtered films was studied by Wickersham and Greene [171] and Morris *et al* [172]. The base pressure in their system was typically 2×10^{-6} Torr and Ar pressure was maintained at 1.5×10^{-2} Torr during deposition. The RF target bias was kept at 750 V while the applied RF substrate bias was varied between 0 and 150 V. Figure 2.53 shows the variation of deposition rate as a function of applied substrate bias. The deposition rate initially increases with increase of substrate bias due to the injection of additional secondary electrons from the substrate into the discharge. However, as substrate bias is further increased, the deposition rate starts decreasing due to resputtering from the substrate. X-ray diffraction and electron microscope studies indicate that the effect of substrate bias on the microstructure of the films is quite significant. The degree of (1 0 0) preferred orientation increases and the density of structural defects, such as voids, decreases with the increase of negative value of substrate bias up to 60 V and remains constant thereafter. The grain size of these films also increases with increasing substrate bias. Morris *et al* [172] extended the RF magnetron sputtering technique to deposit indium oxide films on polyester substrates. The films were polycrystalline but the degree of the preferred orientation was a strong function of the

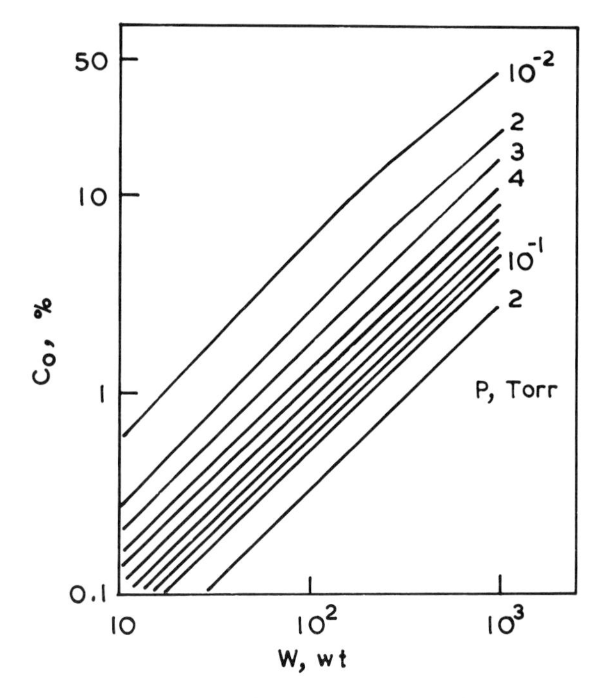

Figure 2.52 The dependence of the values of the critical oxygen concentration in the gas mixture on the discharge power for various working pressures (from [170]).

RF sputtering bias. Under optimum bias conditions, the films had (2 2 2) preferred plane and showed excellent electrical and optical properties [114, 173]. It should be noted that In_2O_3 films produced by other techniques have a variety of preferred orientations, but (2 2 2) orientation is normally produced only by annealing the films at high temperatures (400 °C). The

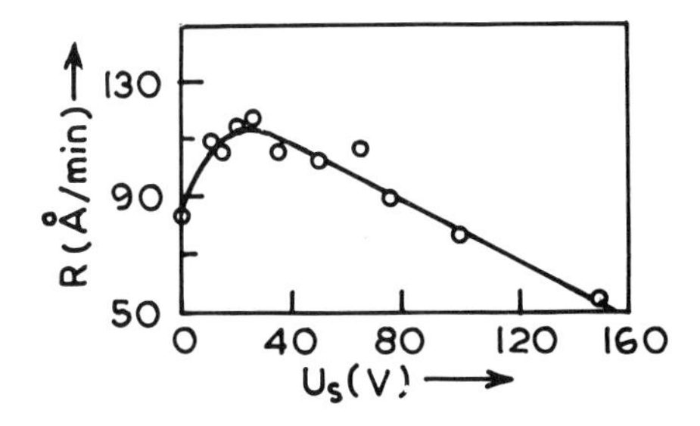

Figure 2.53 In_2O_3 deposition rate (R) at 750 V target bias and 15 mTorr Ar sputtering pressure as a function of applied substrate bias (from [171]).

results confirm that the magnetron sputtering process can produce films at low temperatures with properties otherwise obtainable only at high temperatures.

2.4.5 In_2O_3: ion-assisted deposition

Ion-beam sputtering of an indium metal target in the presence of oxygen has recently been employed to produce amorphous transparent conducting films of indium oxide at room temperature [181–183]. Though the films are amorphous, the electrical and optical properties of these films are comparable to those of polycrystalline indium oxide grown on heated substrates. Typical values of resistivity and transmittance in these films are $5 \times 10^{-4}\ \Omega$ cm and 90%, respectively. The fact that excellent quality films can be produced on unheated substrates implies that indium oxide films prepared by this process have enormous potential for device applications where coatings are required on heat sensitive substrates.

The reactive ion plating technique has been exploited to grow fluorine-doped In_2O_3 films on glass substrates at room temperature (25 °C) [184]. The substrates are cleaned by etching in 30% fluorocarbon–70% O_2 gas mixture by applying an RF glow discharge for 2 min at 100 W RF power (13.56 MHz) and 1 mTorr gas pressure. After the substrates have been etched clean, the O_2 supply is switched off and the partial pressure of CF_4 gas is adjusted to a value between 0.5 and 2.5 mTorr without switching off the discharge. When a dynamic equilibrium is established and the CF_4 pressure is stabilized, the oxygen supply is turned on and the O_2 flow into the chamber is adjusted so that the total pressure is 4 mTorr. The amount of fluorine incorporated into the films depends on the fluorine partial pressure during film deposition. Figure 2.54 shows the variation of fluorine content in

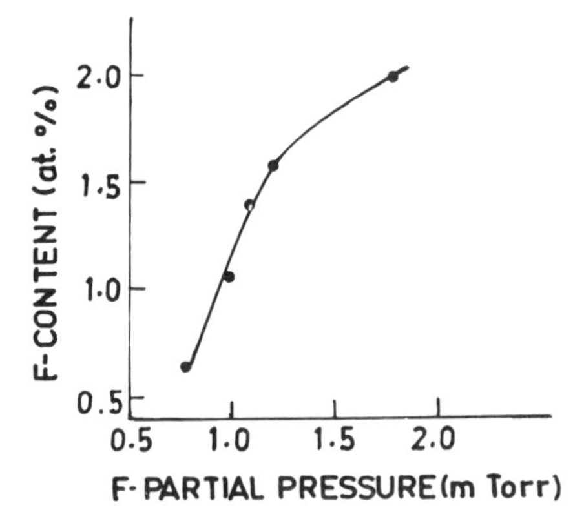

Figure 2.54 Variation of fluorine content in the film versus fluorine partial pressure (from [184]).

the films as a function of fluorine partial pressure. It can be observed that the amount of fluorine incorporated into the films increases linearly with increasing fluorine partial pressure up to 1.2×10^{-3} Torr. Thereafter, the incorporation of fluorine is much less. Typically, the films grown at partial pressure 1.8×10^{-3} Torr exhibit sheet resistivity of 171 $\Omega\,\square^{-1}$ and transmittance of 70%. Annealing of these films in air at 450 °C for one hour results in sheet resistivity of 53 $\Omega\,\square^{-1}$ and transmittance >80%.

2.4.6 In_2O_3: other techniques

Donnelly *et al* [37] deposited In_2O_3 films on quartz, GaAs and InP substrates using excimer laser induced photo-decomposition of $(CH_3)_3InP(CH_3)_3$. A 1930 Å Ar–F excimer laser irradiates the substrates at a glancing angle of 5° with 15 ns, 10 mJ pulses at a 5 Hz repetition rate. The substrate temperature is kept around 300–400 °C. The system is operated under reduced pressure. The reactions involved are

$$(CH_3)_3InP(CH_3)_3 + nh\nu \rightleftarrows (CH_3)_3In + P(CH_3)_3$$

$$(CH_3)_3In + nh\nu \rightarrow In + 3CH_3.$$

The indium is converted to indium oxide at high substrate temperature.

2.4.7 In_2O_3: technique comparison

Table 2.2 presents typical results on indium oxide films grown using different

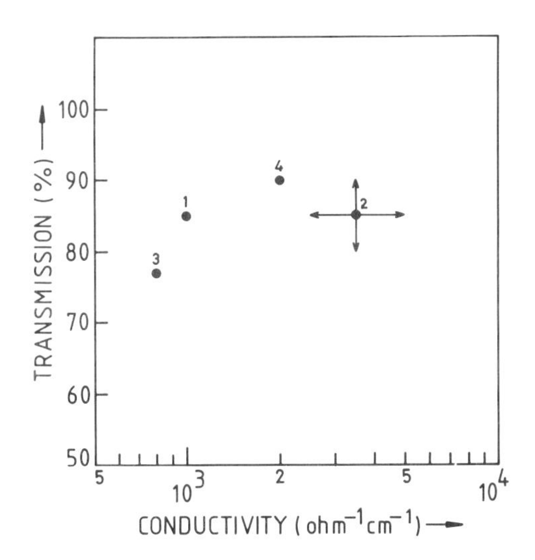

Figure 2.55 Comparison of different growth techniques for indium oxide films. (1) CVD, (2) evaporation, (3) sputtering, (4) ion-assisted. The bars show the variation in data reported by various workers.

Table 2.2 Properties of In_2O_3 films grown using different techniques.

Process	Substrate temperature (°C)	Rate (Å min^{-1})	Resistivity (Ω cm)	Transmission (%)	Remarks	Reference
In_2O_3						
CVD	450	—	$\sim 10^{-3}$	~85	—	153
CVD	527	7980	9.3×10^{-4}	89	—	149
Evaporation	400	—	4×10^{-4}	>80	$O_2 = 5 \times 10^{-5}$ Torr	166
Reactive evaporation	350	400	3.2×10^{-3}	96	Ar–15% O_2	20
Reactive evaporation	350	60	4×10^{-4}	>80	$O_2 = 10^{-4}$ Torr	165
Reactive evaporation	250	—	3×10^{-4}	>90	$O_2 = 5 \times 10^{-4}$ Torr	167
Reactive evaporation	—	60	2.5×10^{-3}	85	Deposited on 10 Å indium film	150
e-beam evaporation	320–350	15–30	2×10^{-4}	>90	In_2O_3 + In $O_2 = 0.5–2 \times 10^{-4}$ Torr	161
e-beam evaporation	300	—	3×10^{-3}	80	$O_2 = 1 \times 10^{-4}$ Torr	169
Sputtering	—	—	1.3×10^{-3}	77	Sputtering in Ar	174
Ion-beam sputtering	—	60	5×10^{-4}	90	$O_2 = 2 \times 10^{-4}$ Torr amorphous	181
In_2O_3:F						
CVD	400	—	2.89×10^{-4}	>85	—	155
In_2O_3:S						
CVD	400	—	5.2×10^{-4}	90	99% Ar–1% O_2 carrier gas	156

techniques. In general, it can be seen that, as compared to SnO_2 films, these films usually have not only better electrical and optical properties but also can be grown at relatively low substrate temperatures. The ion-assisted technique can be used to grow In_2O_3 films of comparable properties even at room temperature. A comparison of different growth techniques is shown in figure 2.55. All the deposition methods seem to be more or less equally suitable.

2.5 Growth of Indium Tin Oxide Films

Numerous deposition techniques have been used to deposit indium tin oxide (ITO) films. Since the electrical and optical properties of these films depend strongly on their structure, tin content, etc., each deposition technique yields films with different properties. In this section, we shall discuss the effect of various deposition parameters, involved in different techniques, on the properties of the films.

2.5.1 ITO: CVD

The CVD technique has not been fully exploited in growing ITO films in spite of the fact that this process does not require a high vacuum and as such it is easier to use for large-scale applications. However, a few workers [153, 154, 185] have used this technique to grow ITO films. Ryabova *et al* [185] reported low resistance (10^{-4} ohm cm) In_2O_3:Sn (ITO) films grown using a CVD technique. The setup used is shown in figure 2.56. The apparatus consists of a quartz tube of diameter 18 mm, in which two separate

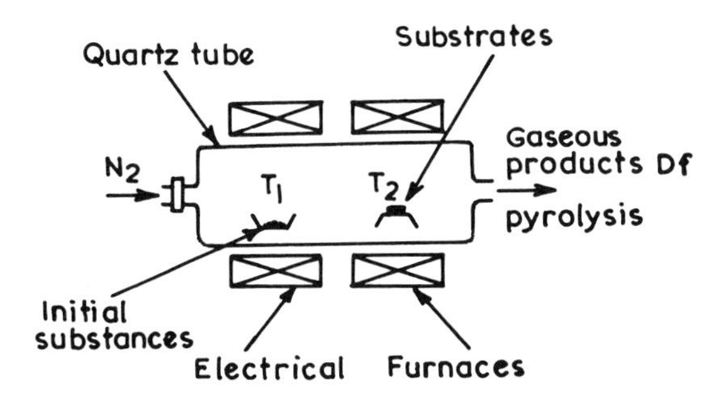

Figure 2.56 Schematic diagram of the experimental setup for the deposition of In_2O_3:Sn films (from [185]).

temperature zones are maintained using resistance furnaces: the evaporation zone ($T_1 = 150$–200 °C) where quartz boats containing reactants are placed, and the reaction zone ($T_2 = 350$–450 °C) where the substrates are located on a rotary holder. Vapours from the reactants are transported into the reaction zone by the carrier gas (N_2) at a flow rate of 450 ml min^{-1}. Indium and tin acetyl acetonates mixed in the ratio 43:57 wt% are used as reactants. The films grown under optimum conditions have an atomic ratio of tin to indium of 0.08. The growth rate is maintained at 10–15 Å min^{-1}. The films have been deposited on various substrates including glass, mica and sapphire. It is observed that the films grown on sapphire are of the best quality, having resistivity 1.6–1.8×10^{-4} Ω cm and transparency of 90–95%. The films annealed in a vacuum of $\sim 10^{-2}$ Torr at 400 °C for 30–45 min exhibit 2 to 3 times higher conductivity, thus making the films more useful.

Recently Maruyama and Fukui [153, 154] prepared ITO films by a CVD technique using indium 2-ethyl hexanoate and tin chloride as source materials. The ratio of tin to indium was varied by varying the flow rate ratio, i.e. the ratio of the flow rate of carrier gas for the tin source to that for the indium source. Figure 2.57 shows the variation of atomic ratio of Sn/In in the films as a function of flow rate ratio of the carrier gas. It can be observed from this figure that the atomic ratio Sn:In increases from 0 to 0.1 with increasing flow rate ratio from 0 to 3. Thus the tin content in the film can be adjusted by changing the flow rate ratio.

Like indium oxide films, ITO films grown by the CVD technique are always polycrystalline in nature. A typical XRD pattern for ITO films grown using indium acetylacetonate and tin acetylacetonate as source materials is given in figure 2.58. The diffraction pattern shows a preferred orientation of the (2 2 2) plane. It should be noted that this preferred orientation is different in the case of ITO films grown using indium 2-ethylhexanoate and

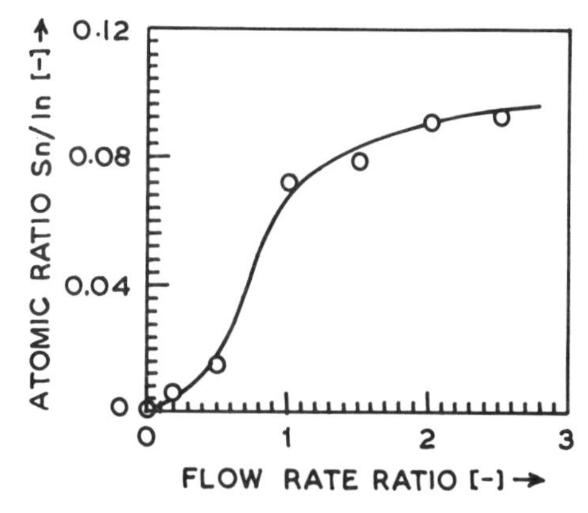

Figure 2.57 Atomic ratio of Sn:In as a function of flow rate of the carrier gas (from [154]).

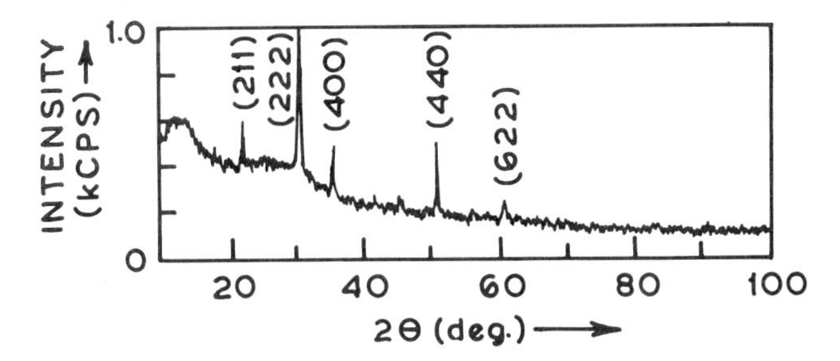

Figure 2.58 X-ray diffraction patterns of ITO films grown using CVD techniques (from [153]).

tin chloride as the starting material. In the latter case the preferred orientation is the (4 0 0) plane [154].

2.5.2 ITO: spray pyrolysis

In the case of ITO films grown by the spray pyrolysis technique, tin doping is performed by adding stannic chloride hydrated in a solution of $InCl_3$. Typical composition [72] of the solution is $InCl_3$: 0.0817; H_2O: 0.4204; CH_3, CH_2OH: 0.4204; $SnCl_4$: $5H_2O$: 0.0024; HCl: 0.0751. The gas flow rate is of the order of $3\,l\,min^{-1}$. The substrate temperature is usually in the range 400–700 °C. The nature of the substrate significantly influences the incorporation of tin in the films. Such effects are shown in figure 2.59. It can be observed that in films grown on silicon substrates the tin concentration is

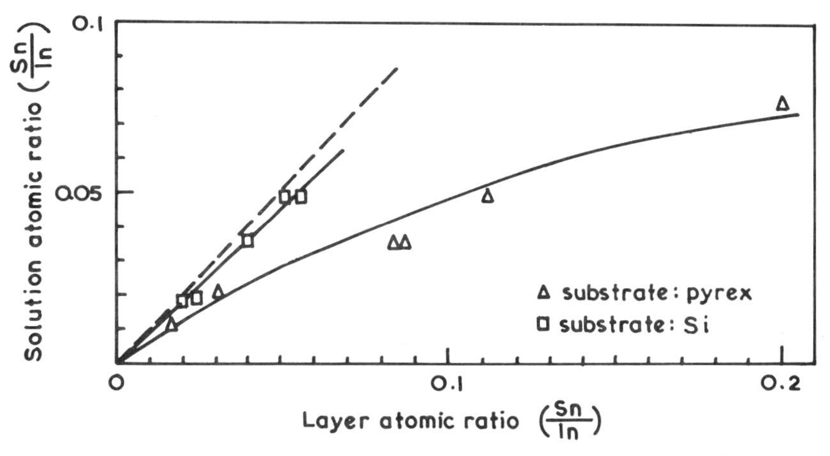

Figure 2.59 Sn/In atomic ratio in the In_2O_3:Sn layers versus Sn/In atomic ratio in the solution (from [72]).

almost identical with that in the solution. However, the concentration of tin in films grown on pyrex is significantly higher than that in the solution. SIMS studies on films grown on borosilicate glass indicate the presence of various other oxides besides In_2O_3, such as $InSiO_3$, AlO, SiO_3 and SiO_2. The depth profiling results on these layers are given in figure 2.60. It is quite evident that the alkali ions and silicon accumulate at the interface of the film and substrate.

Substrate temperature and thickness influence the structural properties of ITO films markedly. Pommier *et al* [16] and Saxena *et al* [186] studied the structural properties of In_2O_3:Sn films as a function of thickness and substrate temperature. A solution of 4 g $InCl_3$ in 100 cm^3 methanol in which Sn was added in the form of $SnCl_4$, such that Sn/In ratio was of the order of 2.5%, was used by Pommier *et al* [16] for depositing In_2O_3:Sn films. Saxena *et al* [186] also used an alcoholic solution of $InCl_3$ with $SnCl_4$ such that the tin concentration was 10 wt%. Spraying was done for a period of 8 s, followed by a pause of 1 min to avoid excessive cooling of the hot substrates. XRD studies and Scanning electron microscopic studies [16, 186] indicate that the

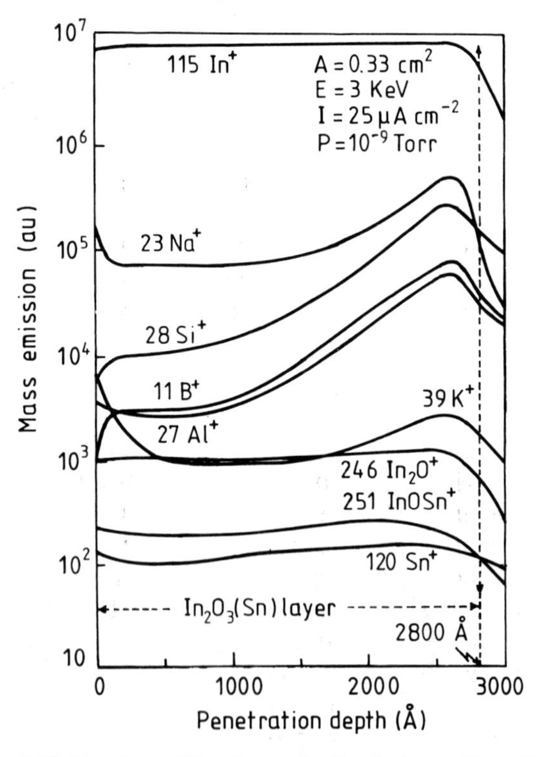

Figure 2.60 Depth profiling for an In_2O_3–Sn layer (from [72]).

grain size and thus the crystallinity of the films improves with increasing thickness. Typically thin films (500 Å) are amorphous in nature whereas thicker films (4000 Å) grown under similar conditions are polycrystalline [186]. XRD studies [16, 79, 83, 100, 186–190] also reveal that an increase in substrate temperature improves the crystallinity significantly. Figure 2.61 shows typical XRD patterns for ITO films grown at different substrate temperatures. Kulaszewicz and co-workers [79, 83, 100, 188] reported that the lattice constants increase linearly as the temperature of spraying is increased. These results are shown in figure 2.62. Over the temperature

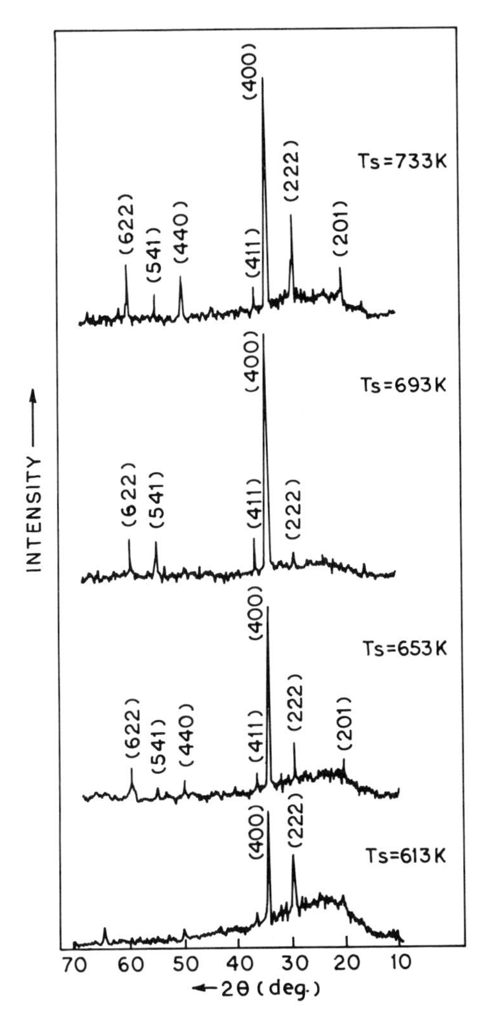

Figure 2.61 XRD patterns for ITO films grown by spray pyrolysis (from [187]).

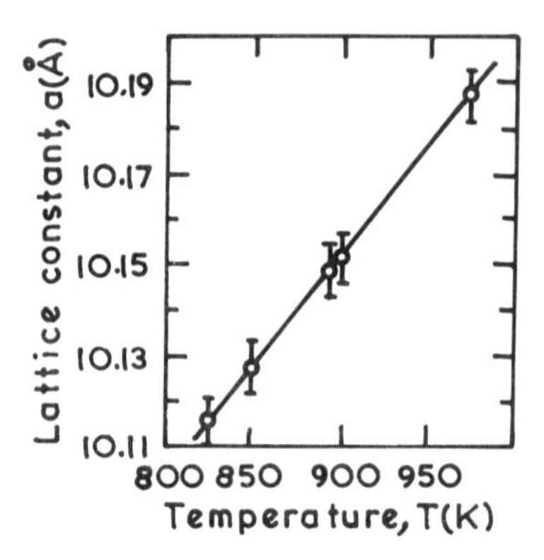

Figure 2.62 Dependence of the lattice constant of In_2O_3:Sn films on the spraying temperature (from [188]).

range 823–973 K, the dependence of lattice constants on the temperature can be represented by the equation:

$$a = 9.7237 + 4.761 \times 10^{-4} T \text{ Å}$$

where T is the temperature.

2.5.3 ITO: vacuum evaporation

The growth techniques for vacuum evaporation of tin oxide or indium oxide films can also be used for the growth of indium tin oxide films. Most workers have deposited these films by reactively evaporating either metallic alloy or an oxide mixture [20, 166, 191, 193–202]. Christian and Shatynski [191] grew ITO films using a reactive evaporation technique from alloys of In–5 wt% Sn. The substrate temperature was varied in the range 25–300 °C. Scanning Auger microscopy indicated the presence of indium, tin, oxygen with small amounts of other impurities such as carbon, sodium, sulphur, etc. The typical composition of 700 Å thick ITO films evaporated at an oxygen partial pressure of 5×10^{-4} Torr was found to be 42 wt% In, 30 wt% Sn, 9 wt% O_2, 0.04 wt% carbon, 0.007 wt% sulphur, 0.003 wt% Na and 0.008 wt% Cl. TEM studies indicated that the ITO films grown in a vacuum of 1×10^{-7} Torr (without oxygen) were discontinuous and had poor electrical and optical properties. Films grown in a vacuum 5×10^{-4} Torr (with oxygen) were found to be continuous and had good electrical and optical properties. Instead of using the In–Sn alloy, Noguchi and Sakata [192] grew ITO films by co-evaporation of In and Sn from two beryllia crucibles. The growth rate was maintained at 3 Å s^{-1}. Yao *et al* [193] used sequential reactive evaporation of indium and tin followed by air annealing. The advantage of this method is that the composition of the films can be controlled accurately. Electrical resistivity $4.0 \times 10^{-3}\ \Omega$ cm and transparency >90% have been achieved.

Films of ITO grown by vacuum evaporation using oxide mixtures are generally deficient in oxygen. An oxygen partial pressure of 10^{-4} Torr is required in order to have transparent films. Mizuhashi [166] vacuum deposited ITO films from two independently controlled beryllia crucibles containing SnO_2 and In_2O_3. Evaporation was carried out in the presence of oxygen and the partial pressure of the oxygen was 10^{-4} Torr. These independently controlled crucibles were used to grow ITO films of controlled composition. A tantalum foil heater was used to avoid contamination of the indium oxide films because tantalum oxide has a fairly low vapour pressure compared with tungsten oxide or molybdenum oxide. The best ITO films can be produced under evaporation conditions: T_S, 400 °C; $p(O_2)$, 3×10^{-4} Torr; rate of deposition, 0.8 Å s^{-1}; Sn, 5 wt%. Instead of using two separate crucibles, Agnihotri *et al* [194] used a single crucible with a pallet of ITO with different concentration of SnO_2 (3–7 mol%). They deposited ITO films at substrate temperatures as low as 200 °C. X-ray studies of vacuum evaporated ITO films revealed no trace of any oxide phase of Sn even for highly doped films [20, 194]. (1 1 1) was the preferred orientation in ITO films. In general, it is observed that ITO films with higher Sn content (>8–10 wt%), have a lower degree of preferred orientation.

2.5.4 ITO: sputtering

Highly transparent and conducting films of ITO have been prepared by sputtering processes [21, 174, 176, 178–180, 203–225]. Targets of In–Sn alloy and In_2O_3–SnO_2 have been used to grow these films. Usually the tin concentration in the target is 10–15%. Doping of indium oxide films with tin generally deteriorates the crystalline properties. Typically the grain size decreases from 1500 Å to 800 Å when indium oxide films are doped with tin [176]. When the films are grown using metallic alloy targets, the deposition rate is a strong function of oxygen partial pressure and sputter power, whereas it is a weak function of total sputtering pressure [143, 153]. The insensitivity of deposition rate to the total sputtering pressure is an inherent property of this process, where plasma intensity is determined by the electromagnetic fields in the system. Figure 2.63 shows the dependence of deposition rate on the sputter power for DC magnetron sputtered ITO films. For low sputtering power levels P, the target surface is oxidized non-stoichiometrically. The sputtered particles can oxidize further during their transport or during growth of the film. If the power P is increased, the number of sputtered species also increases. This effect is translated into a higher consumption of oxygen atoms needed to oxidize the target surface

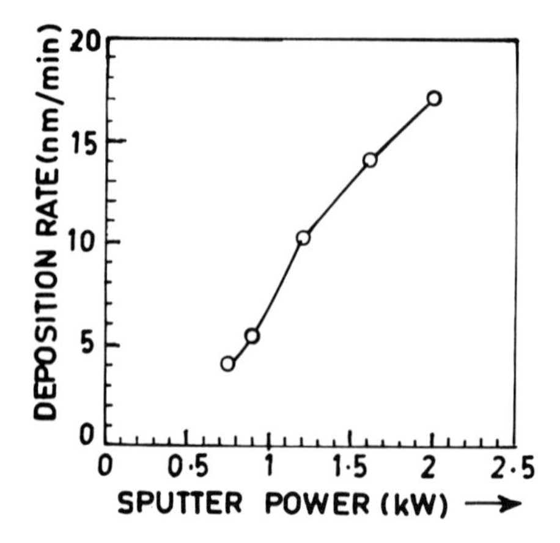

Figure 2.63 Deposition rate as a function of sputter power for ITO films (from [203]).

and the sputtered atoms. With increasing power P the number of oxygen atoms available to oxidize the target decreases and the surface of the target becomes more metallic. Annealing in forming gas at 485 °C for 30 min is found to enhance the quality of the films and is independent of the oxygen partial pressure during sputter deposition.

Smith *et al* [21] used this process for mass-scale production of ITO films. In their system a moving deposition mode is employed in which substrates are introduced through a small volume load lock and are coated as they pass under the target. The films are grown at room temperature and then annealed in N_2:H_2 (95:5) mixture at 300–400 °C. Annealing improves the electrical and optical properties significantly. The sheet resistivity and transmission of ITO films grown under optimum conditions are ~40 $\Omega \, \square^{-1}$ and 90%, respectively.

Instead of using an In–Sn alloy, Fujinaka and Berezin [180] deposited tin-doped In_2O_3 films by RF sputtering using a 0.8 mm thick disc-shaped target consisting of two unequal segments made up of indium and tin. A gas mixture of argon and oxygen flowing at rates of 100 cm^3 min^{-1} and 20–26 cm^3 min^{-1}, respectively was used. The deposition time was 18 min at an RF power level of 50 W, and the films obtained had thicknesses of the order of 1500–2000 Å. The concentration of Sn and In in the films was varied by changing the substrate position laterally with respect to the target. X-ray studies revealed that the as-deposited films were poorly crystalline, whereas annealing at 300 °C in 40 mTorr for 30 min improved the crystalline nature significantly, with a preferred (2 2 2) orientation.

A modified DC reactive sputtering technique, in which an $In_{0.88}$ $Sn_{0.12}$ alloy target was first oxidized in oxygen atmosphere and then ITO films were deposited by sputtering the oxidized target in an argon–oxygen atmosphere, was used successfully by Jachimowski *et al* [179]. First the DC diode sputtering system was pumped down to 6.6×10^{-2} Pa (4.95×10^{-4} Torr), and oxygen then admitted to a pressure of 15 Pa (1.125×10^{-1} Torr). Next, the oxidation of the target surface was carried out for 2 min by glow discharge at a DC voltage of 1100 V and discharge current of 120 mA. After target oxidation, the system was again pumped to 6.6×10^{-2} Pa (4.95×10^{-4} Torr), and the deposition chamber filled with oxygen and argon up to a combined pressure of 5.3 Pa (3.97×10^{-2} Torr). Finally, the oxidized target was sputtered at a constant voltage of 1800 V in a argon–oxygen atmosphere. The substrates were heated to about 413 K. During deposition at constant voltage, a fall in discharge current was observed. This is due to the decrease of secondary electron emission when the oxide is removed from the target surface. The decrease in discharge current depends mainly on the oxygen partial pressure in the O_2–Ar atmosphere. A typical plot of discharge current against time for three oxygen partial pressures at a constant voltage of 1800 V is shown in figure 2.64. It can be observed from the figure that the current decreases rapidly for the lowest value of oxygen partial pressure (1.3×10^{-2} Pa) (9.75×10^{-5} Torr). In this case, metal-rich films were obtained. For 2×10^{-2} Pa (1.5×10^{-4} Torr) oxygen partial pressure, samples are polycrystalline but optical quality is poor. A further increase of oxygen partial pressure to 5×10^{-1} Pa (3.75×10^{-3} Torr) results in highly transparent ITO films.

As already discussed, the oxygen flow rate and deposition rate play crucial roles in determining the properties of reactively sputtered ITO films. The dependence of resistivity on deposition rate and oxygen flow rate of

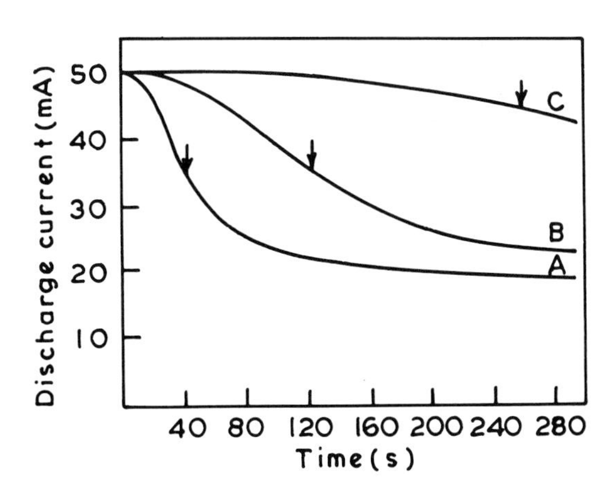

Figure 2.64 Dependence of the discharge current on time during sputtering of the oxidized $In_{0.88}Sn_{0.12}$ target at a DC voltage of 1800 V with different partial pressures (Pa): (A) 1.3×10^{-2}; (B) 2×10^{-1}; (C) 5×10^{-1}. The arrows indicate the ends of deposition (from [179]).

reactively sputtered ITO films form a family of straight lines of constant resistivity. Kawada [204] established such lines of constant resistivity for sputtered ITO films on glass substrates subsequently air baked at 490 °C for one hour. An In(90%)–Sn(10%) metal alloy target was used to deposit ITO films on moving substrates. The substrate transport speed was adjusted to keep the thicknesses of the films approximately the same. Figure 2.65 shows the variation of film thickness times substrate transport speed (proportional to deposition rate) with oxygen flow rate for films of different resistivities. It can be observed from the figure that lines of constant resistivity all emanating from the origin fit the data well. Such constant resistivity lines are possible due to the fact that ITO films with the same resistivity have the same chemical composition [205]. It may be further observed from figure 2.65 that for the production of low resistance ITO films, one requires a low deposition rate/oxygen flow rate ratio. Contour plots with lines of constant resistivity are very useful for depositing films with reproducible properties. Figure 2.65 also illustrates that for preparing films with high deposition rates, the oxygen partial pressure has to be increased in proportion. Such observations have also been made by Howson and Ridge [28] in their studies on ITO films. They [28] grew ITO films using a 10%-Sn-doped In target, using a gas mixture of 3.5 mTorr argon and 1.5 mTorr oxygen. When the power to the sputtering target was increased in order to increase the rate of deposition, the oxygen partial pressure was observed to fall to zero. The stoichiometry of the films produced can only be controlled by balancing the oxygen input rate against the sputtering rate of the metal. Rates as high as 1000 Å min^{-1} have been achieved.

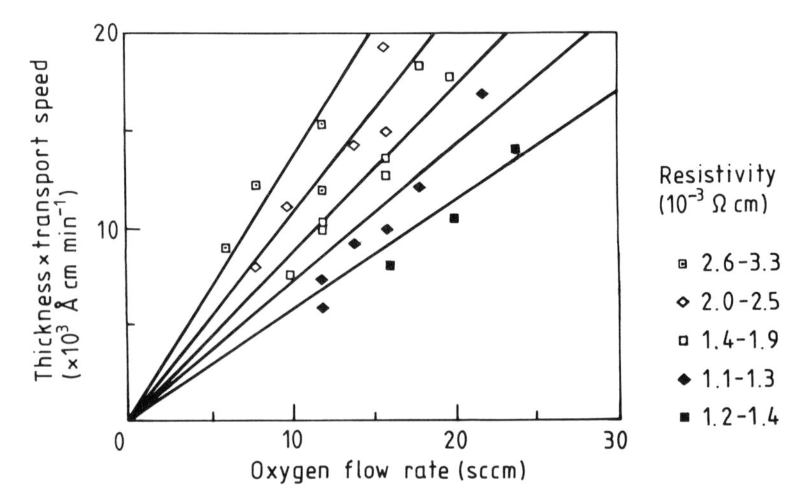

Figure 2.65 Film thickness × transport speed versus O_2 flow rate for ITO films with different resistivities (from [204]).

For mass production of reproducible quality ITO films, the required oxygen flow rate is very critical. Low sheet resistance combined with high transparency occurs for a small range of oxygen partial pressures. This is particularly so in films grown at low substrate temperature (<100 °C). The use of ITO films for many applications requires substrates like polyester, plastic, etc., the thermal properties of which preclude use of elevated temperatures. Consequently, a deposition technique which results in a large process window for good quality ITO films deposited at low temperatures is of considerable interest. The inclusion of H_2 or H_2O in the sputtering gas mixture is reported [206–209] to facilitate the production of transport conductivity oxide films. Naseem and Coutts [206] produced ITO films of resistivity less than 2×10^{-3} Ω cm by sputtering an oxide target in argon containing 10% H_2. Ishibashi *et al* [207] deposited ITO films by introducing H_2O or H_2 in an in-line production magnetron sputtering system. ITO films having resistivity as low as 6×10^{-4} Ω cm could be deposited at room temperature with reproducible properties. Harding and Window [208] demonstrated that reproducible quality ITO films can be deposited over a wide range of oxygen partial pressures if hydrogen is added to the sputtering gas mixture. Figure 2.66 shows the dependence of resistivity on oxygen flow rate for ITO films deposited at room temperature with and without the addition of hydrogen in the gas mixture. It can be observed from this figure that the low resistivity region is very narrow when the films are grown without adding hydrogen to the gas mixture, whereas this region is broadened considerably when hydrogen is added to the gas mixture. Thus the introduction of hydrogen greatly broadens the process window and thereby facilitates mass-scale production of low resistivity transparent films. It may

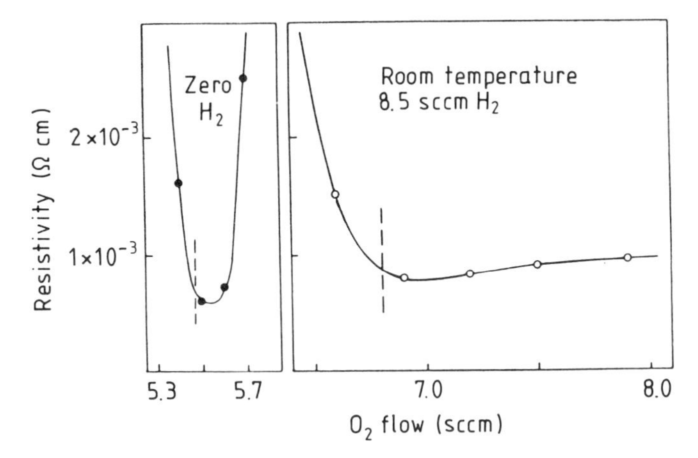

Figure 2.66 Resistivity versus oxygen flow rate for ITO films deposited at room temperature with 0 and 8.5 sccm hydrogen. Transparent films are produced to the right of the dashed line (from [208]).

be noted that the broadening of the process window by the introduction of hydrogen in the gas mixture is comparatively less when films are deposited at higher temperatures.

Transparent conducting ITO films have also been prepared by a sputtering process using an In_2O_3–SnO_2 target [178, 207, 209–212]. The use of oxide targets in place of metallic targets helps in controlling the stoichiometry of the films more precisely. Generally post-deposition annealing is not essential when ITO films are grown from oxide targets. Usually oxygen is added to the sputtering gas mixture in order to improve the structural, electrical and optical properties. However, such an improvement is possible only when the oxygen partial pressure is low and is within a narrow range, typically $(2–4) \times 10^{-5}$ Torr. Films deposited at low temperatures are amorphous in nature, whereas films deposited at higher temperatures (250–400 °C) are polycrystalline. A typical XRD pattern for ITO films grown using an oxide target by DC magnetron sputtering is shown in figure 2.67. The SnO_2 dopant content in the oxide target is 10 wt% and the sputtering gas is a mixture of argon and 1% O_2 at volume ratio. Water vapour with partial pressure less than 10^{-5} Torr has been added to the gas mixture in order to deposit reproducible films. Films have been deposited at 400 °C. All the films exhibit only In_2O_3 peaks, while the peaks of SnO_2 are absent. It can be observed that the most predominant planes are (4 0 0) and (2 2 2). The intensity of these two planes depends on substrate temperature, deposition rate, sputtering voltage and sputtering power [178, 209–211]. For films grown at a

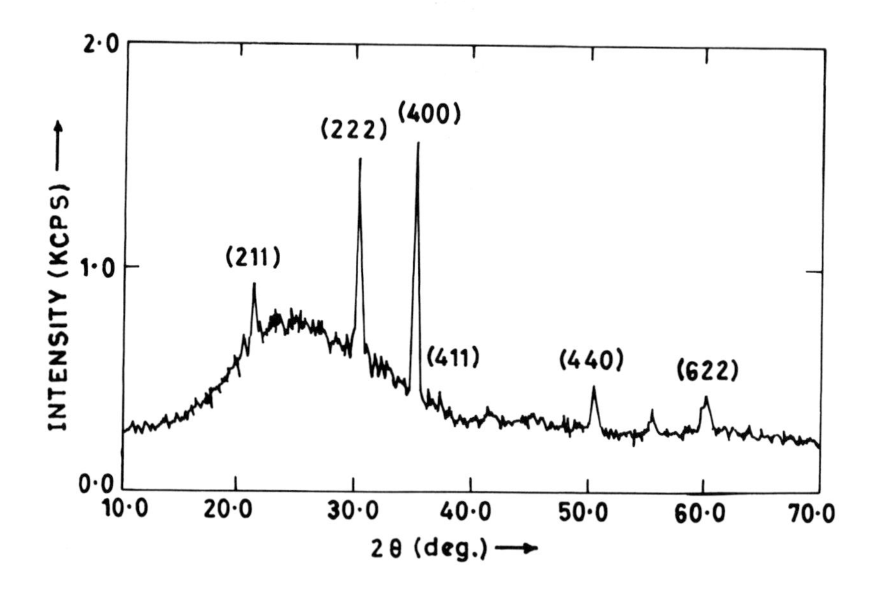

Figure 2.67 X-ray diffraction pattern of sputtered ITO films (from [209]).

substrate temperature of 100 °C, mainly (2 2 2) orientation is present, with little indication of the (4 0 0) plane. At substrate temperatures above 200 °C, the intensity of the (4 0 0) peak increases rapidly and at temperature >300 °C, exceeds that of (2 2 2) orientation [210]. The studies by Itoyama [178] on sputtered ITO films indicate that for films deposited at low deposition rate (below 2000 Å min^{-1}), the (4 0 0) peak is the most prominent. All other peaks are either weak or absent, indicating that the films have strong (1 0 0) texture. In the range of intermediate rates of deposition (2000–3000 Å min^{-1}) two strong peaks of (2 2 2) and (4 0 0) appear. At higher deposition rates (over 3000 Å min^{-1}), the (2 2 2) peak is the most prominent, indicating the existence of (1 1 1) texture. Shigesato and coworkers [209, 211] observed a marked effect of sputtering voltage and sputtering power on the texture of the sputtered ITO films. With an increase of sputtering voltage or sputtering power the preference for the (2 2 2) plane becomes stronger, which is similar to the behaviour observed with increasing substrate temperature [210].

Like SnO_2 films [76], an improvement in structural, electrical and optical properties of ITO films has been observed [213] as a result of laser irradiation. The laser treatment annihilates dislocations and promotes grain growth.

2.5.5 ITO: ion-assisted deposition

The ion-assisted deposition technique has been successfully applied to deposit excellent quality ITO films at low deposition temperatures [27–29, 173, 181, 182, 184, 226–229].

Fan [29, 226] prepared ITO films by an ion-beam sputtering technique at deposition temperatures below 100 °C. An Ar–ion beam was directed to a hot pressed target of composition In_2O_3–12 mol% SnO_2. Typical conditions were: target to substrate distance about 6 cm, Ar beam current 50 mA, target presputtering 10 min followed by deposition for 30 min, deposition rate 170–200 Å min^{-1} and O_2 partial pressure $1–7 \times 10^{-5}$ Torr. X-ray diffractometer measurements made on these films showed very diffuse diffraction patterns, which indicate that the films are either amorphous or weakly polycrystalline. The best and reproducible films were obtained only when the oxygen pressure was maintained over a very narrow range $2–3 \times 10^{-5}$ Torr.

An ion-plating process for the growth of high quality ITO films on glass and plastic substrates at room temperature was used by Howson *et al* [173] and Machet *et al* [227]. Howson *et al* [173] used both inductive heating and DC planar magnetron sputtering for the evaporation of the metal alloy. Typical values of resistivity and transmission were 1×10^{-3} Ω cm and 80%, respectively. TEM studies revealed that films grown on polyester were crystalline with a grain size of 100 Å. Ridge *et al* [27] further extended this technique for the ion plating of large-area plastic sheets with ITO coatings. Highly transparent (~90%) and conducting ($R_s \sim 400\,\Omega\,\square^{-1}$) films have been grown on 250 m long plastic sheets. Recently Oyama *et al* [228] used an ion plating technique to grow resistive ITO films on large-area glass substrates (1.8 m × 2.3 m). Film growth was carried out at a high deposition rate of more than 4000 Å min^{-1}. Films grown at a substrate temperature of 125 °C have a mixture of amorphous and polycrystalline phases, and show a hazy appearance, whereas transparent polycrystalline films with a minimum resistivity of $1.7 \times 10^{-4}\,\Omega$ cm have been obtained at a substrate temperature of 180 °C.

Machet *et al* [227] deposited ITO films by separately evaporating indium and tin in oxygen from resistively heated vapour sources. The cathode supporting the substrate was biased by an RF generator (13.56 MHz). It was observed that the best films were produced when an RF power of about 20 W was applied to the cathode. Application of higher power resulted in an increase in the negative direct potential at the cathode, which prevented the growth of films with low resistivities. The optimum oxygen partial pressure and substrate temperature were found to be 0.6–1.3 Pa (4.5–9.75 × 10^{-3} Torr) and 300–350 °C, respectively.

2.5.6 ITO: other techniques

Other than the above-discussed growth techniques, electroless chemical growth and sol–gel dip coating techniques have also been used for the growth of ITO films. Goyal *et al* [230] used an electroless technique to grow both In_2O_3 and ITO films. A typical starting solution for In_2O_3 was: 1 mol $InCl_5$ – 5 cm^3; 6 mol $Na_3C_6H_5O_7$ – 7 cm^3; 0.5 mol $AgNO_3$ – 0.01 cm^3; distilled water – 100 cm^3; 1 mol NaOH was added to the solution to adjust its pH value to 7.5. For ITO coatings, Sn was added in the form of 1 mol $SnCl_4$ (0.5 cm^3) to give films having optimum properties. Figure 2.68 shows a typical variation of film thickness with time. It can be observed that the film thickness saturates after 25–30 min. The pH value of the solution also significantly affects the ultimate thickness achieved in the films; for example, a change of pH value from 7.5 to 8.5 reduces the thickness from about 3400 Å to 2600 Å. X-ray studies indicate that In_2O_3 and ITO films are polycrystalline and the average grain size is of the order 250 and 540 Å, respectively. Although no peak corresponding to Ag is observed, a small intensity peak due to Cl is always present in these films. The resistivity and transmission of ITO films grown using this technique are $10^{-4}\,\Omega$ cm and 80%, respectively.

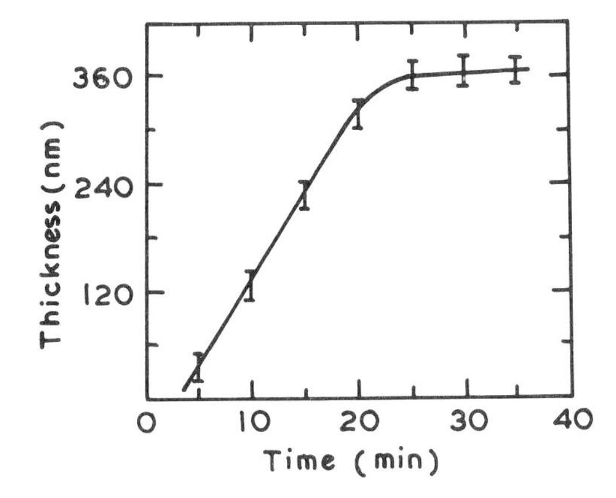

Figure 2.68 Variation of film thickness with time (from [230]).

Sol–gel derived transparent conducting films of ITO have been developed, which require air baking at 500–600 °C [141, 231, 232]. Mattox [141] used a solution of SnO_2 and In_2O_3 as precursors made by dissolving tin isopropoxide (Sn $(OC_3H_7)_4$) and indium isopropoxide ($In(OC_3H_7)_3$) powders in anhydrous ethanol. The solution was heated at about 60 °C for 16 h. There were hydrolysed with 1.9 mol H_2O per mol of alkoxide. The substrates were coated using a spin coating technique. After coating, the substrates were heated in a furnace at 500 °C. The resistivity and the transmission in these coatings were 10^{-1}–10^{-2} Ω cm and 90%, respectively. Yamamoto *et al* [231] deposited ITO films with resistivity 1.1×10^{-4} Ω cm with 7 at% Sn using this technique. Precursors have been prepared by dissolving indium and tin chloride into ethylene glycol. Well-oriented crystalline films have been obtained by heat treatment at 600 °C.

2.5.7 ITO: technique comparison

A brief description of growth parameters and electrical and optical properties of ITO films grown by different techniques is given in table 2.3. In general, highly transparent and conducting ITO films can be grown by any of the techniques discussed in section 2.5. Figure 2.69 depicts graphically the transparent conducting properties of ITO films in terms of transmission and resistivity. Sputtering seems to emerge as the most suitable technique. However, it requires either elevated growth temperatures or annealing at high temperature in a suitable ambient. For applications where coating is required on heat sensitive substrates, ion-assisted techniques can be employed.

Table 2.3 Properties of ITO films grown using different techniques.

Process	Substrate temperature (°C)	Rate (Å min^{-1})	Resistivity (Ω cm)	Transmission (%)	Remarks	Reference
ITO						
CVD	350–450	—	$1.5–1.8 \times 10^{-4}$	90–95	Annealing in vacuum at 400 °C	185
CVD	450	—	1.8×10^{-4}	90	Sn/In = 0.031	153
Spray	400–500	—	1.3×10^{-1}	84	—	89
Spray	400	—	2×10^{-4}	92	Sn/In = 0.025	16
Spray	450	—	2.6×10^{-2}	86	—	186
Spray	677	—	10^{-4}	82	—	100
Spray	420	—	3×10^{-3}	90	—	187
Spray	—	—	2.2×10^{-4}	88	Ultrasonic	15
Reactive evaporation	440	120	1.2×10^{-3}	~80	$O_2 = 10^{-3}–10^{-1}$ Pa	196
Reactive evaporation	400	—	2×10^{-4}	80	$O_2 = 5 \times 10^{-5}$ Torr	166
Reactive evaporation	300	—	—	90	$O_2 = 10^{-3}–10^{-4}$ Torr	191
Reactive evaporation	350	400	8.8×10^{-5}	88	15% O_2–Ar, activated reactive evaporation	20
Reactive evaporation	400	400	7×10^{-5}	90	Activated reactive evaporation	197
Reactive evaporation	(Room)	—	3×10^{-4}	90	Air annealed at 475 °C	198
e-beam evaporation	—	—	$\sim 3 \times 10^{-4}$	90	—	198
e-beam evaporation	300	—	3×10^{-4}	90	$O_2 = 5 \times 10^{-4}$ Torr	199
e-beam evaporation	200	96	2.4×10^{-4}	~90	$O_2 = 1.5 \times 10^{-5}$ Torr	194
e-beam evaporation	—	4000	1.7×10^{-4}	80	Large-area applications	228
e-beam evaporation	300	—	5.2×10^{-4}	90	$O_2 = 8 \times 10^{-5}$ Torr Reactive evaporation	200
Sequential evaporation of Sn, In	—	—	4×10^{-3}	>90	Evaporated in O_2, air annealed	193
Sputtering	—	—	6.2×10^{-4}	87	Sputtering in O_2 Annealed in N_2 at 300 °C	174
Sputtering	453	—	6.5×10^{-4}	84	In–Sn target oxidized first and then	179

[illegible]			1.17 × [illegible]	[illegible]	[illegible] wt% Sn	[illegible]
Sputtering	450	—	1.8×10^{-4}	95	—	217
Sputtering	460	800	1.2×10^{-4}	>85	—	219
Sputtering	130	—	4×10^{-4}	85	$O_2 = 5 \times 10^{-5}$ Torr	212
Sputtering	450	—	3×10^{-4}	90	$O_2 = 3\text{–}4 \times 10^{-5}$ Torr	214
Sputtering	(Room)	—	5.46×10^{-4}	>80	$O_2/H_2O/Ar$	207
Sputtering	(Room)	—	1.6×10^{-2}	>70	$O_2 = 2.6 \times 10^{-3}$ Torr Annealing in O_2	220
Sputtering	(Room)	250	3×10^{-4}	90	$Ar/O_2 = 50/50$ Vacuum annealed	215
Sputtering	(Room)	—	3×10^{-4}	>90	Ar–50% O_2 Vacuum annealed	215
Sputtering magnetron	—	—	5×10^{-4}	~90	Annealing in 10% H_2–N_2 at 485 °C	203
Sputtering magnetron	300	1200	1.4×10^{-4}	~87	$O_2 = 6 \times 10^{-5}$ Torr Large-area applications	210
Sputtering magnetron	450	—	1.8×10^{-4}	~90	2% O_2	216
Sputtering magnetron	(Room)	116–182	7×10^{-4}	90	O_2 = 9%	225
Sputtering magnetron	(Room)	—	$40\ \Omega\ \square^{-1}$ [a]	~90	Annealed in 5% H_2–95% N_2	21
Sputtering magnetron	23	—	5.0×10^{-4}	85	Double-ended magnetron sputtering	201
Ion-beam sputtering	100	—	5.5×10^{-4}	>80	$O_2 = 2\text{–}3 \times 10^{-5}$ Torr	29
Ion-beam sputtering	450	—	3×10^{-4}	>80	Glass	226
	100	—	2×10^{-3}	>80	Myler	
Reactive ion-plating	300–350	660–1000	10^{-3}	>85	—	227
Reactive ion-plating	(Room)	—	$400\ \Omega\ \square^{-1}$ [a]	—	Large-area applications	27
Reactive ion-plating	(Room)	—	$10\ \Omega\ \square^{-1}$ [a]	>80	—	229
Reactive ion-plating	(Room)	500	5×10^{-3}	80	—	28
Sol–gel	500	—	4.23×10^{-4}	90	—	232
Sol–gel	—	—	10^{-2}	~90	Baking temp. = 500 °C	141
ITO:F						
Sputtering	80	—	6.5×10^{-4}	>80	Annealing in Ar	221
Reactive ion-plating	(Room)	—	2.1×10^{-4}	—	Improvement on annealing in air	184

[a]Sheet resistance.

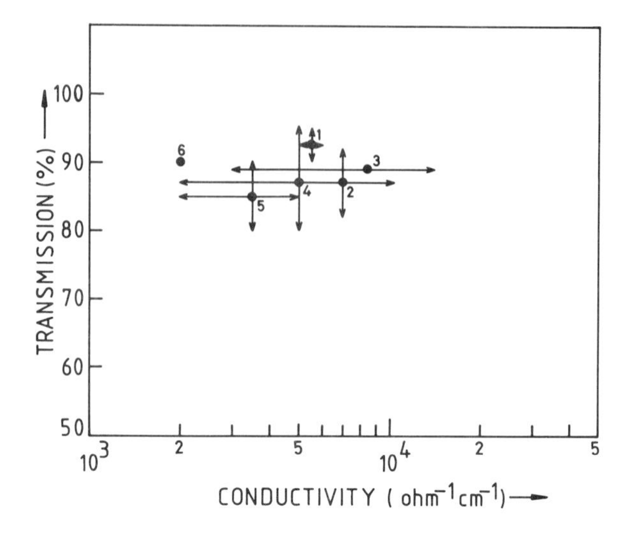

Figure 2.69 Graphical representation of transparent conducting properties of ITO films grown by different techniques. (1) CVD, (2) spray, (3) evaporation, (4) sputtering, (5) ion-assisted, (6) sol-gel. The bars show the variation in the data reported by various workers.

2.6 Growth of Cadmium Stannate Films

Although cadmium stannate thin films, like doped tin oxide and indium tin oxide coatings, have promising electrical and optical properties, very little work has been carried out on these films. Most researchers have used a sputtering process for the growth of these films. However, spray pyrolysis has also been tried because of the large-area applications and low cost of production.

2.6.1 Cd_2SnO_4: spray pyrolysis

A solution containing $CdCl_2$ and $SnCl_4$ dissolved in HCl is generally used for the growth of cadmium stannate films. Haacke [233] observed that if the substrate temperature is in the range 400–500 °C, only translucent coatings are obtained. In the temperature range 500–700 °C, the coatings consist of $CdSnO_3$ phase, whereas films deposited at temperatures higher than 800 °C result in Cd_2SnO_4 phase. However, these results are in disagreement with the results of other workers [234, 235] who have successfully deposited Cd_2SnO_4 films at low temperatures in the range 300–450 °C. In order to improve the quality of films, Agnihotri *et al* [234] annealed all their samples at about 300 °C under high vacuum. X-ray and electron diffraction studies indicate that the films were polycrystalline in nature. It has been observed [234] that with an increase in substrate temperature to 400 °C, the film appearance shows better uniformity and the clusters usually present in films grown at lower temperatures tend to disappear.

2.6.2 Cd_2SnO_4: sputtering

Initial attempts to grow cadmium stannate films using sputtering were not very encouraging. Nozik [236] prepared amorphous films of only very poor optical quality. However, later studies by other workers [237–250] showed that highly transparent polycrystalline and conducting films of cadmium stannate could be produced by sputtering techniques.

Films have been grown using both metallic alloy targets and mixed oxide targets. Films, in general, are amorphous if grown at room temperature, whereas films grown at elevated substrate temperatures are polycrystalline in nature. The minimum temperature for film crystallization depends not only on the substrate material but also on the composition of the sputtering gas. For example, films grown in pure oxygen become crystalline at a lower substrate temperature than those grown in an argon/oxygen mixture. Enoki *et al* [237], Haacke and co-workers [238, 240] and Miyata *et al* [241, 242] used mixed oxide targets to prepare cadmium stannate films. The film texture strongly depended on the initial composition of the oxide target and the substrate temperature. Films grown at substrate temperatures lower than 400 °C were either amorphous or contained secondary phases of CdO and $CdSnO_3$, irrespective of the target composition. Crystalline Cd_2SnO_4 phase was never present at these temperatures. Figure 2.70 depicts the XRD patterns for CTO films grown using different SnO_2 content in the targets for various deposition temperatures. It can be observed that single-phase Cd_2SnO_4 film can be prepared only at high temperature (400 °C) and using a 10% SnO_2–CdO target. However, it should be noted that growth of films from a 20% or higher SnO_2–CdO target never resulted in Cd_2SnO_4 phase even at elevated temperatures [237].

In extensive work, Haacke *et al* [238, 240] studied the effect of cumulative sputtering time, bias and post-deposition annealing on the composition and structure of Cd_2SnO_4 sputtered films. As-sputtered films usually have secondary phases such as $CdSnO_3$ and CdO. The intensity of these phases, however, depends on the deposition conditions. For example, when films are prepared from Cd_2SnO_4 ceramic targets under conditions of poor RF tuning (off-resonance) Cd_2SnO_4 is the predominant phase, but CdO and $CdSnO_3$ are always present. $CdSnO_3$ can be avoided by keeping the RF system tuned at full resonance and minimum reflected power. However,

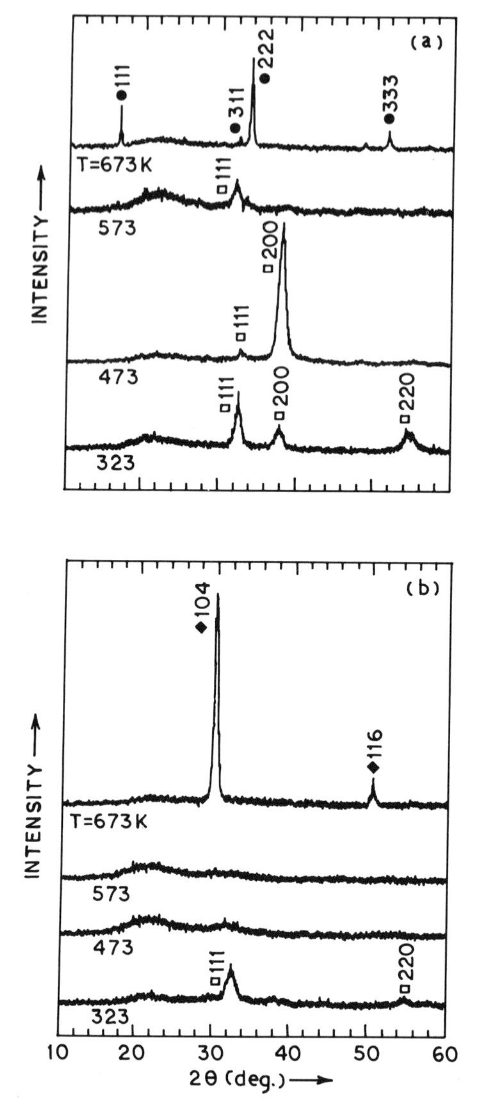

Figure 2.70 X-ray diffraction patterns of sputtered films using (a) 10% SnO_2–CdO, (b) 20% SnO_2–CdO, and (c) 33% SnO_2–CdO targets: ●, Cd_2SnO_4; □, CdO; ◆, $CdSnO_3$; ○, SnO_2 (from [237]).

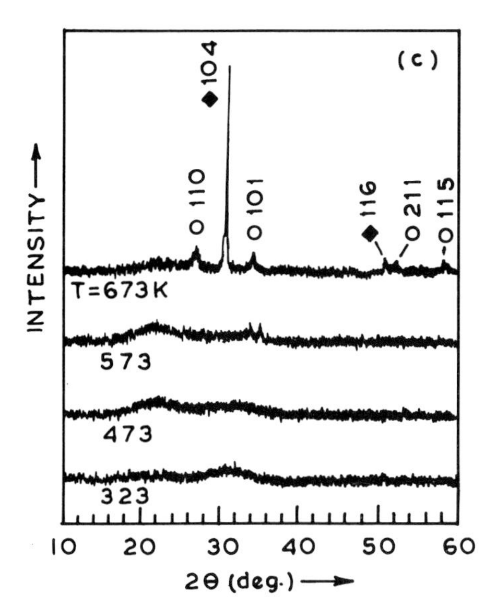

Figure 2.70 *Continued.*

under these conditions, cadmium oxide is still formed. Heat treatment of the as-sputtered film in an Ar–CdS atmosphere at different temperatures reduces CdO phase with increasing annealing temperature up to 700 °C. Annealing at still higher temperatures 750–800 °C, results in the occurrence of $CdSnO_3$ and SnO_2 phases while Cd_2SnO_4 phase is very much reduced. Cadmium stannate films free from CdO can, however, be prepared by applying a negative DC bias voltage to the substrate during deposition. An increasing bias reduces CdO phase, which completely disappears for bias values of approximately −200 V. Although CdO phase disappears with increasing DC bias, $CdSnO_3$ phase develops significantly. Typical parameters for the growth of these films can be summarized as: sputtering in pure O_2, substrate temperature 500 °C, RF power 600 W, deposition rate 150 Å min^{-1}, post-deposition annealing in Ar/CdS for 10 min at 690 °C. These films are found to contain mainly single-phase Cd_2SnO_4.

Cd_2SnO_4 films have also been prepared using a DC reactive sputtering technique from Cd–Sn alloy targets [240, 243–246]. Targets of various composition ranging from 50% Cd–Sn to 75% Cd–Sn have been used to grow Cd_2SnO_4 films. Films sputtered from targets with Cd content 60–70% have, in general, been reported to contain mainly Cd_2SnO_4 phase. The film quality generally improves after post-deposition annealing in a suitable atmosphere of Ar, O_2 or H_2–Ar at 400–500 °C. Leja *et al* [243] prepared Cd_2SnO_4 films using a DC reactive sputtering technique from Cd–Sn targets of various compositions, namely 50% Cd–50% Sn; 60% Cd–40% Sn; 66% Cd–34% Sn; 70%Cd–30%Sn and 75%Cd–25%Sn. Ar–O_2 was used as the sputtering gas mixture and typical parameters were: pumping rate 0.002 $m^3 s^{-1}$; total pressure in the chamber 6 Pa; discharge voltage 2000 V. The discharge current density at constant voltage and constant total pressure depends on the composition of the Ar–O_2 mixture; an increase in discharge current density results from an increase of O_2 concentration. For low oxygen concentration, sputtering assumes a metallic nature, i.e. the sputtering rate of the target is greater than the oxidation rates. In this range, the deposition rate exhibits a maximum. With increasing oxygen concentration in the mixture, the oxidation rate increases, which results in a drop in the deposition rate. In this range of oxygen concentration, sputtering assumes a reactive nature, i.e. the target is covered with an oxide layer. Typical curves of growth rate versus oxygen concentration in the mixture for various types of target are shown in figure 2.71. X-ray studies indicate that films sputtered from targets with 60–70% Cd content and at oxygen concentrations greater than 10% are primarily Cd_2SnO_4 phase, however, small quantities of CdO and SnO_2 secondary phases are also present. SnO_2 is the main phase in films grown from targets with cadmium concentration less than 60%. On the other hand, CdO is the major phase in films grown from targets containing more than 70% Cd. It should be noted that all films obtained at an oxygen concentration of 5–10% are amorphous in nature, irrespective of the target composition. Later studies by Leja and co-workers [244] showed that film quality could be further improved by post-deposition annealing in an atmosphere of O_2 or H_2–Ar at 673 K. Schiller *et al* [245] reported that a reduction in pressure ratio (P_{O_2}/P_{total}) resulted in a reduction in the cadmium to tin ratio (figure 2.72). The reduction is more marked at higher

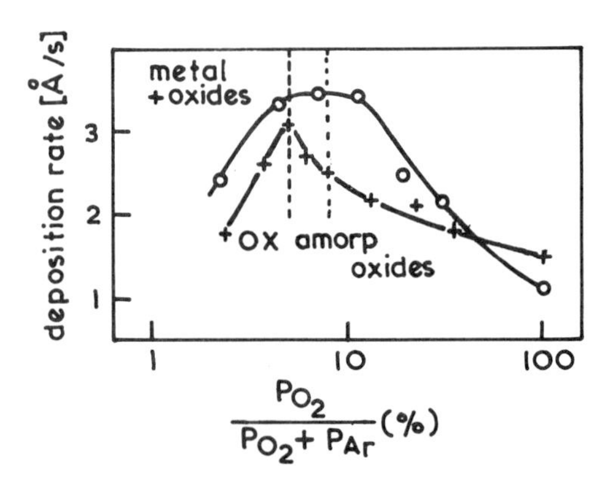

Figure 2.71 Film growth rate versus oxygen concentration in an Ar–O_2 mixture for various target compositions: ○, 50% Cd–50% Sn; +, 67% Cd–33% Sn (from [243]).

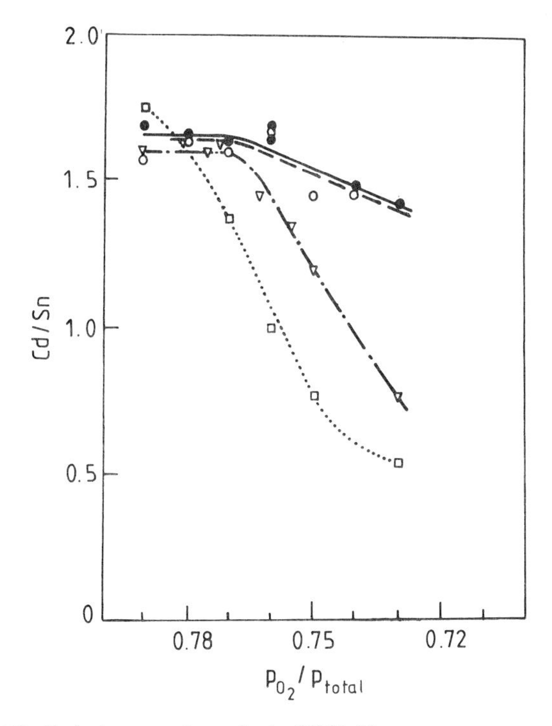

Figure 2.72 Cadmium to tin ratio in CTO films versus pressure ratio P_{O_2}/P_{total} for substrate temperatures 30 °C (●), 100 °C (▽) and 300 °C (□), and films produced at 30 °C and subsequently annealed at 400 °C (○) (from [245]).

substrate temperatures. It has also been observed that at constant argon and O_2 pressure levels, the cadmium to tin ratio in films decreases with an increase in discharge power. Films with optimum properties produced at room temperature have a Cd/Sn ratio of the order of 1.5–1.6, resistivity $4.5 \times 10^{-4}\ \Omega$ cm and transmission 90%.

The texture and composition of CTO films depend strongly on thickness, deposition rate and the target used. Miyata and co-workers [241, 242, 246–249], in their extensive studies of CTO films, observed that films less than 800 Å thick were always amorphous irrespective of the rate of deposition. Thicker films grown at a rate of 20 Å min^{-1} showed strong Cd_2SnO_4, (0 0 1) and (1 3 0) peaks, and those grown at 40 Å min^{-1} exhibited a strong $CdSnO_3$ phase (2 0 0) peak. Further, Haacke *et al* [240] observed that the composition (Cd:Sn ratio) of films grown using reactive sputtering with metal alloy targets changed with sputtering duration. This value, however, stabilizes after about 250 min. Typically, a target of composition Cd/Sn = 2:1 at% resulted in films with Cd/Sn ratio 1.78 after a period of about

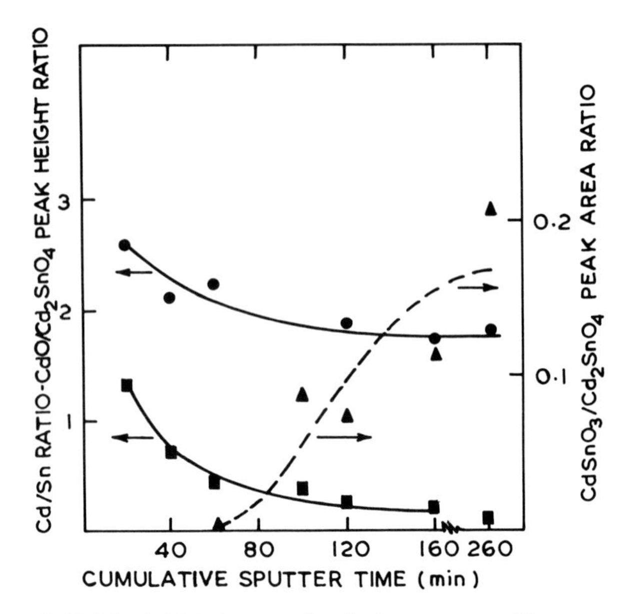

Figure 2.73 The initial change of cadmium stannate film composition during reactive sputtering from a 2 Cd:1 Sn target; ●, Cd:Sn ratio; ■, $CdO:Cd_2SnO_4$ peak height ratio; ▲, $CdSnO_3:Cd_2SnO_4$ peak area ratio (from [246]).

250 min. Figure 2.73 shows the change in film composition with increasing target usage. The change in target surface composition is because the rate of sputtering of one component is higher than that of the other. Under constant deposition conditions, the target surface eventually acquires a new equilibrium composition which determines the composition of the growing film. It can be observed from figure 2.73 that a reduction of Cd in the target reduces the CdO phase, which is quite prominent in films grown from a virgin target. However, $CdSnO_3$ phase increases with increased sputtered time.

2.6.3 Cd_2SnO_4: technique comparison

Various growth, electrical and optical parameters for cadmium stannate films grown by different techniques are listed in table 2.4. Figure 2.74 shows these results graphically. Although the electrical and optical properties of cadmium stannate films grown by a sputtering technique are comparable to those of doped tin oxide and ITO films, these films are non-reproducible and as such are not as suitable for sophisticated device applications.

Table 2.4 Properties of CTO films grown using different techniques.

Process	Substrate temperature (°C)	Rate (Å min^{-1})	Resistivity (Ω cm)	Transmission (%)	Remarks	Reference
Spray	430	—	9×10^{-3}	83	—	235
Sputtering	500	180	1.5×10^{-5}	85	Annealing in Ar/CdS	238
Sputtering	—	5–30	6.2×10^{-4}	93	Ar	242
			4.4×10^{-3}	87	Ar/2% O_2	
			1.8×10^{-2}	86	Ar/13% O_2	
Sputtering	400	5–70	4×10^{-4}	90	Ar–O_2	248
Sputtering	500–600	150–200	$1\text{–}3 \times 10^{-4}$	90	$O_2 = 5 \times 10^{-3}$ Torr	240
Sputtering	200	20–40	4×10^{-4}	90	Ar/O_2	246
Sputtering	—	—	6×10^{-4}	>80	Heating in air	243
Sputtering	400	—	10^{-3}	>80	Ar	237
Sputtering magnetron	Room	—	4.5×10^{-4}	>80	Ar/O_2	245
Atom-beam sputtering	120	—	9×10^{-4}	80	Annealing in Ar	250
Reactive ion-plating	Room	500	10^{-3}	~90	—	28

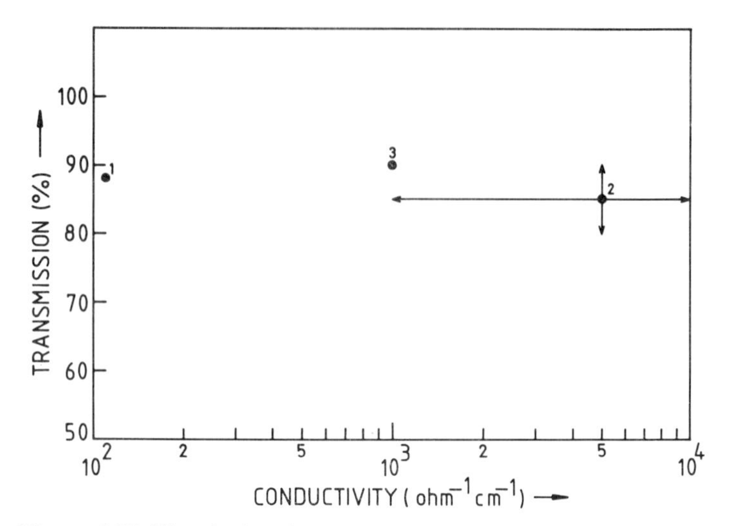

Figure 2.74 Electrical and optical parameters of cadmium stannate films grown by different techniques: a comparison. (1) Spray, (2) sputtering, (3) ion-assisted. The bars represent the variation in the data reported by various workers.

2.7 Growth of Zinc Oxide Films

Zinc oxide based coatings have recently received much attention because they have advantages over the more commonly used indium and tin-based oxide films. Indium oxide and tin oxide films are usually more expensive than zinc oxide films. Pure zinc oxide films, although transparent, are usually highly resistive. Non-stoichiometric and doped zinc oxide films, however, have high conductivities, but non-stoichiometric films are not very stable at high temperatures. For practical applications, therefore, doped ZnO films are more suitable. The most commonly employed techniques for the growth of ZnO films are spray pyrolysis and sputtering.

2.7.1 ZnO: CVD

ZnO films have been grown using a CVD technique for piezoelectric, electro-optic and guided wave device applications [251–255]. Aoki *et al* [251] fabricated single-crystal ZnO films onto sapphire substrates. The growth reactions used are

$$3ZnO + 2NH_3 \rightarrow 3Zn(v) + N_2 + 3H_2O \quad \text{(at source)}$$

$$Zn(v) + H_2O \rightarrow ZnO + H_2 \quad \text{(at substrate).}$$

In conventional CVD systems H_2 is normally used as the reducing gas for the ZnO source, but Aoki *et al* used NH_3 in order to slow down the growth rate and improve the film quality. A typical layout of the CVD apparatus used by these authors is shown in figure 2.75. The optimum growth parameters were as follows: source temperature 870–880 °C, substrate temperature 730–740 °C, NH_3 flow rate 0.6 l min^{-1}, and distance between the source and the substrate 20–30 cm. The growth rate of the films was observed to be ~2–3 $\mu m\ h^{-1}$. The plasma-enhanced metallo-organic chemical vapour deposition (PEMOCVD) technique has been used for the growth of highly oriented ZnO films at substrate temperature as low as 150–350 °C. The setup is shown in figure 2.76, and consists of a gas supply system and

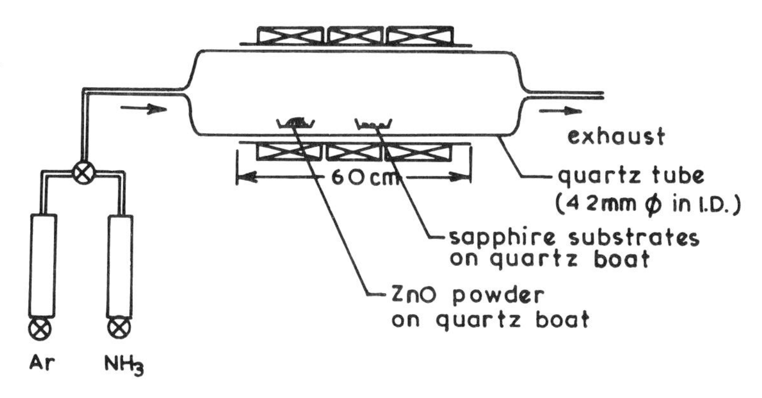

Figure 2.75 Schematic representation of the CVD apparatus (from [251]).

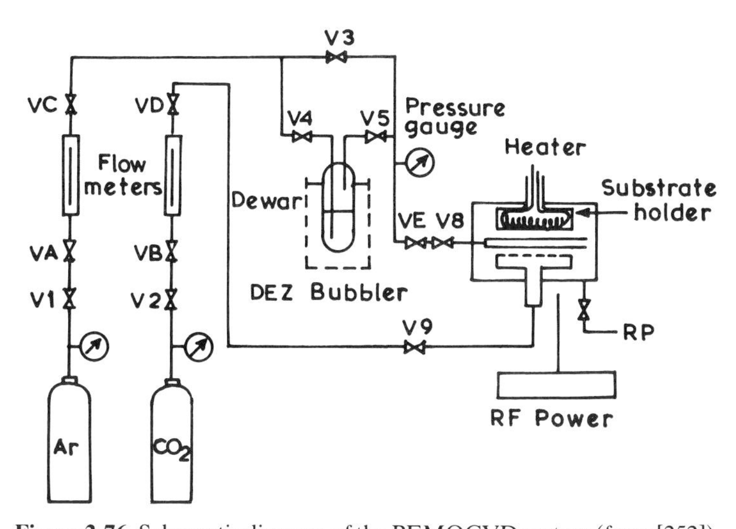

Figure 2.76 Schematic diagram of the PEMOCVD system (from [252]).

plasma-enhanced CVD system. CO_2 gas has been used along with diethyl zinc to produce ZnO films. Diethyl zinc is introduced into the reaction chamber using argon as a carrier gas. This technique is similar to that of Ghandhi *et al* [54], used for the growth of SnO_2 films. The authors [252] grew highly *c*-axis oriented films on glass substrates, and epitaxial films on sapphire substrates. However, the adhesion of these films to sapphire substrates is stronger than that on glass substrates. Recently the CVD technique has been exploited for the growth of transparent conducting doped ZnO films. Resistivity as low as $3 \times 10^{-4}\ \Omega$ cm and transmission ~85% have been observed in aluminium-doped ZnO films [257].

2.7.2 ZnO: spray pyrolysis

Transparent conducting films of zinc oxide have been successfully prepared using spray pyrolysis techniques [258–266]. An aqueous solution of zinc acetate is usually used as the spray solution. This precursor is selected due to its high vapour pressure at low temperature. The addition of a few drops of acetic acid prohibits the precipitation of zinc hydroxide thereby making the spray solution clear. This helps in producing better quality optically transparent films. The substrate temperature is normally varied between 350 and 550 °C. Tomar and Garcia [258] sprayed a 0.05 mol solution of zinc acetate in a water–alcohol mixture at a rate of 15 $cm^3\ min^{-1}$ onto heated substrates. Dry air was used as carrier gas at a rate of 10 $l\ min^{-1}$, and the spray solution was maintained at 80 °C. It was observed that the as-grown films were highly resistive, but annealing in N_2 at 350 °C for 45 min reduced the resistivity. The optimum substrate temperature for minimum resistivity was 400 °C. Aranovich *et al* [259] produced ZnO films using an aqueous solution of 0.1 mol $ZnCl_2$ plus H_2O_2 and 0.1 mol zinc acetate. The rates of solution flow and air flow were varied between 1.8 and 3.5 $cm^3\ min^{-1}$ and 8.6–10.6 $l\ min^{-1}$, respectively. It was observed that substrate temperature and air flow rate significantly affect the crystalline nature and chlorine content of the films. Figure 2.77 shows chlorine content in ZnO films as a function of substrate temperature for various flow rates. It can be seen that chlorine concentration decreases with increasing substrate temperature and air flow rates. Figure 2.78 shows the effect of substrate temperature and flow rates on the thickness of the films. The correlation between chlorine content and film thickness is quite obvious from these figures, i.e. films are thicker when the chlorine content is increased for any given substrate temperature or air flow rate. This indicates that the presence of chlorine in the films possibly favours the thermodynamic reaction leading to the formation of ZnO. It should, however, be noted that higher flow rates yielded better films in all cases; films deposited at low flow rates had a large number of inhomogeneities and pin holes. For a given flow rate, the film thickness is found to decrease with an increase of substrate temperature, which is unlike

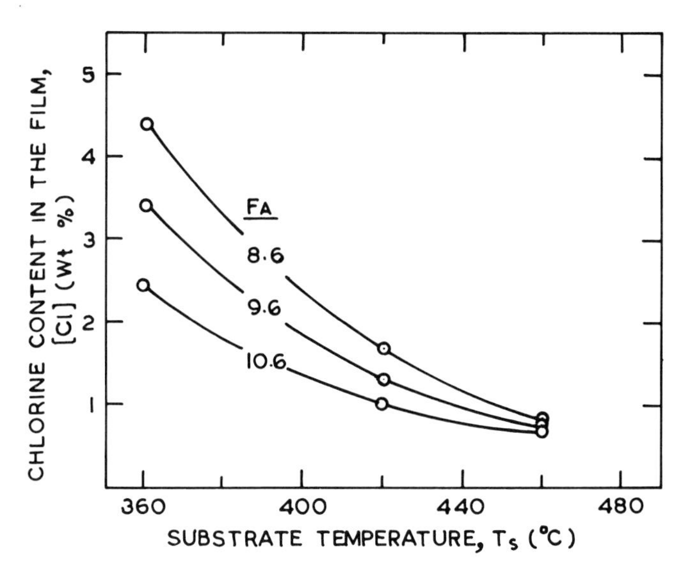

Figure 2.77 Chlorine contents in ZnO films as a function of substrate temperature and air flow rate F_A in $1\,min^{-1}$ as indicated (from [259]).

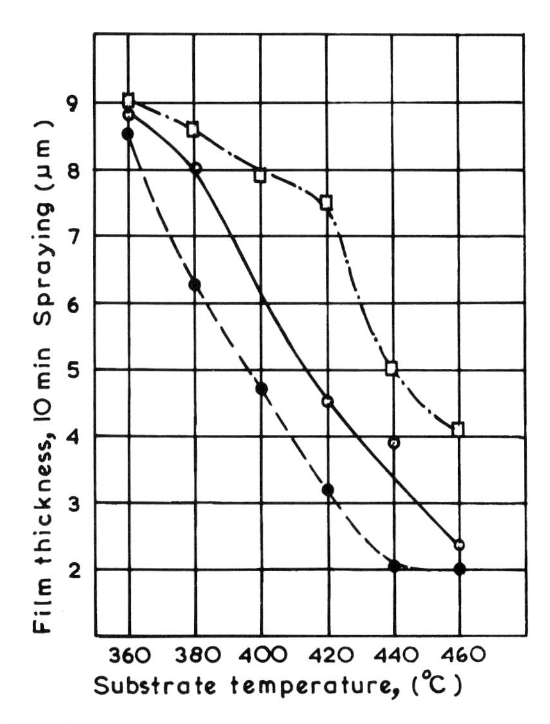

Figure 2.78 Film thickness of sprayed ZnO films as a function of substrate temperature with air flow rates of $8.6\,l\,min^{-1}$ (□), $9.6\,l\,min^{-1}$ (○), $10.6\,l\,min^{-1}$ (●) (from [259]).

the case of SnO_2 films where film thickness is found to increase with an increase in substrate temperature due to the enhanced reactivity at higher temperatures. Such a decrease in film thickness with an increase in substrate temperature has also been observed by Tomar and Garcia [258].

X-ray diffraction studies [260–263] indicate that at deposition temperatures lower than 300 °C, (1 0 1) and (1 0 0) are the most dominant orientations. Although the (0 0 2) plane is also present at these deposition temperatures the intensity of this plane is significantly less. However, at an increased temperature (300 °C), the (0 0 2) orientation becomes progressively more prominent and the intensity of the (1 0 1) and (1 0 0) peaks starts decreasing. The variation in relative intensity of some important planes as a function of deposition temperature is shown in figure 2.79.

Film morphology is very sensitive to substrate temperature and film thickness. Usually, homogeneous crystalline films are obtained at deposition temperatures of 400–450 °C [258, 260]. At very low temperatures, crystallite size and shape are not uniform. Nobbs and Gillespie [264] observed that films thinner than 400 Å were amorphous, whereas thicker films were polycrystalline. Non-stoichiometric ZnO films, although having good electrical and optical properties, in general are unstable at high temperatures, irrespective of thickness. Films thinner than 400 Å are unstable even at room temperature. These thinner films tend to develop translucent areas, which in some cases cover 25% of the total film area. For practical applications, therefore, undoped ZnO films are unsuitable.

Doping of ZnO films not only improves their electrical and optical properties, but also makes them highly stable. Aktaruzzaman *et al* [265]

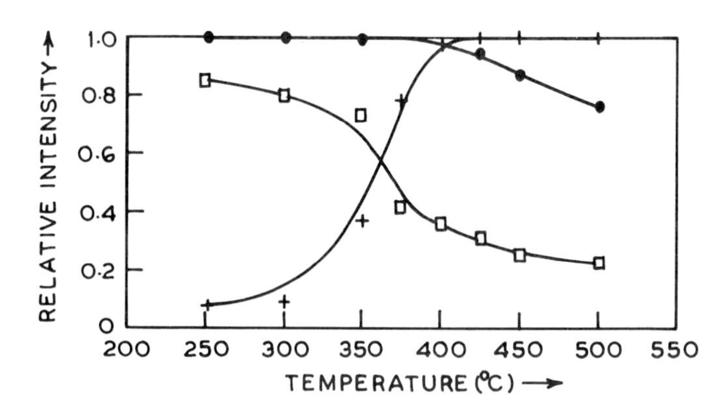

Figure 2.79 Relative intensity of (1 0 0), □; (1 0 1), ●; (0 0 2), + diffraction peaks as a function of deposition temperature for spray-deposited ZnO films (from [260]).

prepared transparent conducting thin films of aluminium-doped ZnO using a spray pyrolysis technique. The aluminium to zinc ratio was varied from 0 to 6 at% by adding aluminium chloride to zinc acetate. As-deposited aluminium doped ZnO films are highly resistive. Annealing of these films in a hydrogen atmosphere at 400 °C resulted in a significant reduction in resistivity, without affecting optical transmission. Major *et al* [263] and Ortiz *et al* [261] deposited indium-doped ZnO films using a spray solution of 0.1 mol zinc acetate in a mixture of isopropyl alcohol and water (volume ratio 3 to 1). These films were doped with indium by adding $InCl_3$ to the spray solution. The substrate temperature was maintained in the range 250–450 °C and the carrier gas, either O_2 or N_2, had a flow rate of $10\,l\,min^{-1}$.

These doped ZnO films (Al or In) grown under optimum conditions have resistivity $\sim 10^{-3}\,\Omega$ cm and transmission 85%. X-ray diffraction studies on these doped films reveal that the diffraction spectra are similar to those for undoped ZnO films. The films are polycrystalline with hexagonal wurtzite structure and the grain size is usually of the order 200–500 Å. Figure 2.80 shows a typical XRD pattern for Al-doped ZnO films grown at 350–450 °C. It can be observed that, like undoped films, the (0 0 2) orientation is the most dominant plane. The same orientation has also been observed in In-doped ZnO films [263] grown at 375 °C. However, Ortiz *et al* [261] observed the predominance of (1 0 0) and (1 0 1) orientations in their In-doped ZnO films grown at 280 °C. This is expected in accordance with figure 2.79, since the deposition temperature is less than 300 °C.

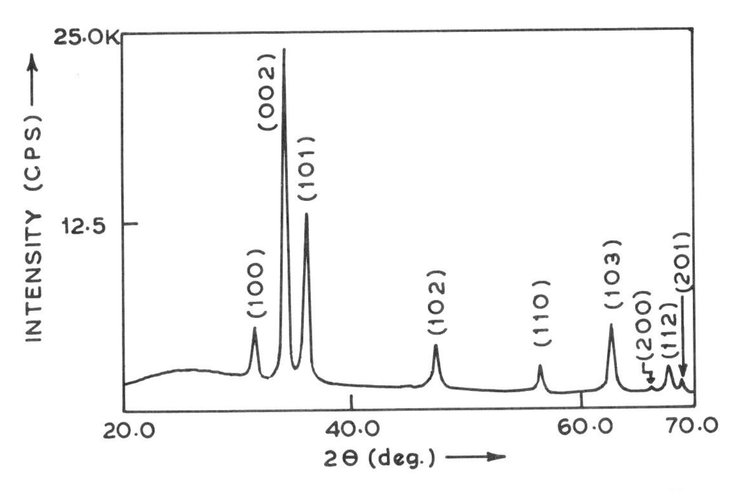

Figure 2.80 XRD patterns for aluminium-doped ZnO films (from [265]).

2.7.3 ZnO: sputtering

In the past few decades many workers have used a sputtering technique to grow ZnO thin films. The main aim of these studies has been to prepare highly resistive ZnO films for acoustic wave transducers. However, the sputtering technique has also been used to grow the low resistance ZnO films required for transparent conducting applications, either by adjusting the film stoichiometry or by doping ZnO with Al, Sn or In [22, 241, 243, 267–295, 297].

Non-stoichiometric ZnO films can be prepared by sputtering either a metallic zinc target in the presence of an oxygen–argon atmosphere, or an oxide target, usually in a gas mixture of hydrogen and argon. In general, the structural properties and growth rates of sputtered films are strongly influenced by various processing conditions, such as gas phase composition, plasma conditions, deposition temperatures and deposition geometry. In general, crystallinity improves with increase in substrate temperature [267–269]. Further, if the distance between the substrate and the target is less than the mean free path of the zinc atoms, films of better texture are observed [268]. Typical sputtering conditions [267] for growth of ZnO films using a zinc target are: sputtering power, 100 W; total pressure, 5×10^{-3} Torr; oxygen concentration 0–29.4% in volume; substrate temperature, 300 to 600 K. Figures 2.81 and 2.82 show XRD patterns for ZnO films sputtered using a zinc target at different oxygen concentrations for two substrate temperatures. It can be observed that the phase and composition of films grown at room temperature depend on oxygen content; without oxygen, the films are metallic in nature, whereas films have both zinc and zinc oxide peaks when the oxygen concentration is 1%. However, single-phase ZnO films can be prepared at room temperature only if the oxygen concentration is 2%. As the deposition temperature increases, the amount of oxygen required to form ZnO phase decreases. Typically, for films grown at 573 K, 1% oxygen is sufficient to produce crystalline ZnO films. Further addition of oxygen gives a significant improvement of crystallinity and grain size.

The growth rate of reactively sputtered ZnO films depends markedly on oxygen concentration and deposition temperature. Figure 2.83 shows the dependence of growth rate on oxygen concentration for different deposition temperatures. Growth rate increases initially with an increase of oxygen concentration, then reaches a plateau at which it is independent of oxygen concentration. Growth rate decreases drastically for higher oxygen concentrations. This behaviour is characteristic of the reactive sputtering process [243, 270, 271]. The growth rate is governed by the sputtering rate of the metallic target (in this case zinc), the fraction of sputtered metal atoms that can reach the substrate and the fraction of metallic atoms that are oxidized to form metallic oxide. The growth rate plateau may be caused by an oxygen flux on to the substrate, sufficient to oxidize all the arriving metallic atoms.

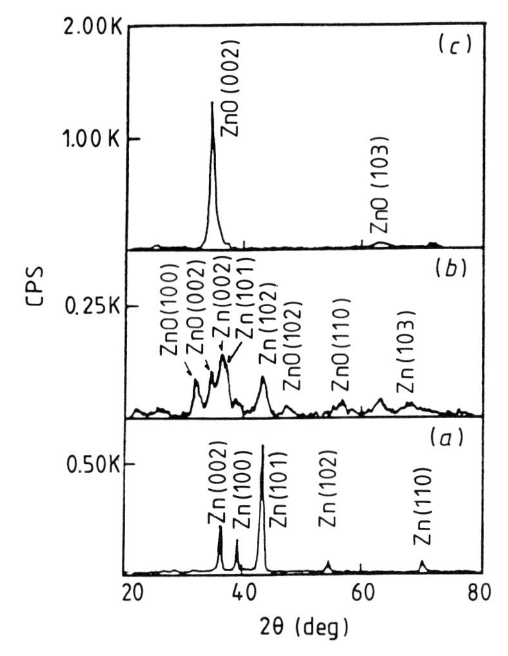

Figure 2.81 XRD patterns of ZnO films grown at room temperature with oxygen concentration (a) 0%, (b) 1%, and (c) 2% (from [267]).

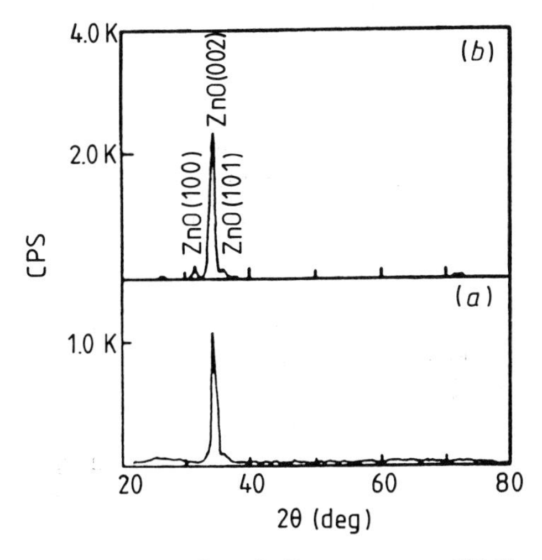

Figure 2.82 XRD patterns of ZnO films grown at 573 K with oxygen concentration (a) 1% and (b) 2% (from [267]).

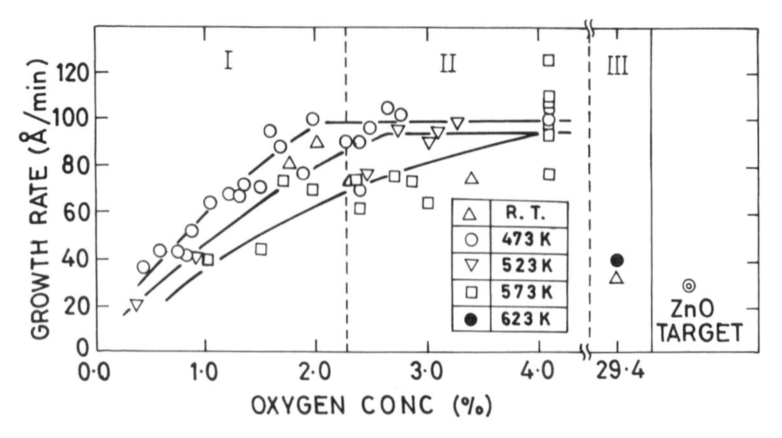

Figure 2.83 Growth rate as a function of oxygen concentration for sputtered ZnO films (from [267]).

The decrease in growth rate for higher oxygen concentrations is due to the fact that the rate of oxidization of the metallic target is much higher than the rate of sputtering. This is further confirmed by the fact that the growth rates obtained using a zinc target at high oxygen concentrations are comparable to those observed using a ZnO target (also shown in figure 2.83). It can further be observed from figure 2.83 that the growth rate decreases with increase of substrate temperature in the low oxygen concentration region. This is expected because the vapour pressure of zinc, which is quite high and also increases rapidly with increase of temperature, results in re-evaporation of zinc atoms from the substrate before being oxidized. A similar decrease in deposition rate with increase of substrate temperature has been observed by Sundaram and Garside [269]. Such results are shown in figure 2.84. Typically the rate of deposition varies from 38 Å min^{-1} to 30 Å min^{-1} for a change of substrate temperature from 50 to 300 °C.

Sputtering from a ZnO + Zn target in an Ar–O_2 atmosphere, or from a ZnO target using a mixture of hydrogen and oxygen has been widely used to grow transparent conducting ZnO films. Barnes *et al* [272] prepared ZnO films using a target made from hot pressed ZnO powder. The substrate temperature was maintained at 500 °C. The growth rate of the films depends on the oxygen concentration and the applied RF power. Figure 2.85 shows film growth rate versus applied RF power. The addition of 10% oxygen is seen to reduce the growth rate by 36%. This is because in atmospheres containing less than 10% oxygen, individual zinc atoms are sputtered, while in an atmosphere containing more than 10% oxygen, ZnO is sputtered in addition to Zn. Since the binding energy of ZnO is greater than that of Zn, this results in a reduction in sputtering rate when oxygen is present. X-ray studies have revealed that the sputtered films are polycrystalline with an

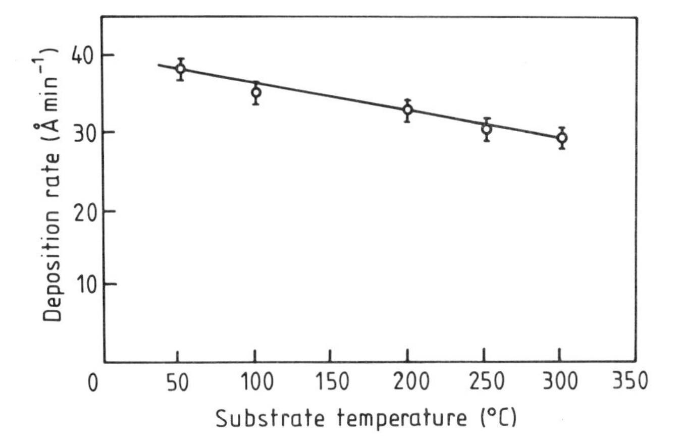

Figure 2.84 Variation of deposition rate with substrate temperature (from [269]).

average grain size of ~200–900 Å, depending on the substrate temperature; higher substrate temperatures result in larger grain size.

It is possible to grow low resistance ZnO films at low substrate temperatures by using either bias sputtering or magnetron sputtering systems [273–275]. The resistivity of ZnO films can be decreased by several orders of magnitude by varying the magnitude of bias voltage applied to the substrate. Caporaletti [273] fabricated films with resistivity $\sim 10^{-2}\,\Omega$ cm and transmission ~80% using a ZnO target in a plasma consisting of Ar/5% H_2 and at a substrate bias of −200 V. Webb *et al* [274] were able to grow ZnO films at relatively low substrate temperatures (40–275 °C) using magnetron sputtering. Hydrogen was introduced into the argon sputter gas to change the

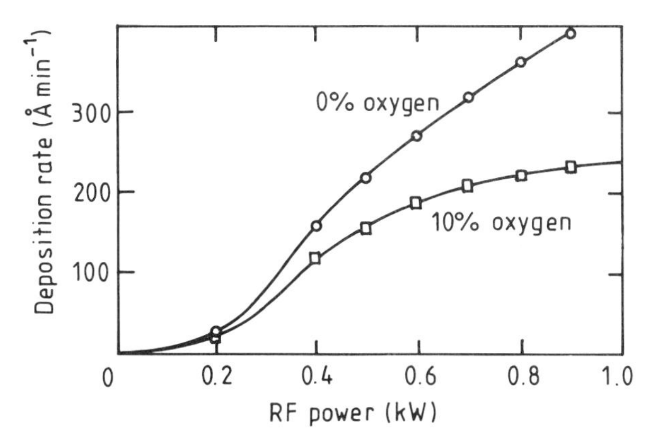

Figure 2.85 Dependence of deposition rate of undoped ZnO on applied RF sputtering power (from [272]).

zinc/oxide ratio in order to produce low resistance films. The optimum value of hydrogen partial pressure was observed to be 1×10^{-5} Torr.

As mentioned earlier, undoped ZnO films are unsuitable for practical applications as the films are unstable at high temperatures; doping of ZnO films is thus essential to make them a viable alternative for transparent conducting applications. For doping purposes, oxides of different metals such as In, Al, Ga, Sn, etc. are added to zinc oxide powder and the target is prepared from this mixture powder. Usually a dopant concentration in the range 2–10 wt% is sufficient to achieve good quality films. Highly transparent (>80%) doped ZnO films with low resistivity ($\sim 10^{-4}\ \Omega$ cm) have been prepared using this technique. XRD studies reveal that, like undoped films, the grains are strongly oriented along the *c*-axis [276–280]. Figure 2.86 shows XRD patterns for undoped and 7.5 wt% Ga_2O_3-doped ZnO films. The improvement in grain size and crystallinity is evident. Typically, grain size increases from 500 Å in undoped films to 1000 Å in doped films. Qiu and Shih [276, 277] observed that doping with indium or tin introduces an increment in the lattice constant Δc, and the value of Δc increases with an increase in the concentration of the dopant. The value of Δc for Sn-doped films is, however, larger than that for In-doped films for the same level of doping.

As the sputtering rate of different materials varies, it is expected that the dopant concentration in films will be different from that in the target. Qiu

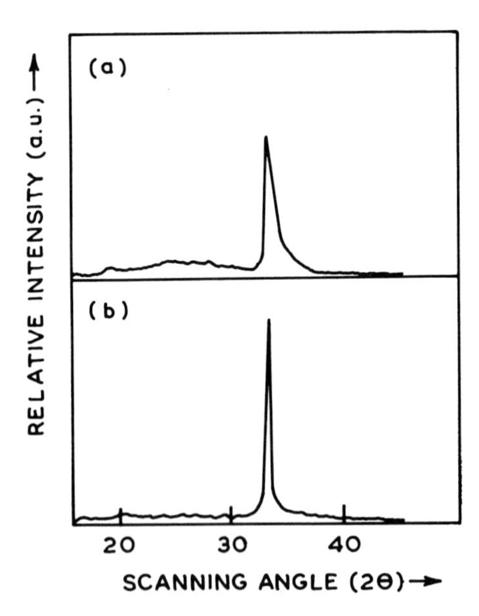

Figure 2.86 XRD patterns for (a) undoped and (b) 7.5 wt% Ga_2O_3-doped ZnO films (from [278]).

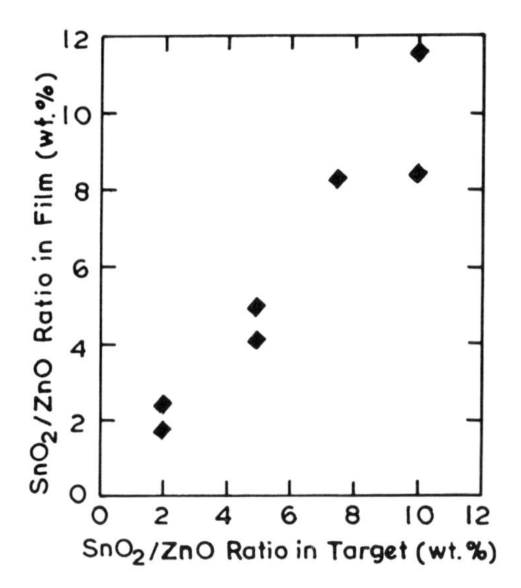

Figure 2.87 SnO_2/ZnO ratio in the film versus that in the target (from [276]).

and Shih [276, 277] studied such effects for Sn as well as In-doped ZnO films. Figures 2.87 and 2.88 depict the results of electron probe microanalyser studies performed on Sn-doped and In-doped films, respectively. These studies indicate that the Sn contents in the films are roughly equal to that in

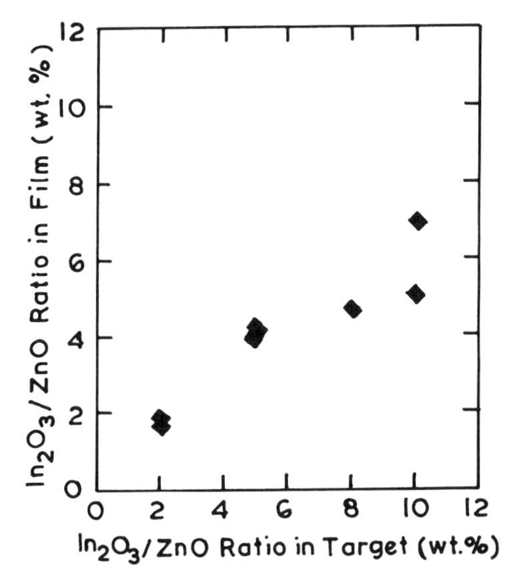

Figure 2.88 In_2O_3/ZnO ratio in the film versus that in the target (from [276]).

the target. The results for In-doped films have, however, revealed that In contents in the films are less than those in the targets, especially in those films grown from targets with higher In_2O_3/ZnO weight percentages.

2.7.4 ZnO: technique comparison

Table 2.5 gives growth parameters, resistivity and transmission of ZnO films prepared by various techniques. Undoped ZnO films are always unstable, irrespective of the growth process. A graphical comparison of electrical and optical properties of ZnO films as a function of process is given in figure 2.89. It is quite evident that aluminium-doped ZnO prepared by the sputtering technique has the best properties and can be an alternative to doped SnO_2 and ITO films.

2.8 Conclusions

Various deposition techniques such as CVD, spray pyrolysis, sputtering, evaporation, etc. can be employed for the growth of transparent conducting

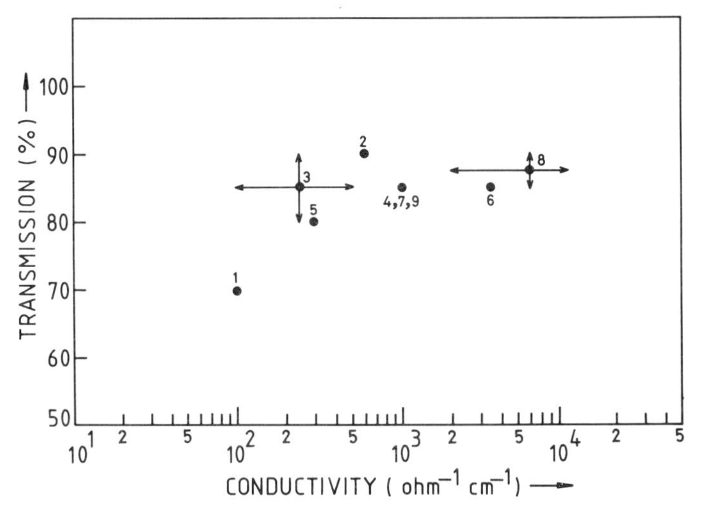

Figure 2.89 Graphical comparison of electrical and optical properties of doped and undoped ZnO films as a function of growth process. Undoped: (1) spray, (2) evaporation, (3) sputtering. Doped: (4) In–ZnO: spray, (5) In–ZnO: sputtering, (6) Al–ZnO: CVD, (7) Al–ZnO: spray, (8) Al–ZnO: sputtering, (9) Ga–ZnO: sputtering. The bars show the variations in data reported by different workers.

Table 2.5 Properties of ZnO films grown using different techniques.

Process	Substrate temperature (°C)	Rate (Å min^{-1})	Resistivity (Ω cm)	Transmission (%)	Remarks	Reference
ZnO						
Spray	400	—	10^{-2}	70	Annealing in N_2	258
Reactive evaporation	150–200	—	1.5×10^{-3}	89	—	296
Sputtering	200–250	—	$\sim 10^{-2}$	>80	O_2 = 1–2%	267
Sputtering	125	—	2×10^{-3}	90	Sputtering in hydrogen	274
Bias sputtering	Room	—	7×10^{-3}	80	Plasma 5% H_2/Ar	241
ZnO:In						
Spray	375	—	$\sim 10^{-3}$	85	—	263
Sputtering	Room	90–120	3.72×10^{-3}	>80	Ar/5%O_2; annealing in H_2	290
Sputtering magnetron	—	72–120	4.4×10^{-2}	>80	O_2 = 5%; Ar = 95% Annealing in hydrogen	295
ZnO:Al						
CVD	367–444	—	3×10^{-4}	85	—	257
Spray	500	—	2×10^{-2}	>80	—	266
Spray	300–500	—	10^{-3}	85	Annealing in H_2	265
Sputtering	—	—	10^{-4}	85	—	286
Sputtering	200	7–274	$1.4–3 \times 10^{-4}$	90	Ar	280
Sputtering	Room	90–120	9.12×10^{-3}	90	Ar/5% O_2 annealing in H_2	290
Sputtering magnetron	>250	—	2.7×10^{-4}	85	Plasma controlled using external magnetic field	279
Sputtering magnetron	<100	72	5×10^{-4}	~85	O_2/Ar	289
ZnO:Ga						
Sputtering	—	—	10^{-3}	85	Ar; 5 wt% Ga_2O_3	278

Table 2.6 A comparison of various growth techniques employed for the deposition of semiconducting transparent thin films.

Deposition technique	Substrate temperature	Rate of growth	Uniformity	Reproducibility	Cost	Electrical conductivity	Transmission
CVD	High	High	High	High	Moderate	Moderate–excellent	Moderate–excellent
Spray	High	High	Poor	Moderate	Low	Moderate–excellent	Moderate–excellent
Sputtering	Low	Low	Excellent	Excellent	High	Excellent	Excellent
Ion plating	Room	Low	Excellent	Excellent	High	Excellent	Excellent
Evaporation	High	High	Moderate	Moderate	Moderate	Moderate–excellent	Moderate

oxide films. The properties of the resulting films depend markedly on the deposition parameters of each technique. Important results have been summarized in tables 2.1 to 2.5. A broad comparison of different growth techniques in relation to various growth parameters and characteristics of transparent conducting oxide films has been made and is depicted in table 2.6. The important features related to various techniques are as follows:

(i) Spray pyrolysis can be employed for the growth of low-cost films for large-area applications where uniformity is not the primary requirement.

(ii) The ion-assisted growth technique is particularly suitable for deposition on polystyrene-like materials where substrate heating is not possible.

(iii) For the growth of reproducible device quality films, CVD and sputtering have been extensively used in one form or another. However, deposition rates of CVD methods are usually greater than those of sputtering techniques. The sputter deposition technique, although more complex and more expensive, is preferred as it permits better control of film composition and thickness.

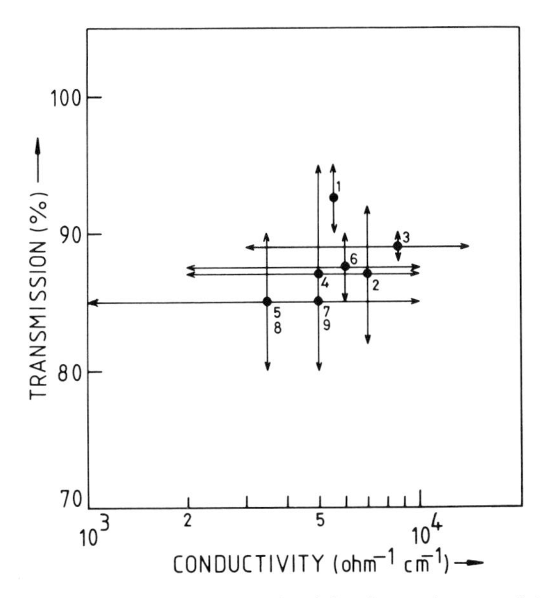

Figure 2.90 Transmission and conductivity for various useful conducting oxide films prepared by different techniques under optimum conditions. (1) ITO: CVD, (2) ITO: spray, (3) ITO: evaporation, (4) ITO: sputtering, (5) ITO: ion-assisted, (6) Al–ZnO: sputtering, (7) F–SnO_2: spray, (8) In_2O_3: evaporation, (9) Cd_2SnO_4: sputtering. The bars show the variations in data reported by various workers.

(iv) Techniques such as vacuum evaporation, dip coating, electroless deposition, etc. are of academic interest only and are not suitable for the production of device-grade films.

Figure 2.90 compares the transparent conducting properties of thin films of various oxide materials, grown under optimum conditions. It can be seen that ITO coatings are always highly transparent and conducting, irrespective of the growth technique employed. Spray-deposited fluorine-doped SnO_2 films also exhibit comparable properties. In the case of In_2O_3, it is possible to obtain useful transparent conducting films only if grown using reactive evaporation or ion-assisted deposition. Recent trends suggest that aluminium-doped ZnO films grown using a sputtering technique can be a cost-effective alternative to ITO films. Although CTO films have excellent properties, their use has been limited because it is difficult to control their chemical composition and crystal structure using any of the deposition techniques.

References

[1] Maissel LI and Glang R (eds) 1970 *Handbook of Thin Films Technology* (New York: McGraw-Hill)
[2] Holland L 1956 *Vacuum Deposition of Thin Films* (New York: Wiley)
[3] Chopra KL 1969 *Thin Film Phenomena* (New York: McGraw-Hill)
[4] Anderson JC (ed) 1966 *The Use of Thin Films in Physical Investigations* (New York: Academic)
[5] Dawar AL and Joshi JC 1984 *J. Mater. Sci* **19** 1
[6] Chopra KL, Major S and Pandya DK 1983 *Thin Solid Films* **102** 1
[7] Vossen JL 1977 *Phys. Thin Films* **9** 1
[8] Jarzebski M 1982 *Phys. Status Solidi* a **71** 13
[9] Manifacier JC 1982 *Thin Solid Films* **90** 297
[10] Mooney JB and Reading SB 1982 *Ann. Rev. Mater. Sci.* **12** 81
[11] Sundaram KB and Bhagavat GK 1981 *Thin Solid Films* **78** 35
[12] Fiest WM, Steele SR and Ready DW 1969 *Physics of Thin Films* vol 5 ed G Hass and RE Thun (New York: Academic) p 237
[13] Blocher JM Jr 1981 *Thin Solid Films* **77** 51
[14] Gottlieb B, Koropecki R, Arce R, Crisalle R and Ferron J 1991 *Thin Solid Films* **199** 13
[15] Blandenet G, Court M and Lagarde Y 1981 *Thin Solid Films* **77** 81
[16] Pommier R, Gril C and Maruchhi J 1981 *Thin Solid Films* **77** 91
[17] Grosse P, Schmitte FJ, Frank G and Kostlin H 1982 *Thin Solid Films* **90** 309
[18] Siefert W 1984 *Thin Solid Films* **121** 275
[19] Siefert W 1984 *Thin Solid Films* **120** 267
[20] Nath P and Bunshah RF 1980 *Thin Solid Films* **69** 63
[21] Smith JF, Aronson AJ, Chen D and Class WH 1980 *Thin Solid Films* **72** 469
[22] Matsuoka M, Hoshi Y and Naoe M 1987 *J. Vac. Sci. Technol.* A **5** 52

[23] Onishi S, Eschwel M, Wang WC 1981 *Appl. Phys. Lett.* **38** 419
[24] Bunshah RF 1985 *J. Vac. Sci. Technol.* A **3** 553
[25] Thornton JA 1983 *Thin Solid Films* **107** 3
[26] Matthews A 1985 *J. Vac. Sci. Technol.* A **3** 2354
[27] Ridge MI, Stentake M, Howson RP and Bishop CA 1981 *Thin Solid Films* **80** 31
[28] Howson RP and Ridge MI 1981 *Thin Solid Films* **77** 119
[29] Fan JCC 1979 *Appl. Phys. Lett.* **34** 515
[30] Cuomo JJ 1986 Synthesis by reactive ion-beam deposition *Proc. Conf. on the Applications of Ion-Plating and Implantation to Materials (1985)* ed RF Hochman (American Society for Metals) p 25
[31] Raviendra D and Sharma JK 1985 *J. Phys. Chem. Solids* **46** 945
[32] Roas B, Schultz C and Endres G 1988 *Appl. Phys. Lett.* **53**, 1557
[33] Kwok HS, Mattocks P, Shi L, Wang XW, Witanachchi S, Ying QY, Zheng JP and Shaw DT 1988 *Appl. Phys. Lett* **52** 1825
[34] Cheung JT, Niizawa C, Moyle J, Ong NP, Paire BM and Vrecland T 1986 *J. Vac. Sci. Technol.* A **4** 2086
[35] Kwok HS, Zheng JP, Witanachchi S, Mattocks P, Shi L, Ying QY, Wang XW and Shaw DT 1988 *Appl. Phys. Lett.* **52** 1095
[36] Dai CM, Su CS and Chuu DS 1990 *Appl. Phys. Lett.* **57** 1879
[37] Donnelly VM, Geva M, Long J and Karlicek RF 1984 *Appl. Phys. Lett.* **44** 951
[38] Kunz RR, Rothschild M and Ehrlich DJ 1989 *Appl. Phys. Lett.* **54** 1631
[39] Ghoshtagore RN 1978 *J. Electrochem. Soc.* **125** 110
[40] Livesay RG, Lyford E and Moore H 1968 *J. Phys. E.* **1** 947
[41] Gromer R 1960 *Rev. Sci. Instrum.* **31** 992
[42] Bartholomew RF and Garfinkel HM 1969 *J. Electrochem. Soc.* **116** 1205
[43] Baliga BJ and Ghandhi SK 1976 *J. Electrochem. Soc.* **123** 941
[44] Tabata O, Tanaka T, Waseda M and Kinuhara K 1979 *Surf. Sci.* **86** 230
[45] Kane J, Schweizer HP and Kern W 1976 *J. Electrochem. Soc.* **123** 270
[46] Kane J, Schweizer HP and Kern W 1975 *J. Electrochem. Soc.* **122** 1144
[47] Srinivasamurty N and Jawalekar SR 1983 *Thin Solid Films* **102** 283
[48] Lou JC, Lin MS, Ghyi JI and Shieh JH 1983 *Thin Solid Films* **106** 163
[49] Advani GN, Jordon AG, Lupis CHP and Longini RL 1979 *Thin Solid Films* **62** 361
[50] Muranoi T and Furukoshi M 1978 *Thin Solid Films* **48** 309
[51] Kalbskopf R 1981 *Thin Solid Films* **77** 65
[52] Upadhyay JP, Vishwakarma SR and Prasad HC 1989 *Thin Solid Films* **169** 195
[53] Reich S, Suhr H and Waimer B 1990 *Thin Solid Films* **189** 293
[54] Ghandhi SK, Siviy R and Borrego JM 1979 *Appl. Phys. Lett.* **34** 833
[55] Aboaf JA, Marcotte VC and Chou NJ 1973 *J. Electrochem. Soc.* **120** 701
[56] Kojima M, Kato H, Imai A and Yoshida A 1988 *J. Appl. Phys.* **64** 1902
[57] Jarzebski ZM and Morton JP 1976 *J. Electrochem. Soc.* **123** 199C, 299C, 333C
[58] Borman CG and Gordon RG 1989 *J. Electrochem. Soc.* **136** 3820
[59] Hsu YS and Ghandhi SK 1979 *J. Electrochem. Soc.* **126** 1434
[60] Wan CF, McGrath RB, Keenan WF and Frank SN 1989 *J. Electrochem. Soc.* **136** 1459
[61] Vetrone J and Chung YW 1991 *J. Vac. Sci. Technol.* A **9** 3041
[62] Zawadzki AGJ, Giunta CJ and Gordon RG 1992 *J. Phys. Chem.* **96** 5364

[63] Mani A, Karuppiah N and Mahalingam R 1990 *Mater. Res. Bull.* **25** 799
[64] Sanz Maudes J and Rodriguez T 1980 *Thin Solid Films* **69** 183
[65] Yueguan X, Lennard WN and Akano U 1992 *Appl. Phys. Lett.* **60** 335
[66] Proscia J and Gordon RG 1992 *Thin Solid Films* **214** 175
[67] Chambouloyron I and Saucedo E 1979 *Solar Energy Mater.* **1** 299
[68] Shanthi E, Dutta V, Banerjee A and Chopra KL 1980 *J. Appl. Phys.* **51** 6243
[69] Fung T, Ghosh A and Fishman C 1979 *Appl. Phys. Lett.* **35** 266
[70] Badawy W, Decker F and Doblohofer K 1983 *Solar Energy Mater.* **8** 363
[71] Karlsson T, Roos A and Ribbing CG 1985 *Solar Energy Mater.* **11** 469
[72] Manifacier JC, Szepessy L, Bresse JP, Perotin M and Stuck R 1979 *Mater. Res. Bull.* **14** 109, 163
[73] Islam MN and Hakim MO 1986 *J. Phys. D: Appl. Phys.* **19** 615
[74] Shanthi E, Banerjee A, Dutta V and Chopra KL 1980 *Thin Solid Films* **71** 237
[75] Jagadish C, Dawar AL, Sharma S, Shishodia PK, Tripathi KN and Mathur PC 1988 *Mater. Lett.* **6** 149
[76] Dawar AL, Kumar A, Sharma S, Tripathi KN and Mathur PC 1993 *J. Mater. Sci.* **28** 639
[77] Chambouleyron I, Constantino C, Pantini M and Parias M 1983 *Solar Energy Mater.* **9** 127
[78] Carlson DE 1975 *J. Electrochem. Soc.* **122** 1334
[79] Kulaszewicz S, Lasocka I and Michalski Cz 1978 *Thin Solid Films* **55** 283
[80] Kaneko H and Miyake K 1982 *J. Appl. Phys.* **63** 3629
[81] Shanthi E, Banerjee A and Chopra KL 1982 *Thin Solid Films* **88** 93
[82] Hoflund GB, Cox DF, Woodson GL and Laitinen HA 1981 *Thin Solid Films* **78** 357
[83] Kulaszewicz S 1980 *Thin Solid Films* **74** 211
[84] Ishiguro K, Sasaki T, Arai I and Imai I 1958 *J. Phys. Soc. Japan* **13** 296
[85] Afify HH, Momtaz RS, Badawy WA and Nasser SA 1991 *J. Mater. Sci.: Mater. Electron.* **2** 40
[86] Shanthi E, Banerjee A, Dutta V and Chopra KL 1982 *J. Appl. Phys.* **53** 1615
[87] Agashe C, Marathe BR, Takawale MG and Bhide VG 1988 *Thin Solid Films* **164** 261
[88] Abass AK and Mohammad MT 1986 *J. Appl. Phys.* **59** 1641
[89] Demichellis F, Mezzetti EM, Smurro V, Tagliaferro A and Tresso E 1985 *J. Phys. D: Appl. Phys.* **18** 1825
[90] Antonaia A, Aprea S and Menna PT 1990 Proc. IEEE 21st Photovoltaic Specialist Conf. p 1601
[91] Carroll AF and Slack LH 1976 *J. Electrochem. Soc.* **123** 1889
[92] Vincent CA 1972 *J. Electrochem. Soc.* **19** 515
[93] Sears WM and Gee MA 1988 *Thin Solid Films* **165** 265
[94] Kato H, Yoshida A and Arizume T 1976 *Japan. J. Appl. Phys.* **15** 1819
[95] Vasu V and Subrahmanyam A 1990 *Thin Solid Films* **189** 217
[96] Manifacier JC, De Murcia M, Fillard JP 1975 *Mater. Res. Bull.* **10** 1215
[97] Omar OA, Ragaie HF and Fikry WF 1990 *J. Mater. Sci.: Mater. Electron.* **1** 79
[98] Unaogu AL and Okeke CE 1990 *Solar Energy Mater.* **20** 29
[99] Onyia AI and Okeke CE 1989 *J. Phys. D: Appl. Phys.* **22** 1515
[100] Kulaszewicz S, Jarmoc W and Turowska K 1984 *Thin Solid Films* **112** 313

[101] Chaudhuri UR, Ramkumar K and Satyam M 1990 *J. Phys. D: Appl. Phys.* **23** 994
[102] Caillaud F, Smith A and Baumard JF 1992 *Thin Solid Films* **208** 4
[103] Vasu V and Subrahmanyam A 1990 *Thin Solid Films* **193/194** 973
[104] Vasu V and Subrahmanyam A 1991 *Thin Solid Films* **202** 283
[105] Zhang J and Colbow K 1992 *J. Appl. Phys.* **71** 2238
[106] Geurts J, Rau S, Richter W and Schmitte FJ 1984 *Thin Solid Films* **121** 217
[107] Feng T, Ghosh AK and Fishman C 1979 *Appl. Phys. Lett.* **34** 198
[108] Qazi IA, Akhter P and Mufti A 1991 *J. Phys. D: Appl. Phys.* **24** 81
[109] Manifacier JC, De Murcia M, Fillard JP and Vicario E 1977 *Thin Solid Films* **41** 127
[110] Spence W 1967 *J. Appl. Phys.* **38** 3767
[111] Casey V and Stephenson MI 1990 *J. Phys. D: Appl. Phys.* **23** 1212
[112] Muranaka S, Bando Y and Takada T 1981 *Thin Solid Films* **86** 11
[113] Watanabe M 1970 *Japan. J. Appl. Phys.* **9** 1551
[114] Howson RP, Avaritsiotis JN, Ridge MI and Bishop CA 1979 *Thin Solid Films* **58** 379
[115] Giani E and Kelly R 1974 *J. Electrochem. Soc.* **121** 394
[116] Sinclair WR, Peters FG, Stillinger DW and Koonce SE 1965 *J. Electrochem. Soc.* **112** 1096
[117] Leja E, Pisarkiewicz T and Kolodziej A 1980 *Thin Solid Films* **67** 45
[118] Stedile FC, Leite CVB, Schreiner WH and Baumvol IJR 1990 *Thin Solid Films* **190** 119
[119] Munz WD, Heimbach J and Reineck SR 1981 *Thin Solid Films* **86** 175
[120] Howson RP, Barankova H and Spencer AG 1991 *Thin Solid Films* **196** 315
[121] De A and Ray S 1991 *J. Phys. D: Appl. Phys.* **24** 719
[122] Croitoru N, Seidman A and Yassin K 1984 *Thin Solid Films* **116** 327
[123] Croitoru N, Seidman A and Yassin K 1985 *J. Appl. Phys.* **57** 102
[124] Beensh-Marchwicka G, Krol-Stepniewska L and Misiuk A 1984 *Thin Solid Films* **113** 215
[125] Czapla A, Kusior E and Bucko M 1989 *Thin Solid Films* **182** 15
[126] Stjerna B and Granqvist CG 1990 *Thin Solid Films* **193/194** 704
[127] Howson RP, Barankova H and Spencer AG 1990 *Proc. SPIE* **1275** 75
[128] Stjerna B and Granqvist CG 1990 *Solar Energy Mater.* **20** 225
[129] Soares MR, Dionisio PH, Baumvol IJR and Schreiner WH 1992 *Thin Solid Films* **214** 6
[130] Cavicchi RE, Semancik S, Antonik MD and Lad RJ 1992 *Appl. Phys. Lett.* **61** 1921
[131] Vaynshteyn VM 1967 *Sov. J. Opt. Technol.* **34** 45
[132] Lehmann HW and Widmer R 1975 *Thin Solid Films* **27** 359
[133] Vossen JL 1971 *J. Vac. Sci. Technol.* **8** 512
[134] Sabnis AG and Moldoran AG 1978 *Appl. Phys. Lett.* **33** 885
[135] Suzuki K and Mizuhashi M 1982 *Thin Solid Films* **97** 119
[136] Vossen JL and Poliniak ES 1972 *Thin Solid Films* **13** 281
[137] Hecq M and Portier E 1972 *Thin Solid Films* **9** 341
[138] Tohda T, Wasa K and Hayakawa S 1976 *J. Electrochem. Soc.* **123** 1719
[139] Sabnis AG and Chang KY 1977 *Electron. Lett.* **13** 113

[140] Dislich H and Hussmann E 1981 *Thin Solid Films* **77** 129
[141] Mattox DM 1991 *Thin Solid Films* **204** 25
[142] Maddalena A, Maschio RD, Dire S and Raccanelli A 1990 *J. Non-Cryst. Solids* **121** 365
[143] Arfsten NJ 1984 *J. Non-Cryst. Solids* **63** 243
[144] Cocco G, Enzo S, Carturan G, Orsini PG and Scardi P 1987 *Mater. Chem. Phys.* **17** 541
[145] Borsella E, Padova PD and Larciprete R 1991 *Proc. SPIE* **1503** 312
[146] Koinkar VN and Ogale SB 1991 *Thin Solid Films* **206** 259
[147] Dhar S, Sen S and Biswas D 1985 *J. Electrochem. Soc.* **132** 2030
[148] Yoon KH and Song JS 1993 *Thin Solid Films* **224** 203
[149] B Mayer 1992 *Thin Solid Films* **221** 166
[150] Muranaka S 1992 *Thin Solid Films* **221** 1
[151] Korzo VP and Ryabova LA 1967 *Sov. Phys. Solid State* **9** 745
[152] Korzo VP and Chernyaev VN 1973 *Phys. Status Solidi* a **20** 695
[153] Maruyama T and Fukui K 1991 *J. Appl. Phys.* **70** 3848
[154] Maruyama T and Fukui K 1991 *Thin Solid Films* **203** 297
[155] Maruyama T and Fukui K 1990 *Japan. J. Appl. Phys.* **29** L1705
[156] Nomura R, Konishi K and Matsuda H 1991 *J. Electrochem. Soc.* **138** 631
[157] Groth R 1966 *Phys. Status Solidi* **14** 69
[158] Raza A, Agnihotri OP and Gupta BK 1977 *J. Phys. D: Appl. Phys.* **10** 1871
[159] Viguie JC and Spitz J 1975 *J. Electrochem. Soc.* **122** 585
[160] Mirzapour S, Rozati SM, Takwale MG, Marathe BR and Bhide VG 1992 *Mater. Res. Bull.* **27** 1133
[161] Pan CA and Ma TP 1981 *J. Electron. Mater.* **10** 43
[162] Pan CA and Ma TP 1981 *Appl. Phys. Lett.* **37** 163
[163] Lau WS and Fonash SJ 1986 *J. Electron. Mater.* **15** 117
[164] Sundaram KB and Bhagawat GK 1981 *Phys. Status Solidi* a **63** K15
[165] Laser D 1982 *Thin Solid Films* **90** 317
[166] Mizuhashi M 1980 *Thin Solid Films* **70** 91
[167] Golan A, Bregman J, Shapira Y and Eizenberg M 1990 *Appl. Phys. Lett.* **57** 2205
[168] Muranaka S 1991 *Japan. J. Appl. Phys.* **30** L2062
[169] Lee CH, Kuo CV and Lee CL 1989 *Thin Solid Films* **173** 59
[170] Kagenovich KB, Ovsjannikov VD and Svevhnikov SV 1979 *Thin Solid Films* **60** 335
[171] Wickersham CE and Greene J 1978 *Phys. Status Solidi* a **47** 329
[172] Morris JE, Bishop CA, Ridge MI and Howson RP 1979 *Thin Solid Films* **62** 19
[173] Howson RP, Avaritsiotis JN, Ridge MI and Bishop CA 1979 *Appl. Phys. Lett.* **35** 163
[174] Frasaer DB and Cook HD 1972 *J. Electrochem. Soc.* **119** 1368
[175] Mukherjee A 1989 *Vacuum* **39** 537
[176] Bawa SS, Sharma SS, Agnihotri SA, Biradar AN and Chandra S 1983 *Proc. SPIE* **428** 22
[177] Sczyrbowski J, Dietrich A and Hoffmann 1982 *Phys. Status Solidi* a **69** 217
[178] Itoyama K 1978 *Japan. J. Appl. Phys.* **17** 1191
[179] Jachimowski M, Brudnik A and Czternastex H 1985 *J. Phys. D: Appl. Phys.* **18** L145

[180] Fujinaka M and Berezin AA 1983 *Thin Solid Films* **101** 10
[181] Bellingham JR, Phillips WA and Adkins CJ 1991 *Thin Solid Films* **195** 23
[182] Bellingham JR, Phillips WA and Adkins CJ 1990 *J. Phys.: Condens. Matter* **2** 6207
[183] Graham MR, Bellingham JR and Adkins CJ 1992 *Phil. Mag.* B **65** 669
[184] Avaritsiotis JN and Howson RP 1981 *Thin Solid Films* **77** 351
[185] Ryabova LA, Salun VS and Serbinov IA 1982 *Thin Solid Films* **92** 327
[186] Saxena AK, Singh SP, Thangaraj R and Agnihotri OP 1984 *Thin Solid Films* **117** 95
[187] Vasu V and Subrahmanyam A 1990 *Thin Solid Films* **193/194** 696
[188] Kulaszewicz S 1981 *Thin Solid Films* **76** 89
[189] Haitjema H and Elich JJP 1991 *Thin Solid Films* **205** 93
[190] Mirzapour S, Rozati SM, Takwale MG, Marathe BR and Bhide VG 1992 *Mater. Lett.* **13** 275
[191] Christian KDJ and Shatynski SR 1983 *Thin Solid Films* **108** 319
[192] Noguchi S and Sakata H 1980 *J. Phys. D: Appl. Phys.* **13** 1129
[193] Yao JL, Hao S and Wilkinson JS 1990 *Thin Solid Films* **189** 227
[194] Agnihotri SA, Saini KK, Saxena TK, Nagpal KC and Chandra S 1985 *J. Phys. D: Appl. Phys.* **18** 2087
[195] Patel NG and Lashkari BH 1992 *J. Mater. Sci.* **27** 3026
[196] Habermeier HU 1981 *Thin Solid Films* **80** 157
[197] Nath P, Bunshah RF, Basol BM and Staffsud OM 1980 *Thin Solid Films* **72** 463
[198] Krokoszinski HJ and Oesterlein R 1990 *Thin Solid Films* **187** 179
[199] Hjortsberg A, Hamberg I and Granqvist CG 1982 *Thin Solid Films* **90** 323
[200] Banerjee R, Das D, Ray S, Batabyal AK and Barua AK 1986 *Solar Energy Mater.* **13** 11
[201] Lee WK, Machino T and Sugihara T 1993 *Thin Solid Films* **224** 105
[202] Shigesato Y, Hayashi Y and Haranoh T 1992 *Appl. Phys. Lett.* **61** 73
[203] Theuwissen AJ and Declerck GJ 1984 *Thin Solid Films* **121** 109
[204] Kawada A 1990 *Thin Solid Films* **191** 297
[205] Larsson T, Blom HO, Nender C and Berg S 1988 *J. Vac. Sci. Technol.* A **6** 1832
[206] Naseem S and Coutts TJ 1986 *Thin Solid Films* **138** 65
[207] Ishibashi S, Higuchi Y, Ota Y and Nakamura K 1990 *J. Vac. Sci. Technol.* A **8** 1399
[208] Harding GL and Window B 1990 *Solar Energy Mater.* **20** 367
[209] Shigesato Y, Hayashi Y, Masui A and Haranoh T 1991 *Japan. J. Appl. Phys.* **30** 814
[210] Latz R, Michael K and Scherer M 1991 *Japan. J. Appl. Phys.* **30** L149
[211] Shigesato Y, Takaki S and Haranoh T 1992 *J. Appl. Phys.* **71** 3356
[212] Buchanan M, Webb JB and Williams DF 1981 *Thin Solid Films* **80** 373
[213] Echigoya J, Kato S and Enoki H 1992 *J. Mater. Sci.: Mater. Electron.* **3** 168
[214] Fan JCC, Bachner FJ and Poley GH 1977 *Appl. Phys. Lett.* **31** 773
[215] Tueta R and Braguier M 1981 *Thin Solid Films* **80** 143
[216] Dutta J and Ray S 1988 *Thin Solid Films* **162** 119
[217] Ray S, Dutta J and Barua AK 1991 *Thin Solid Films* **199** 201
[218] Sberveglieri G, Benussi P, Coccoli G, Groppelli S and Nelli P 1990 *Thin Solid Films* **186** 349

[219] Ishibashi S, Higuchi Y, Ota Y and Nakamura K 1990 *J. Vac. Sci. Technol.* A **8** 1403
[220] Hoheisel M, Heller S, Mrotzek C and Mitwalsky A 1990 *Solid State Commun.* **76** 1
[221] Geoffroy C, Campet G, Portier J, Salardenne J, Couturier G, Bourrel M, Chabagno JM, Ferry D and Quet C 1991 *Thin Solid Films* **202** 77
[222] Howson RP and Jafar HA 1992 *J. Vac. Sci. Technol.* A **10** 1784
[223] Martinez MA, Herrero J and Guiterrez MT 1992 *Solar Energy Mater. Solar Cells* **26** 309
[224] Zhang HW and Xu WY 1992 *Vacuum* **43** 835
[225] Karasawa T and Miyata Y 1993 *Thin Solid Films* **223** 135
[226] Fan JCC 1981 *Thin Solid Films* **80** 125
[227] Machet J, Guille J, Saulnier P and Robert S 1981 *Thin Solid Films* **80** 149
[228] Oyama T, Hashimoto N, Shimizu J, Akao Y, Kojima H, Aikawa K and Suzuki K 1992 *J. Vac. Sci. Technol.* A **10** 1682
[229] Howson RP, Ridge MI and Bishop CA 1981 *Thin Solid Films* **80** 137
[230] Goyal RP, Raviendra D and Gupta BRK 1985 *Phys. Status Solidi* a **87** 79
[231] Yamamoto O, Sasamoto T and Inagaki M 1992 *J. Mater. Res.* **7** 2488
[232] Maruyama T and Kojima A 1988 *Japan. J. Appl. Phys.* **27** L1829
[233] Haacke G 1977 *Ann. Rev. Mater. Sci.* **7** 73
[234] Agnihotri OP, Gupta BK and Sharma AK 1978 *J. Appl. Phys.* **49** 4540
[235] Ortiz A 1982 *J. Vac. Sci. Technol.* **20** 7
[236] Nojik AJ 1976 *Phys. Rev.* B **6** 453
[237] Enoki H, Satoh T and Echigoya J 1991 *Phys. Status Solidi* a **126** 163
[238] Haacke G 1976 *Appl. Phys. Lett.* **28** 622
[239] Haacke G 1978 *Thin Solid Films* **55** 55
[240] Haacke G, Mealmaker WE and Siegel LA 1978 *Thin Solid Films* **55** 67
[241] Miyata N, Miyake K, Fukushima T and Koga K 1979 *Appl. Phys. Lett.* **35** 542
[242] Miyata N, Miyake K, Koga K and Fukushima T 1980 *J. Electrochem. Soc.* **127** 918
[243] Leja E, Budzynska K, Pisarkiewicz T and Stapinski T 1983 *Thin Solid Films* **100** 203
[244] Stapinski T, Leja E and Pisarkiewicz T 1984 *J. Phys. D: Appl. Phys.* **17** 407
[245] Schiller S, Beister G, Buedke E, Becker HJ and Schicht H 1982 *Thin Solid Films* **96** 113
[246] Miyata N, Miyake K and Nao S 1979 *Thin Solid Films* **58** 385
[247] Miyata N and Miyake K 1978 *Japan. J. Appl. Phys.* **17** 1693
[248] Miyata N and Miyake K 1979 *Surf. Sci.* **86** 384
[249] Miyata N, Miyake K and Yamaguchi Y 1980 *Appl. Phys. Lett.* **37** 180
[250] Nakazawa T and Ito K 1988 *Japan. J. Appl. Phys.* **27** 1630
[251] Aoki M, Tada K and Murai T 1981 *Thin Solid Films* **83** 283
[252] Shimizu M, Horii T, Shiosaki T and Kawabata A 1982 *Thin Solid Films* **96** 149
[253] Ghandhi SK, Field RJ and Shelly JR 1980 *Appl. Phys. Lett.* **37** 449
[254] Kaufmann T, Fuchs G and Webert M 1988 *Cryst. Res. Technol.* **23** 635
[255] Roth AP and Williams DP 1981 *J. Appl. Phys.* **52** 6685
[256] Hu J and Gordon RG 1992 *J. Electrochem. Soc.* **139** 2014
[257] Hu J and Gordon RG 1992 *J. Appl. Phys.* **71** 880
[258] Tomar MS and Garcia FJ 1982 *Thin Solid Films* **90** 419

[259] Aranovich J, Ortiz A and Bube RH 1979 *J. Vac. Sci. Technol.* **16** 994
[260] Caillaud F, Smith A and Baumard J 1990 *J. Europ. Cer. Soc.* **6** 313
[261] Ortiz A, Garcia M and Falcony C 1992 *Thin Solid Films* **207** 175
[262] Ortiz A, Falcony C, Garcia M and Sanchez A 1987 *J. Phys. D: Appl. Phys.* **20** 670
[263] Major S, Banerjee A and Chopra KL 1983 *Thin Solid Films* **108** 333
[264] Nobbs JMcK and Gillespie FC 1970 *J. Phys. Chem. Solids* **31** 2353
[265] Aktaruzzaman AF, Sharma GL and Malhotra LK 1991 *Thin Solid Films* **198** 67
[266] Goyal D, Solanki P, Marathe B, Takwale M and Bhide V 1992 *Japan. J. Appl. Phys.* **31** 361
[267] Tsuji N, Komiyama H and Tanaka K 1990 *Japan. J. Appl. Phys.* **29** 835
[268] Ievlev VM, Shvedova OG, Belonogov EK and Seleznev AD 1991 *Inorg. Mater.* **27** 430
[269] Sundaram KB and Garside BK 1984 *J. Phys. D: Appl. Phys.* **17** 147
[270] Donaghey LF and Gerghty KG 1976 *Thin Solid Films* **38** 271
[271] Shinoki F and Itoh A 1975 *J. Appl. Phys.* **46** 3381
[272] Barnes JO, Leary DJ and Jordan AG 1980 *J. Electrochem. Soc.* **127** 1636
[273] Caporaletti O 1982 *Solar Energy Mater.* **7** 65
[274] Webb JB, Williams DF and Buchanan M 1980 *Appl. Phys. Lett.* **39** 640
[275] Murti DK and Bluhm TL 1982 *Thin Solid Films* **87** 57
[276] Qiu CX and Shih I 1986 *Solar Energy Mater.* **13** 75
[277] Shih I and Qiu CX 1985 *J. Appl. Phys.* **58** 2400
[278] Choi BH, Im HB, Song JS and Yoon KH 1990 *Thin Solid Films* **193/194** 712
[279] Minami T, Oohashi K, Takata S, Mouri T and Ogawa N 1990 *Thin Solid Films* **193/194** 721
[280] Igasaki Y and Saito H 1991 *J. Appl. Phys.* **70** 3613
[281] Brett MJ and Parsons RR 1986 *J. Vac. Sci. Technol.* **A4** 423
[282] Schropp R and Madan A 1989 *J. Appl. Phys.* **66** 2027
[283] Jin ZC, Hamberg I and Granqvist CG 1988 *J. Appl. Phys.* **64** 5117
[284] Matsuoka M and Ono K 1989 *J. Vac. Sci. Technol.* A **7** 2975
[285] Igasaki Y, Ishikawa M and Shimaska G 1988 *Appl. Surf. Sci.* **33–34** 926
[286] Minami T, Nanto H, Sato H and Takata S 1988 *Thin Solid Films* **164** 275
[287] Minami T, Tamura Y, Takata S, Mouri T and Ogawa N 1994 *Thin Solid Films* **246** 86
[288] Ruth M, Tuttle J, Goral J and Noufi R 1989 *J. Cryst. Growth* **96** 363
[289] Jin ZC, Hamberg I, Granqvist CG, Sernelius BE and Berggren KF 1988 *Thin Solid Films* **164** 381
[290] Ghosh S, Sarkar A, Bhattacharya S, Chaudhuri S and Pal AK 1991 *J. Cryst. Growth* **108** 534
[291] Meng LJ, Andritschky M and dos Santos MP 1994 *Vacuum* **45** 19
[292] Czternastek H, Brudnik A, Jachimowski M and Kolawa E 1992 *J. Phys. D: Appl. Phys.* **25** 865
[293] Nakada T, Ohkubo Y and Kunioka A 1991 *Japan. J. Appl. Phys.* **30** 3344
[294] Kobayashi K, Maeda T, Matsushima S and Okada G 1992 *J. Mater. Sci.* **27** 5953
[295] Sarkar A, Ghosh S, Chaudhuri S and Pal AK 1991 *Thin Solid Films* **204** 255
[296] Swamy HG and Reddy PJ 1990 *Semicond. Sci. Technol.* **5** 980
[297] Minami T, Sato H, Ohashi K, Tomofuji T and Takata S 1992 *J. Cryst. Growth* **117** 370

3. Electrical Properties

3.1 Introduction

Numerous investigations have been made of the electrical properties of transparent conducting oxide films in order to understand the conduction phenomena involved. A marked scattering of results is found mainly due to diversified methods of preparation and the large number of parameters involved, namely nature and temperature of substrate, growth rate, film thickness, post-deposition annealing effects, dopant and its concentration, etc. However, researchers have made a systematic study of the effects of these parameters on the electrical properties of these oxide films in order to optimize growth conditions.

All the semiconductor oxide films have n-type conductivity. The high conductivity of these films results mainly from stoichiometric deviation. The conduction electrons in these films are supplied from donor sites associated with oxygen vacancies or excess metal ions. These donor sites can easily be created by chemical reduction. Sometimes, unintentional doping usually with chlorine when oxide films are grown by chemical vapour deposition or a spray pyrolytic technique, also enhances the electrical conductivity. Intentional doping, e.g. with tin in indium oxide and antimony or fluorine in tin oxide, also helps in increasing their conductivity. Moreover, contamination by alkali ions from substrates of sheet glass may have a marked effect on electrical conductivity, especially for undoped layers deposited at high substrate temperatures greater than 500 °C. The early work on the electrical properties of these films has been reviewed by many workers [1–5]. This chapter starts with a brief introduction to the basic theory of electrical transport phenomena in semiconductor thin films. The electrical properties of transparent conducting films are discussed in detail later in this chapter.

3.2 Transport Phenomena in Semiconductor Films

Transport phenomenon is the term applied to the motion of charge carriers under the action of internal or external fields. In the absence of an electric field, the electron gas in a semiconductor is in an equilibrium state, which is established as a result of the interaction of electrons with lattice defects. Such defects include lattice imperfections, thermal vibrations of the lattice (phonons) and impurity atoms.

If an electric field $\boldsymbol{E}$ is applied to a material, an electric current will flow, whose density $\boldsymbol{J}$ is given by

$$\boldsymbol{J} = \sigma \boldsymbol{E} \tag{3.1}$$

where σ is called the electrical conductivity of the material. The reciprocal of electrical conductivity is known as electrical resistivity ρ. For a rectangular-shaped sample (figure 3.1(a)), the resistance R is given by

$$R = \rho(l/bt) \tag{3.2}$$

where l is the length, b is the width and t is the thickness of the sample. If $l = b$, equation 3.2 becomes

$$R = \rho/t = R_s.$$

The quantity R_s is known as the sheet resistance and it is the resistance of one square of the film and is independent of the size of the square. The sheet resistance is expressed in ohms/square.

The most commonly used method for measuring the sheet resistance R_s is a four-point probe technique. A typical schematic setup is shown in figure 3.1. When the probes are placed on a material of semi-infinite volume, the resistivity is given by [6]

$$\rho = \frac{V}{I}\frac{2\pi}{1/d_1 + 1/d_2 - 1/(d_1 + d_2) - 1/(d_2 + d_3)}.$$

When $d_1 = d_2 = d_3 = d$ (say),

$$\rho = \frac{V}{I} 2\pi d. \tag{3.3}$$

If the material is in the form of an infinitely thin film resting on an insulating support, equation 3.3 leads [6] to

$$\rho = \frac{V}{I}\frac{\pi t}{\ln 2}$$

or

$$\frac{\rho}{t} = R_s = 4.53V/I.$$

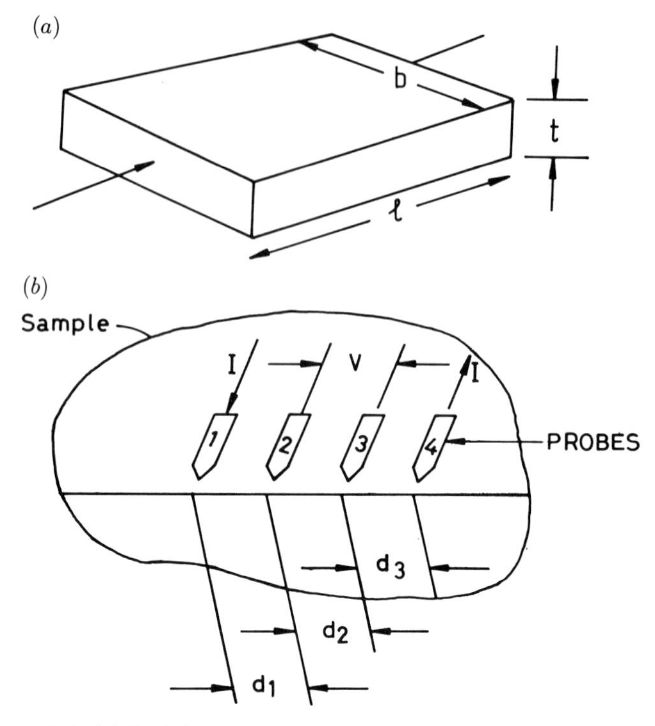

Figure 3.1 (a) Definition of resistivity and sheet resistance. (b) Four-point probe technique.

Under the influence of an electric field, the electrons begin to move in a specific direction and such directional motion is termed drift. The average velocity of this motion is known as drift velocity ($\boldsymbol{v}_{\mathrm{d}}$).

If N is the number density of electrons, the current density is given by

$$\boldsymbol{J} = Ne\boldsymbol{v}_{\mathrm{d}} \tag{3.4}$$

where e is the electron charge.

Combining equations (3.1) and 3.4)

$$\boldsymbol{v}_{\mathrm{d}} = \{\sigma/Ne\}\boldsymbol{E}. \tag{3.5}$$

Here the proportionality factor is called the mobility (μ) of charge carriers, i.e.

$$\mu = \sigma/Ne. \tag{3.6}$$

The charge carrier mobility μ, is related to the effective mass of the charge carriers (m^*) and the relaxation time (τ) according to

$$\mu = \frac{e\tau}{m^*}. \tag{3.7}$$

In the case of thin films, the conductivity is greatly influenced by the thickness of the films. The surface of a thin film affects the conduction of charge carriers by interrupting carrier transit along their mean free path. They may either be diffusely scattered, in which case they emerge from the surface with no memory of their previous velocity; or they may be specularly reflected so that only their velocity component perpendicular to the surface is reversed, their energy remaining constant. Any surface which is not completely specular in its behaviour will result in a decrease in conductivity of the films. In general, when some fraction p_s of the electrons is scattered specularly, while the rest are scattered diffusely, the effective conductivity is given by [7]

$$\sigma = \sigma_0 \left[1 + \frac{3}{8\gamma}(1 - p_s) \right]^{-1} \qquad \text{for } \gamma > 1 \tag{3.8}$$

and

$$\sigma = \sigma_0 \frac{3\gamma}{4}(1 + 2p_s) \ln\left(\frac{1}{\gamma}\right) \qquad \text{for } \gamma < 1 \tag{3.9}$$

where $\gamma = t/\lambda_f$; λ_f being the mean free path. In the case that the scattering at the surface is entirely diffuse, $p_s = 0$, and equations (3.8) and (3.9) reduce to

$$\sigma = \sigma_0 \Big/ \left(1 + \frac{3}{8\gamma}\right) \qquad \text{for } \gamma > 1 \tag{3.10}$$

and

$$\sigma = \sigma_0 \frac{3\gamma \ln 1/\gamma}{4}, \qquad \text{for } \gamma < 1. \tag{3.11}$$

In addition to size effects, the lattice impurity and the enormous number of structural defects in films also affect the conductivity.

It is often necessary to determine whether the sample is n-type or p-type. The conductivity measurement does not give this information since it cannot distinguish between hole and electron conduction. A Hall effect study is usually required to distinguish between the two types of carriers. It also allows determination of the density of the charge carriers.

When a current is passed through a slab of material in the presence of a transverse magnetic field (figure 3.2), a small potential difference, known as the Hall voltage, is developed (between face 1 and face 2), in a direction perpendicular to both the current and the magnetic field. Face 1 is positive for p-type samples, whereas it is negative for n-type samples. Mathematically, it is given by

$$V_H = R_H I (B/t) \tag{3.12}$$

where V_H is the Hall voltage, B is the magnetic field and I is the current

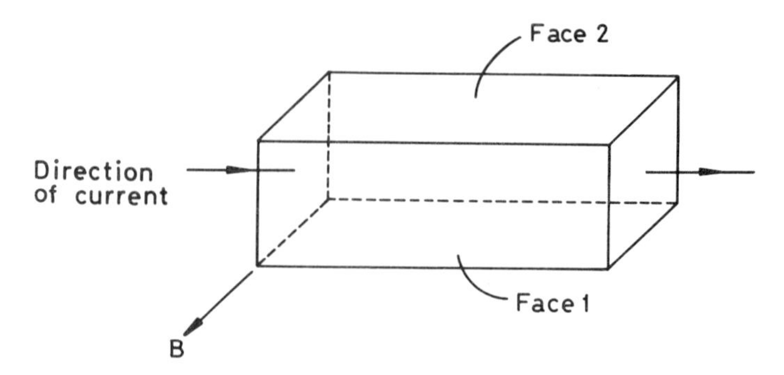

Figure 3.2 The Hall effect phenomenon.

through the sample. R_H is the Hall coefficient and is related to the carrier density according to the relation

$$R_H = r_H(1/Ne) \tag{3.13}$$

where r_H is the Hall scattering factor. The value of r_H depends on the geometry of the scattering surface and the mechanisms by which the carriers are scattered. In general, r_H does not depart very significantly from unity.

For n-type semiconductors, R_H is negative, while for p-type it is positive. Hall effect measurements, in conjunction with the measurement of conductivity, enable calculation of the mobility of charge carriers.

Relations (3.6) and (3.13) lead to

$$\mu = R_H\sigma. \tag{3.14}$$

The value of mobility of charge carriers determined by Hall measurements is known as Hall mobility (μ_H).

3.2.1 Electrical conduction in crystalline semiconducting materials

The expression for mobility of charge carriers (equation (3.7)) depends on the relaxation time, which further depends on the drift velocity and the mean free path of the charge carriers. These parameters, in turn, depend on the mechanism by which the carriers are scattered by the lattice imperfections. A brief account of various scattering mechanisms involved in semiconducting materials is given here.

(a) *Lattice scattering*

In addition to the various stationary imperfections, lattice vibrations also distort perfect lattice periodicity. The degree of distortion is a strong function of temperature. Lattice vibrations are categorized into acoustical and optical modes.

Acoustic deformation potential scattering. When an acoustic wave propagates in a crystal lattice, the atoms oscillate about their equilibrium position. Lattice scattering in non-polar materials is normally discussed in terms of deformation potential models. According to the Bardeen and Shockley model [8], the mobility for this type of scattering is given by

$$\mu_{ac} = \frac{2^{3/2}\sqrt{\pi}}{3} \frac{e\hbar^4 \rho_d U_l^2}{m^{*5/2}(kT)^{3/2} E_{ac}^2} \tag{3.15}$$

where ρ_d is the density of the material, U_l is the longitudinal sound velocity, E_{ac} is the deformation potential constant and k is Boltzmann's constant.

Piezoelectric scattering. The scattering of carriers by the lattice may be due to the strains produced by lattice vibrations. This is more pronounced when a semiconductor crystal consists of dissimilar atoms, where the bonds are partly ionic and the unit cell does not contain a centre of symmetry. The mobility due to this type of scattering is given by [9]

$$\mu_{PZ} = \frac{16\sqrt{2\pi}}{3} \frac{\hbar^2 \varepsilon_s \varepsilon_0}{m^{*3/2} e K_c^2 (kT)^{1/2}} \tag{3.16}$$

where K_c is the electromechanical coupling constant, ε_s is the permittivity of the semiconductor, and ε_0 is the permittivity of free space.

Optical phonon scattering. In vibrations associated with optical phonons, the neighbouring atoms in a crystal vibrate in opposite phase. These vibrations may produce a strain called the optical strain, which is measured in terms of the displacement of the sublattice of one type of atom with respect to the sublattice containing the other type of atom. The perturbing potential arising from such strains depends on the orientations of the constant energy surfaces with respect to the crystallographic axes. This optical phonon scattering, whether of non-polar or polar nature, will be important only when either the lattice temperature or the phonon temperature is higher than the Debye temperature θ_D

In the case of non-polar optical phonon scattering, the mobility is given by [10]

$$\mu_{NPO} = \frac{2^{5/2}\sqrt{\pi} e\hbar^2 \rho_d (k\theta_D)^{1/2}}{3m^{*5/2} E_{NPO}^2} f\left(\frac{T}{\theta_D}\right) \tag{3.17}$$

where E_{NPO} is the non-polar optical deformation potential constant and

$$f\left(\frac{T}{\theta_D}\right) = \left(\frac{\theta_D}{T}\right)^{5/2} \left\{\exp\left(\frac{\theta_D}{T}\right) - 1\right\} \int_0^\infty \frac{y^{3/2} \exp(-\theta_D y/T)\, dy}{\sqrt{y+1} + \exp(\theta_D/T) \mathrm{Re}\sqrt{y-1}} \tag{3.18}$$

where $y = E'/k\,\theta_D$ and E′ is the carrier energy.

In the case of polar optical phonon scattering, the mobility is given by [11]

$$\mu_{PO} = A_{PO} T\chi\left(\frac{\theta_D}{T}\right)\left\{\exp\left(\frac{\theta_D}{T}\right) - 1\right\} y^{1/2} \tag{3.19}$$

where

$$A_{PO} = \frac{\sqrt{\pi}\hbar^2}{\sqrt{2K\theta_D} m^{*3/2} e}\left\{\frac{1}{\varepsilon_\infty} - \frac{1}{\varepsilon_s}\right\} \tag{3.20}$$

and ε_∞ and ε_s are the high frequency and static dielectric constants, respectively. $\chi(\theta_D/T)$ is a function of temperature and has been evaluated by Putley [12].

(b) *Neutral impurity scattering*
The scattering of carriers by a neutral impurity atom in the crystal lattice is similar to the scattering of low energy electrons in a gas. According to Erginsoy [13], the mobility due to neutral impurity scattering is given by

$$\mu_N = \frac{m^* e^3}{20\hbar^3 \varepsilon_0 \varepsilon_s N_n} \tag{3.21}$$

where N_n is the concentration of neutral impurities.

(c) *Ionized impurity scattering*
Of all the impurities that may be present in the crystal, the greatest effect on the scattering of the carriers is produced by ionized impurities. This is because the electrostatic field due to such impurities remains effective even at a great distance. The treatment of Brooks and Herring [14] is most widely used for non-degenerate semiconductors and the corresponding relation is

$$\mu_{IIS} = \frac{2^{7/2}(4\pi\varepsilon_0\varepsilon_s)^2(kT)^{3/2}}{\pi^{3/2} Z^2 e^3 m^{*1/2} N_I\{\ln(1 + y_{BH}) - y_{BH}^2/(1 + y_{BH}^2)\}} \tag{3.22}$$

where

$$y_{BH} = \frac{2m^*}{\hbar}\left(\frac{2}{m^*} \times 3kT\right)^{1/2} L_D \tag{3.23}$$

and L_D is the Debye length and N_I is the concentraton of ionized impurities.

However, in the case of degenerate semiconductors, the contribution of ionized impurity scattering is given by [15]

$$\mu_{IIS} = \frac{4e}{h}\left(\frac{\pi}{3}\right)^{1/3} N^{-2/3}. \tag{3.24}$$

(d) *Electron–electron scattering*
Electron–electron scattering has little influence on mobility because in this process the total momentum of the electron gas is not changed. However, it

is always combined with another scattering mechanism, which is influenced by it. Typically, for a non-degenerate semiconductor dominated by ionized impurity scattering, the mobility is reduced by 60%, whereas in the case of degenerate semiconductors, there is no reduction [16, 17].

3.2.2 Electrical conduction in polycrystalline thin films

In addition to the scattering mechanisms discussed above, grain boundary scattering is another important scattering mechanism in polycrystalline semiconductors. In polycrystalline thin films, the conduction mechanism is dominated by the inherent inter-crystalline boundaries (grain boundaries) rather than the intra-crystalline characteristics. These boundaries generally contain fairly high densities of interface states which trap free carriers from the bulk of the grain and scatter free carriers by virtue of the inherent disorders and the presence of trapped charges. The interface states result in a space charge region in the grain boundaries. Due to this space charge region, band bending occurs, resulting in potential barriers to charge transport. The most commonly used model to explain the transport phenomenon in polycrystalline films is due to Petritz [18]. According to this model, the current density is given by the relation

$$\boldsymbol{J} = \left\{ e\mu_0 N \exp\left(\frac{-e\phi_\mathrm{b}}{kT}\right) \right\} \boldsymbol{E} \tag{3.25}$$

where $\mu_0 = (M/n_\mathrm{c}kT)$, ϕ_b is the height of the potential barrier, n_c is the number of crystallites per unit length along the film, and M is a factor that is barrier dependent.

The grain boundary potential barrier ϕ_b is related to N_1 and N_2, the number of carriers in the grain and the grain boundary, respectively, by

$$\phi_\mathrm{b} = kT \ln\left(\frac{N_1}{N_2}\right). \tag{3.26}$$

The quantity in the bracket in equation (3.25) is nothing other than the conductivity of charge carriers dominated by grain boundaries (σ_g).

$$\sigma_\mathrm{g} = eN\mu_0 \exp\left(\frac{-e\phi_\mathrm{b}}{kT}\right). \tag{3.27}$$

Thus the grain boundary limited mobility can be written as

$$\mu_\mathrm{g} = \mu_0 \exp\left(\frac{-e\phi_\mathrm{b}}{kT}\right). \tag{3.28}$$

Seto [19] modified the pre-exponential term in equation (3.28) on the assumption that (i) current flows between grains by thermionic emission and

(ii) conduction in the crystallites is much higher than that through the grain boundary. The resulting relation of mobility is

$$\mu_g = el'(2\pi m^* kT)^{-1/2} \exp\left(\frac{-e\phi_b}{kT}\right) \tag{3.29}$$

where l' is the grain size.

Orton *et al* [20] showed that the pre-exponential term can be written in a more generalized way as

$$\mu_0 = el' \{8/(\pi\beta^2 kTm^*)\}^{1/2} \tag{3.30}$$

where β is a numerical constant.

In general, the grain boundary mobility can be written as

$$\mu_g = \mu_0' T^{-1/2} \exp\left(\frac{-e\phi_b}{kT}\right) \tag{3.31}$$

where $\mu_0 = \mu_0' T^{-1/2}$.

Later work [21–26] has shown that the conductivity term in equation (3.27) should be written more generally as

$$\sigma_g = \sigma_0 \exp\left(\frac{-E_\sigma}{kT}\right) \tag{3.32}$$

where E_σ is the conductivity activation energy.

Similarly,

$$N = N_0 \exp\left(-\frac{E_n}{kT}\right) \tag{3.33}$$

where E_n is the carrier activation energy.

Further

$$E_\sigma = E_n + e\phi_b. \tag{3.34}$$

However, it should be noted that for degenerate samples, the effect of grain boundary potential barrier is not so significant [27].

3.2.3 Conduction mechanism in amorphous materials

In amorphous materials, the hopping process is the most dominant conduction mechanism. Conduction by hopping results in conductivity of the form

$$\sigma = \frac{\sigma_0'}{T^{1/2}} \exp\left[-\left(\frac{T_0}{T}\right)^x\right] \tag{3.35}$$

where the value of x depends on the dimensionality and nature of the hopping process. When conduction is three-dimensional, variable range hopping (VRH) gives $x = 0.25$ for a constant density of states and $x = 0.5$ for a parabolic density of states [28].

In amorphous materials, VRH conduction occurs at temperatures at which the phonons do not have sufficient energy for transfer to a nearest-neighbour atom, and the charge carriers hop from a neutral atom to another neutral atom situated at the same energy level, which can be many inter-atomic distances away. The most commonly used model for VRH is due to Mott [29]. Mott's theory is based on a number of assumptions, of which the most important are (i) energy independence of the density of states, (ii) neglect of correlation effects in the tunnelling process, (iii) omission of multiphonon process, and (iv) neglect of electron–phonon interaction. According to Mott's theory

$$\sigma = \frac{\sigma_0'}{T^{1/2}} \exp\left[-\left(\frac{T_0}{T}\right)^{1/4}\right] \tag{3.36}$$

$$\sigma_0' = 3e^2\nu_{\mathrm{ph}}\left[\frac{N(E_{\mathrm{F}})}{8\pi\alpha k}\right]^{1/2} \tag{3.37}$$

$$T_0 = \frac{\lambda_{\mathrm{c}}\alpha^3}{kN(E_{\mathrm{F}})} \tag{3.38}$$

$$R = \left[\frac{9}{8\pi\alpha kTN(E_{\mathrm{F}})}\right]^{1/4} \tag{3.39}$$

$$W = \frac{3}{4\pi R^3 N(E_{\mathrm{F}})} \tag{3.40}$$

where $N(E_{\mathrm{F}})$ is the density of states near the Fermi level, λ_{c} is a dimensionless constant ($\simeq 18$), ν_{ph} is a frequency factor taken here as Debye frequency ($\simeq 3.3 \times 10^{12}$ Hz), and α is the decay constant of the wave function of the localized states near the Fermi level, R is the mean hopping distance and W is the hopping energy.

3.3 Electrical Properties of Undoped Tin Oxide Films

Tin oxide films can be fabricated by means of a number of techniques as discussed in chapter 2. The electrical properties of these films depend on the different growth parameters involved in these techniques. Some of the important parameters which most influence electrical properties are substrate temperature, film thickness, post-deposition annealing, gas flow rate, composition of the spray solution and sputtering power. Many workers [30–46, 48, 50–73] have investigated in detail the effect of these parameters on the growth of tin oxide films in order to optimize the electrical properties of these films.

(a) *Effect of deposition temperature*
Several workers [30–37] have tried to optimize the deposition substrate temperature in order to grow highly conductive SnO_2 films. Figure 3.3 shows the variation of sheet resistance R_s with deposition temperature for SnO_x ($x \simeq 2$) films deposited by a CVD technique. It may be observed that the films deposited at a temperature of 600 °C have sheet resistance as low as $10^3\ \Omega\ \square^{-1}$, while films deposited at temperatures below 600 °C have significantly higher sheet resistance. Scanning electron micrograph studies on these films revealed the presence of sharp cornered pits on the smooth surface, whose density decreased with increase of deposition temperature. The presence of these pits results in a decrease in mobility and thus affects sheet resistance. Croitoru and Bannett [31] observed a change in resistivity of three to four orders of magnitude due to change in substrate temperature (figure 3.4). It may be further seen from figure 3.4 that the variation in resistivity with temperature is less in films deposited at higher substrate temperatures than it is in those grown at lower substrate temperatures. These observations suggest that the contribution of grain boundary scattering potential decreases as the substrate temperature increases. Shanthi *et al* [32] reported the effect of substrate temperature on tin oxide films grown on glass substrate by a spray pyrolysis technique. Typically, resistivity as low as $2 \times 20^{-2}\ \Omega$ cm was observed at a substrate temperature of 540 °C. Figure 3.5 shows the variation of resistivity ρ, carrier concentration N and mobility μ with substrate temperature. It is observed that films grown at a substrate temperature of 540 °C have a minimum value of resistivity and maximum value of mobility. The increase in mobility with substrate temperature can

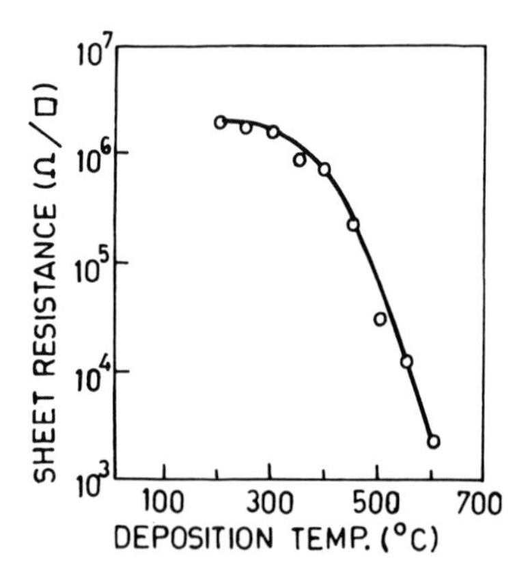

Figure 3.3 Variation of sheet resistance with deposition temperature for CVD grown SnO_x ($x \simeq 2$) films (from [30]).

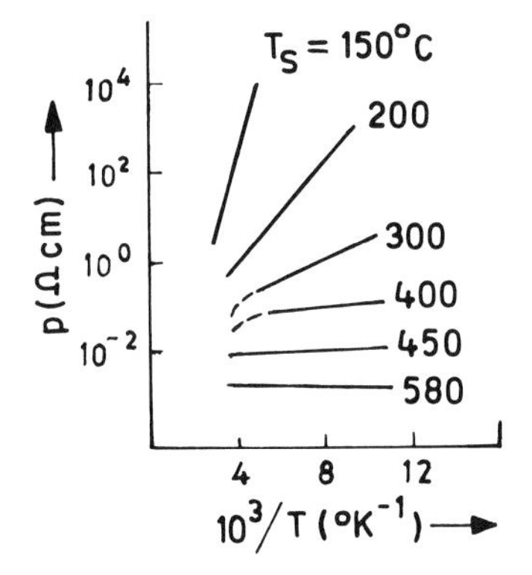

Figure 3.4 Temperature dependence of resistivity of SnO_2 films deposited at various substrate temperatures (from [31]).

be explained on the basis of grain boundary scattering. With an increase in substrate temperature, the grain size increases [33] causing a decrease in grain boundary potential and hence an increase in mobility. The decrease in grain boundary potential is also responsible for an increase in carrier concentration with substrate temperature in accordance with relation [3.33]

$$N = N_0 e^{-E_n/kT} \tag{3.33}$$

Vasu and Subrahmanyam [34,35] studied the effect of water content in the initial solution $SnCl_4$ in addition to the effect of substrate temperature on the electrical properties of spray-deposited tin oxide films. The starting material

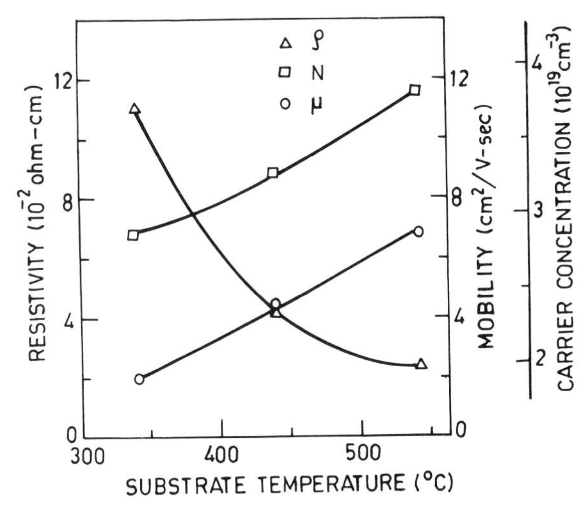

Figure 3.5 Variation of ρ, μ and N with substrate temperature for spray-deposited SnO_2 films (from [32]).

was anhydrous $SnCl_4$ as well as pentahydrate $SnCl_4 : 5H_2O$. It was observed that with an increase of substrate temperature, mobility and carrier concentration increased while sheet resistance decreased. However, for any given substrate temperature, the mobility and sheet resistance were much higher for films grown from the anhydrous solution of $SnCl_4$. On the other hand, the carrier concentration was significantly lower in these films. Typical values of R_s, N and u_H for films grown at substrate temperature 713 K from hydrous and anhydrous solutions were $29\,\Omega\,\square^{-1}$, $2.4 \times 10^{20}\,cm^{-3}$, $11.4\,cm^2\,V^{-1}\,s$ and $167\,\Omega\,\square^{-1}$, $1.26 \times 10^{20}\,cm^{-3}$, $13.4\,cm^2\,V^{-1}\,s$, respectively. A possible reason for the higher value of carrier concentration and lower value of mobility in films prepared using hydrous solution is the presence of impurities [36].

(b) *Effect of film thickness*

The resistivity ρ of SnO_2 films is reported [38] to be almost independent of film thickness (t) if it is greater than 200 nm; for lower values of t, ρ decreases significantly with increase of t. The decrease in ρ with t is more pronounced for film deposited at low substrate temperatures. A typical variation of resistivity ρ and sheet resistance R_S with film thickness is shown in figure 3.6. The rapid increase of resistivity with decrease of film thickness in the case of very thin films is due to the size effect as discussed earlier in section 3.2. Kuznetsov [39] also observed that in SnO_2 films grown by hydrolysis of $SnCl_4$, electrical conductivity is independent of film thickness between 500 and 2500 Å. Results on the nature of the variation of sheet resistance with film thickness by Omar *et al* [40] are in agreement with those of Murty *et al* [38]. However, they observed two distinct zones (figure 3.7) wherein (i) the sheet resistance decreased rapidly when the thickness varied between 50 and

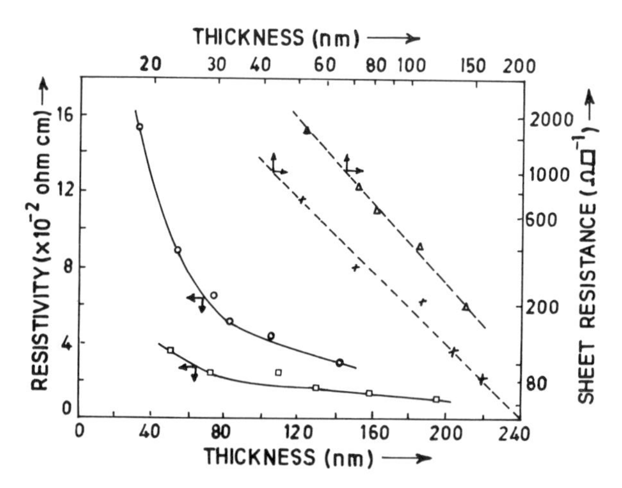

Figure 3.6 Variation of resistivity (———) and sheet resistance (– – – –) as a function of film thickness for CVD grown SnO_2 films deposited at 400 °C (○, △) and at 500 °C (□, ×) (from [38]).

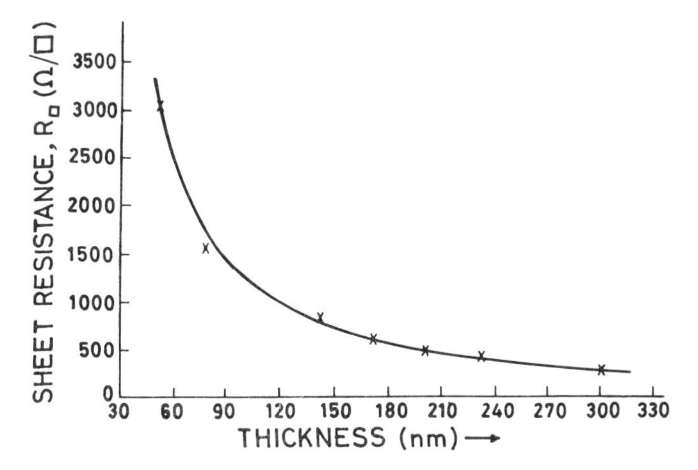

Figure 3.7 Variation of sheet resistance versus film thickness for SnO_2 films (from [40]).

120 nm, and (ii) the sheet resistance decreased moderately when the thickness varied from 120 to 300 nm.

(c) *The effect of oxygen partial pressure*

The oxygen gas flow rate is an important deposition parameter in CVD and sputtering techniques, and influences the properties of these films [30, 38, 41–46]. The effect of oxygen flow rate on the sheet resistance of SnO_2 films grown by a CVD technique is shown in figure 3.8. It can be observed that the sheet resistance initially decreases, reaches a minimum and then increases again with an increase in oxygen flow rate $F(O_2)$. The relatively high value of

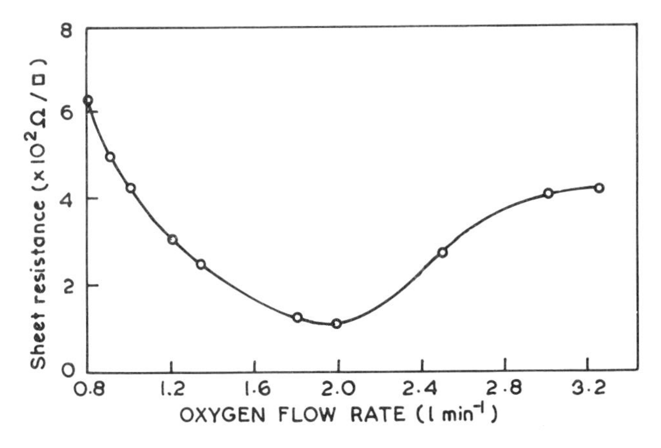

Figure 3.8 Variation of sheet resistance as a function of oxygen flow rate for CVD deposited SnO_2 films grown at 500 °C (from [41]).

R_S for low values of $F(O_2)$ may be due to the presence of highly resistive SnO and/or Sn_3O_4 phases. The decrease in concentration of these phases with initial increase in $F(O_2)$ results in a decrease of R_s. The minimum is reached when SnO_2 films have the optimum concentration of oxygen vacancies. A further increase in $F(O_2)$ results in filling of oxygen vacancies in SnO_2 and hence an increase in R_S. However, Lou *et al* [30] observed that the sheet resistance of SnO_x ($x \approx 2$) films prepared by a CVD technique is mainly controlled by the flow rate of $SnCl_4$ vapour and not by the flow rate of O_2 as long as the flow rate of O_2 is less than 4.1×10^{-2} mol min^{-1} (figures 3.9 and 3.10). For a given deposition duration, the sheet resistance decreases with increase in $SnCl_4$ flow rate. The sheet resistance is also found to decrease as the deposition duration is increased, indicating that SnO_x films prepared at a high $SnCl_4$ flow rate and long deposition time have better electrical properties. However, the sheet resistance is limited to about 10^4 $\Omega\,\square^{-1}$ for flow rates of O_2 of 4.1×10^{-2} mol min^{-1} and deposition time more than 5 min. This saturation of sheet resistance has been attributed to the amorphous nature of the films.

Many researchers [42–46] have studied the effect of O_2 partial pressure on the electrical properties of SnO_2 sputtered films. Figure 3.11 illustrates the

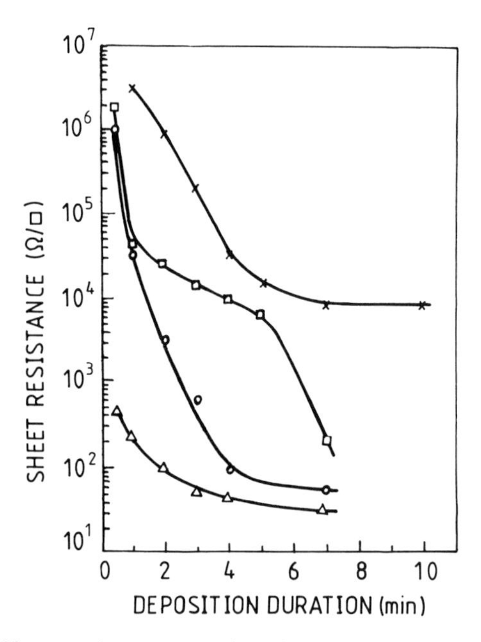

Figure 3.9 Sheet resistance as a function of deposition duration for CVD grown SnO_2 films with oxygen flow rate 4.1×10^{-2} mol min^{-1} and $SnCl_4$ vapour flow rate (×) 2.32×10^{-3} mol min^{-1}, (□) 4.64×10^{-3} mol min^{-1}, (○) 6.96×10^{-3} mol min^{-1}, (△) 9.28×10^{-3} mol min^{-1} (from [30]).

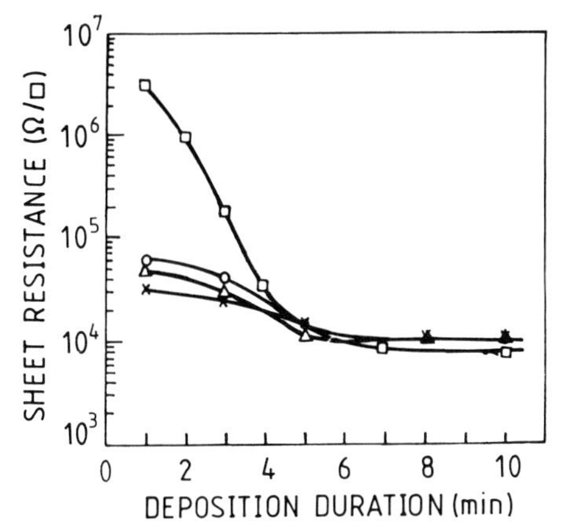

Figure 3.10 Sheet resistance as a function of deposition duration for CVD grown SnO_2 films with $SnCl_4$ vapour flow rate 2.32×10^{-3} mol min^{-1} and O_2 flow rate (□) 4.1×10^{-2} mol min^{-1}, (△) 2.03×10^{-3} mol min^{-1}, (○) 3.05×10^{-2} mol min^{-1}, (×) 1.02×10^{-2} mol min^{-1} (from [30]).

dependence of ρ, N and u_H on oxygen partial pressure $p(O_2)$ during sputtering. The films were grown at 300 °C using a hot pressed SnO_2 target. It can be seen from the figure that the resistivity passes through a minimum at a partial pressure of ~1.25 mTorr. On the other hand, mobility and carrier concentration attain a maximum value at this partial pressure. For low values of oxygen partial pressure, the presence of SnO and Sn_3O_4 phases are responsible for high resistivity and low carrier concentration. With an

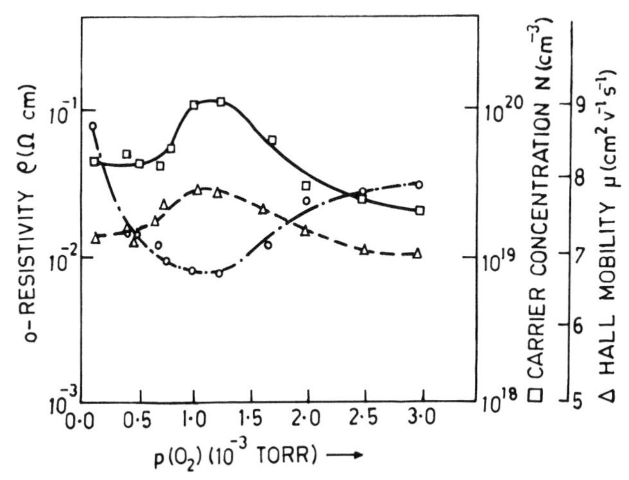

Figure 3.11 Variation of resistivity (○), carrier concentration (□) and Hall mobility (△) with oxygen partial pressure (from [42]).

increase in oxygen partial pressure, the carrier concentration becomes maximum and resistivity becomes minimum due to the formation of shallow donor levels. Above the optimized value of oxygen partial pressure, the carrier concentration decreases because of the incorporation of oxygen atoms, thereby reducing the oxygen vacancies. It should be noted that the nature of the resistivity curves for films grown by a CVD or sputtering technique is similar (figures 3.8 and 3.11).

Stjerna and Granqvist [43, 44] also reported similar results for the variation of sheet resistance with oxygen partial pressure in SnO_2 films grown by RF magnetron sputtering using a Sn target instead of a SnO_2 target. Further, the optimum value of O_2/Ar flow ratio as well as the corresponding minimum value of sheet resistance are strong functions of the sputtering power. Typically the optimum values of the O_2/Ar flow ratio are ~4.2 and 26.4% for 10 W and 150 W sputtering power, respectively. Figure 3.12 shows the variation of minimum sheet resistance with sputtering power for tin oxide films grown using an optimized O_2/Ar gas flow ratio. It can be noticed that the minimum value of sheet resistance, R_{sm} increases linearly with increase of sputtering power P_{rf} in accordance with the relation.

$$R_{sm} = 90 + 0.9P_{rf} \tag{3.41}$$

Croitoru *et al* [45,46] also reported an increase in the value of resistivity with sputtering power. This effect has been attributed to compositional change due to the variation of x in SnO_x films. Typically, the value of x varies from 1.35 to 1.8 when the RF power increases from 300 to 800 W.

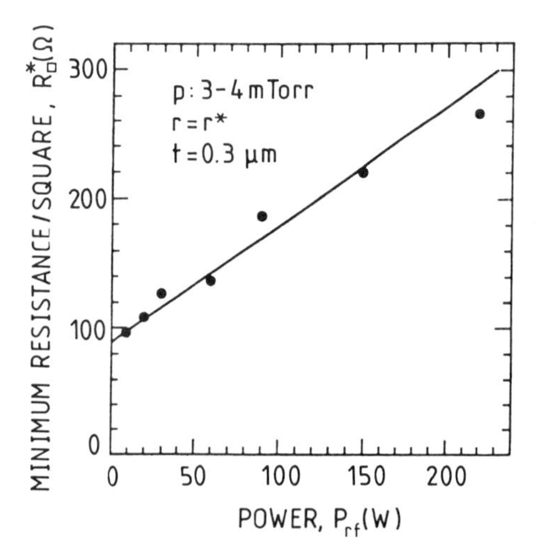

Figure 3.12 Minimum resistance/square versus RF power for SnO_2 films (from [43]).

(d) *Conduction mechanisms*

The scattering mechanisms involved in the electron conduction process in tin oxide films are not well established. Workers have explained their results on the basis of different scattering processes. Figure 3.13 shows a typical variation of Hall mobility with temperature for SnO_2 films prepared by a spray pyrolysis technique. It may be observed that the mobility increases with increase of temperature in the temperature range 200–300 K. The increase in mobility can be attributed to the presence of grain boundary scattering. The results have been analysed [32] in accordance with Seto's model [19]:

$$\mu_g \simeq \mu_H = \mu_0' T^{-1/2} \exp\left(\frac{-e\phi_b}{kT}\right) \tag{3.31}$$

According to this relation ln $\mu_H T^{1/2}$ against $1/T$ gives the value of ϕ_b. Such a plot is shown in figure 3.14 and the value of ϕ_b has been estimated to be about 30 meV. It is worth mentioning that Seto's model is based on the use of Maxwell–Boltzmann statistics for the electron gas and is valid only for the non-degenerate case. For a degenerate gas where Fermi–Dirac statistics [47] are used, the mobility is given by

$$\mu_H \simeq \mu_g = A' T \exp\left(\frac{-\phi_a}{kT}\right) \tag{3.42}$$

where $A' = 4\pi m^* e^2 l'/h^3$ and ϕ_a represents an activation energy related to ϕ_b by the expression

$$\phi_a = \phi_b - (E_f - E_c) \tag{3.43}$$

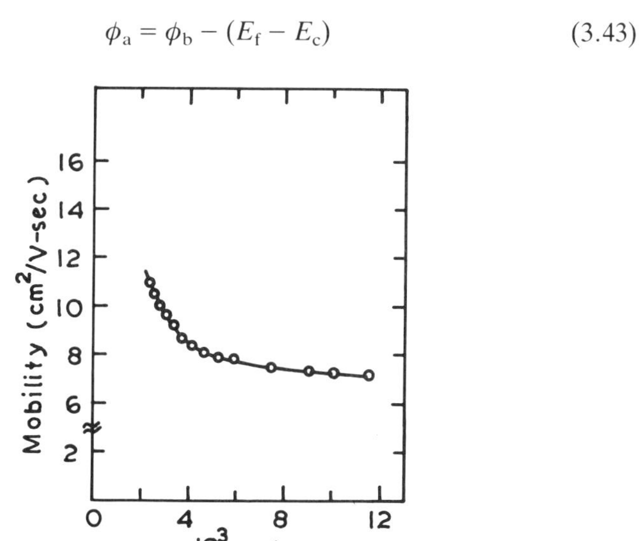

Figure 3.13 Variation of μ_H with temperature for spray-deposited SnO_2 films (from [32]).

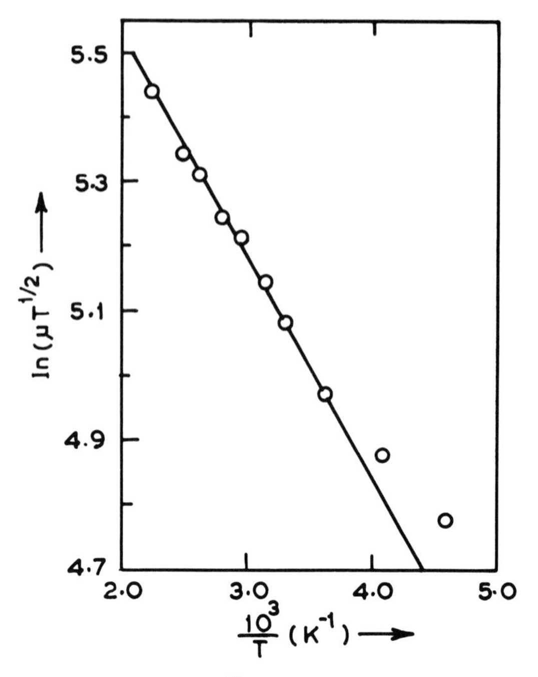

Figure 3.14 Plot of $\ln(\mu_H T^{1/2})$ versus $1/T$ for spray-deposited SnO_2 films (from [32]).

where $(E_f - E_c)$ is the distance between the Fermi level and the bottom of the conduction band.

When the applied voltage across a grain is distributed between the neutral bulk grain and the space charge region created by the grain boundaries, the mobility is the sum of contributions from bulk as well as grain boundaries:

$$\frac{1}{\mu_H} = \frac{1}{\mu_{bulk}} + \frac{1}{\mu_g} \tag{3.44}$$

The change of electron statistics leads to a different temperature dependence for the mobility μ_H, i.e. $\ln(\mu_H T^{1/2})$, or $\ln(\mu_H/T)$ is proportional to $1/T$ for a non-degenerate or a degenerate material, respectively. In the case of tin oxide films, there can be three regions, depending on the value of carrier concentration in the samples on the basis of which the mobility data can be analysed: (i) $N < 10^{18}\ cm^{-3}$, mobility is governed by the barrier effects at grain boundaries and as such Seto's model (equation (3.31)) can be applied; (ii) $N > 10^{19}\ cm^{-3}$, the mobility depends only on bulk transport properties, and the contribution of grain boundaries is almost negligible; (iii) $10^{18}\ cm^{-3} < N < 10^{19}\ cm^{-3}$, the two contributions, namely bulk properties and grain boundary effects based on film degeneracy, should be considered. In this case μ_g should be estimated using the relation (3.42).

Figure 3.15 shows the mobility temperature variation for spray-deposited undoped SnO_2 films with carrier concentration $4 \times 10^{18}\ cm^{-3}$. The curves (a), (b) and (c) represent observed μ, observed μ/T and μ/T corrected for bulk contribution, respectively. It may be observed that the mobility increases with temperature in the range 220–300 K. As the sample is degenerate, the data have been analysed in accordance with relations (3.42) and (3.44). The activation energy is found to be 0.028 eV using curve (b) and 0.042 eV using curve (c). Since the bulk contribution must be considered in this case, the value of ϕ_a estimated using curve (c) is more realistic. Knowing that the value of $E_F - E_C = 0.018$ eV for SnO_2, the value of ϕ_b is estimated to be ~0.06 eV according to relation (3.43). This value of ϕ_b is greater than $2kT$ indicating that the conduction is governed by thermionic emission [49]. Islam and Hakim [50], on the basis of thermoelectric power studies, also concluded that the grain boundary scattering process dominates other scattering processes in polycrystalline SnO_2 films. However, ionized impurity scattering becomes dominant when the carrier concentration exceeds $10^{20}\ cm^{-3}$.

Dewall and Simonis [37], on the other hand, reported that the mobility of undoped SnO_2 films decreases with an increase in temperature in the temperature range 0 to 300 °C. This observed decrease of mobility with temperature is probably because of the predominance of lattice scattering in this temperature range, in their films.

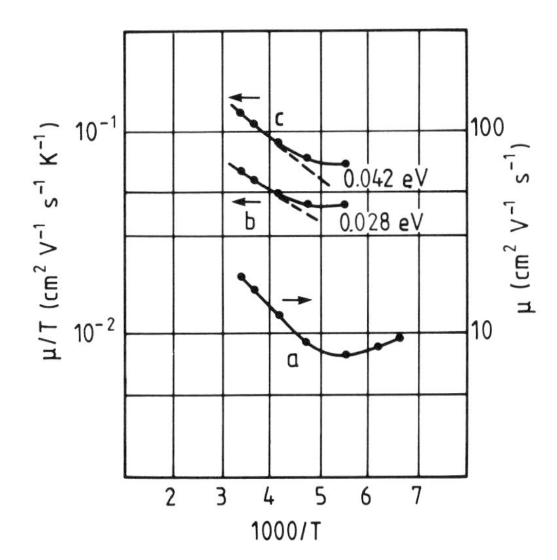

Figure 3.15 Temperature dependence of Hall mobility μ and μ/T for a SnO_2 film with carrier concentration $4.2 \times 10^{18}\ cm^{-3}$ (from [48]).

(e) *Post-deposition annealing*

Post-deposition annealing [30, 50–67] of tin oxide films in different atmospheres such as air, vacuum, oxygen, nitrogen, forming gas (80% H_2 and 20% N_2), hydrogen, etc. further improves their electrical properties. The effects of annealing are complex, and several phenomena may take part in bringing about the observed changes. These include: (a) crystallinity of the film may improve, thereby increasing the grain size; (b) the highly resistive SnO phase which is present [51] in addition to SnO_2 phase in the film, may change to the relatively low resistive SnO_2 phase; (c) chemisorption and desorption of oxygen from the grain boundaries may occur [52–54]. These phenomena may not be equally effective in all cases. Phenomenon (a) may take place in cases where the deposition temperature is much less than the annealing temperature. As SnO phase is less stable than SnO_2 phase, the phase transition from SnO to SnO_2 phase may take place in air. Such a transition is not possible in vacuum. Since oxide films are normally polycrystalline in nature, the process of oxygen chemisorption and desorption is important. As the major source of donors in these films is oxygen vacancies, the incorporation of oxygen states in the grain boundaries is important. At any annealing temperature, the concentration of oxygen states at the grain boundaries relative to the ambient oxygen pressure over the film, determines whether chemisorption or desorption takes place to reach an equilibrium.

Figure 3.16 shows the variation in sheet resistance of SnO_x ($x \sim 2$) films during air annealing at various temperatures between 275 and 550 °C. The films were grown using a CVD technique at a substrate temperature of

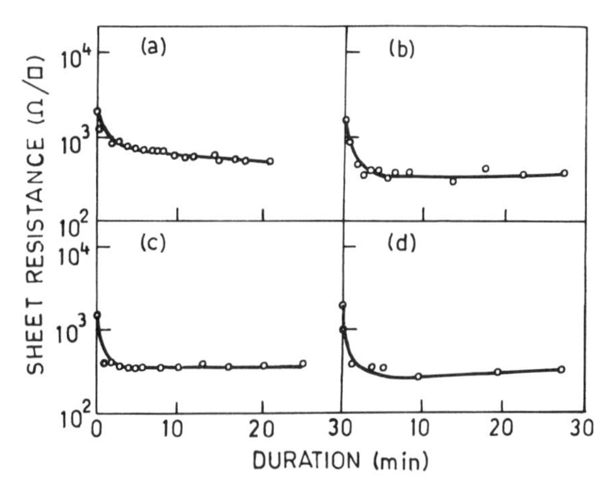

Figure 3.16 Variation of sheet resistance with annealing duration for CVD grown SnO_x, ($x \simeq 2$) films annealed at (a) 275 °C, (b) 325 °C, (c) 350 °C, (d) 550 °C (from [30]).

600 °C. The sheet resistance is found to decrease significantly after air annealing. The decrease in sheet resistance is significant only in the first few minutes of annealing: thereafter, it becomes almost constant. This change in sheet resistance has been explained on the basis of the presence of highly resistive SnO phase in as-deposited SnO_x films. This SnO phase decomposes to less resistive SnO_2 phase at temperatures higher than 300 °C in accordance with [55, 56]

$$2SnO \rightarrow Sn + SnO_2 \tag{3.45}$$

Islam and Hakim [50] studied the effect of post-deposition heat treatment on polycrystalline SnO_2 thin films prepared by a pyrosol process. During post-deposition heat treatment, the films were subjected to a heating cycle from 27 to 250 °C followed by a cooling cycle in the same ambient, i.e. either in air or in vacuum of the order 10^{-1} to 10^{-5} Torr. The studies revealed that post-deposition heat treatment significantly affects the film resistivity, without any appreciable change of carrier concentration. Typically the values of ρ and N are $4 \times 10^{-3}\ \Omega$ cm and 1.3×10^{20} cm^{-3} for as-deposited and $2.27 \times 10^{-3}\ \Omega$ cm and 1.3×10^{20} cm^{-3} for vacuum annealed SnO_2 films. Hence, the change in resistivity is associated with a change in mobility of the carriers. A typical variation of resistivity ρ with temperature T is shown in figure 3.17. AB and CD represent heating and cooling cycles in air,

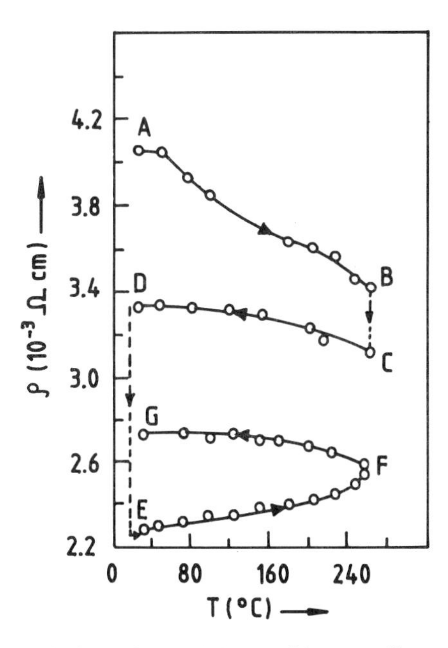

Figure 3.17 Variation of resistivity ρ with annealing temperature for spray-deposited SnO_2 films (from [50]).

BC represents the change of ρ at a constant temperature of 250 °C in one hour. The resistivity drops from point D to E as a result of heat treatment in vacuum. The EFG cycle represents a subsequent heat treatment in air. It can be observed that ρ decreases with increasing temperature in the case of as-deposited films (cycle AB), while it increases with increasing temperature for films which are initially heated in vacuum (cycle EF). Since the films are polycrystalline in nature, oxygen chemisorption/desorption mechanisms at the grain boundary may be considered to be responsible for bringing about the observed changes. It is presumed that as-deposited films have an excess of chemisorbed oxygen and on heat treatment, are subjected to desorption of oxygen from the grain boundaries. This lowers the resistivity by lowering the inter-grain boundary barrier height. However, the increase in resistivity with temperature for the film first subjected to vacuum heat treatment indicates that oxygen chemisorption has taken place. In air at normal pressure, if a heating cycle produces desorption, the cooling cycle will produce chemisorption and vice versa. However, in vacuum, the desorption process is predominant due to low pressure and oxygen deficiency. Figure 3.18 shows the variation of Hall mobility as a function of temperature for as-

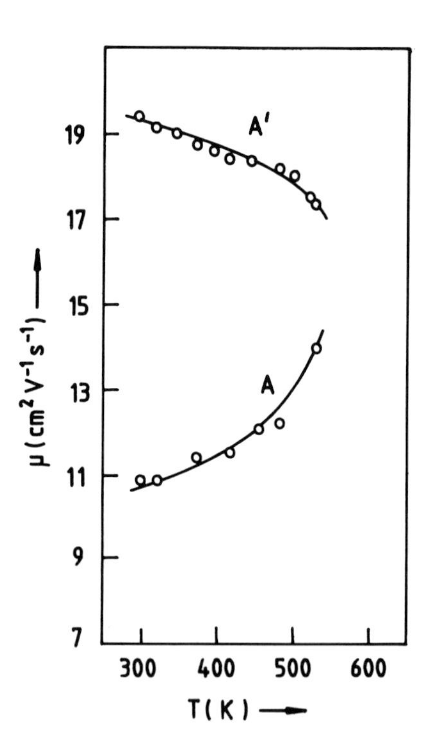

Figure 3.18 Variation of Hall mobility with temperature for spray-deposited films: (A) as grown; (A′) vacuum annealed (from [50]).

grown and vacuum annealed SnO_2 films (as discussed in figure 3.17). The observed variation of Hall mobility with temperature has been analysed in the light of the oxygen diffusion mechanism at the grain boundary by using the grain boundary trapping model due to Orton *et al* [20]. Accordingly,

$$\mu_H \simeq \mu_g = \mu_0' T^{-1/2} \exp\left(\frac{-e\phi_b}{kT}\right) \qquad (3.31)$$

where ϕ_b is given by

$$\phi_b = kT \ln (N_1/N_2) \qquad (3.26)$$

where N_1 and N_2 are the carrier concentration in the grain and boundary regions, respectively. A plot of $\ln(\mu T^{1/2})$, versus T^{-1} (figure 3.19(a) and (b)) suggests that the observed data obey equation 3.31 and that the mobility is thermally activated. Values of ϕ_b obtained from the curves are 0.036 and 0.011 eV for as-deposited and vacuum heat-treated films, respectively. The values of carrier density in the grain boundary region N_2 estimated using relation 3.26 are 3.14×10^{19} cm^{-3} and 8.48×10^{19} cm^{-3} for as-deposited and vacuum annealed SnO_2 films, respectively. It may thus be inferred that oxygen diffusion process modulates the inter-grain boundary barrier height and thus controls the carrier transport process. Shanthi *et al* [57] also suggested that chemisorption or desorption from the grain boundaries changes the potential barrier height at the grain boundaries, thereby

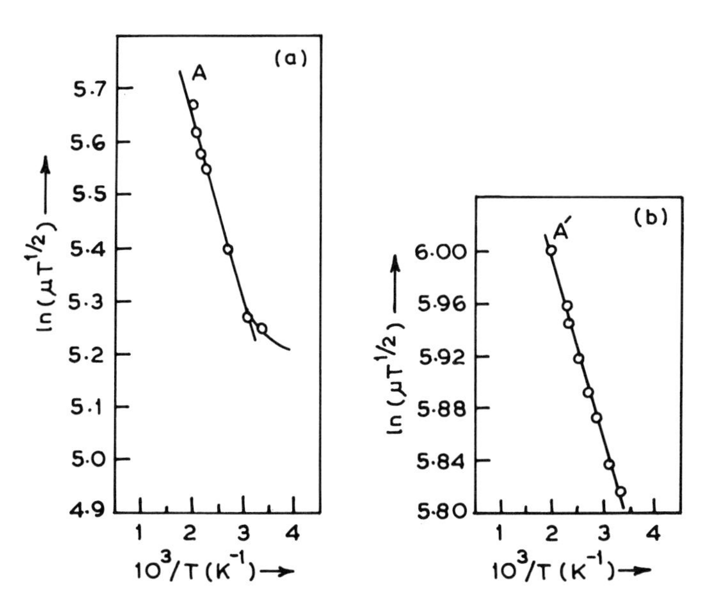

Figure 3.19 Plot of $\mu\ T^{1/2}$ versus $1/T$ for spray-deposited SnO_2 films: (a) as grown; (b) vacuum annealed (from [50]).

governing the electrical properties of tin oxide films. The dominance of any of these mechanisms depends on the temperature of annealing and on the relative concentration of oxygen atoms at the grain boundaries and in the ambient atmosphere.

Beensh-Marchwicka *et al* [58] reported the influence of annealing on electrical resistance of undoped SnO_2 films prepared by DC reactive ion sputtering. The films were annealed in vacuum, air and nitrogen in the temperature range between 570 and 750 K. A decrease in resistance for films annealed in vacuum or dry nitrogen was observed. However, annealing in air or nitrogen containing a small percentage of oxygen, increased the resistance about three-fold. Other workers [66, 67] have also reported an increase in resistivity when SnO_2 films are heated in an environment containing oxygen. The increase in resistivity can be attributed to the reduction of oxygen vacancy concentration in the annealed films.

3.4 Electrical Properties of Doped Tin Oxide Films

Many workers [1, 5, 32, 41, 48, 58, 59, 68–70, 76, 78–85, 88–98] have tried to improve the electrical properties of tin oxide films by doping the films with different dopants such as antimony, fluorine, arsenic, phosphorus, indium, thallium, tellurium, tungsten, molybdenum, etc. However, antimony and fluorine are the most commonly used dopants. The doping of SnO_2 films with antimony or fluorine increases their conductivity. The F^- anion substitutes for an O^{2-} anion in the lattice creating a donor level in the energy band gap, whereas the doping of SnO_2 with antimony gives rise to a donor level at 35 meV [75]. The upper limit of electron density is determined by the solubility of the dopants. In the case of fluorine, the excess amount is volatile, while in antimony the segregated atoms at higher concentration form clusters in the lattice, resulting in a darkening of the films. This phenomenon also explains the higher value of μ_H obtained in fluorine-doped tin oxide films compared to antimony-doped tin oxide films.

3.4.1 Antimony-doped tin oxide films

The electrical properties, namely ρ, N and μ_H of antimony-doped tin oxide films show a marked dependence on the parameters antimony concentration, substrate temperature, thermal annealing and other growth parameters depending on the deposition techniques. It has been observed [41, 76] that the resistivity initially decreases with antimony concentration up to about 3 mol %, goes through a minimum and then increases again for higher Sb concentrations. Figure 3.20 illustrates the typical variation of resistivity,

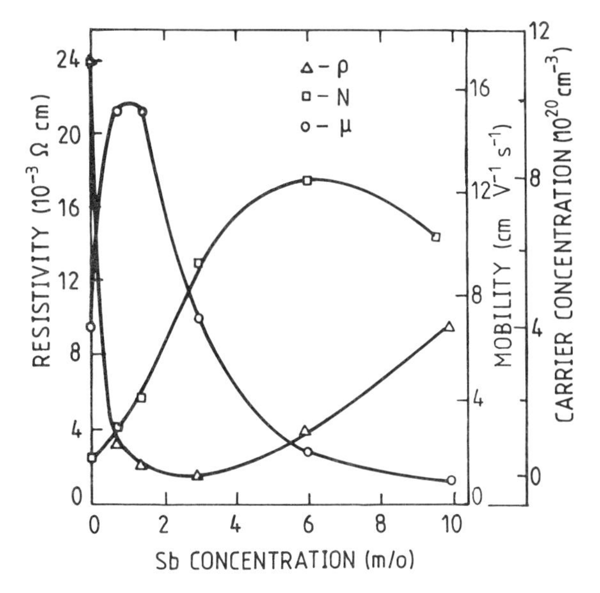

Figure 3.20 Variation of resistivity, mobility and carrier concentration as a function of antimony concentration in spray-deposited SnO_2 films (from [32]).

carrier concentration and mobility with antimony concentration. The initial decrease in resistivity with antimony concentration has been attributed [41] to an increase in the carrier concentration as a result of doping. However, the resistivity, after reaching a minimum, starts increasing again with further antimony doping. This is because the concentration of the traps which result from crystal defects and increase with doping, dominates the carrier concentration above a critical antimony content. As such, antimony atoms act as traps rather than donors. It can also be observed from figure 3.20 that the carrier concentration increases rapidly up to 6 mol% and decreases thereafter. The initial increase in carrier concentration is due to the formation of a shallow donor level [75, 77]. The decrease in carrier concentration above 6 mol% Sb concentration is due to increased disorder causing an increase in the activation energy of donors [77]. The dependence of mobility on Sb concentration shows that the mobility increases sharply, reaches a peak at ~1.4 mol% Sb concentration and then starts decreasing with further increase of Sb concentration. The initial increase in mobility has been attributed [32] to the reduction in grain boundary scattering. This has been confirmed [32] by an increase in the grain size from ~250 Å for undoped tin oxide films to 600 Å for 1.4 mol% Sb doped films. The decrease in mobility with further increase of Sb concentration suggests [32] an increase in the contribution of ionized impurity scattering. Randhawa *et al* [78] also observed a significant decrease in sheet resistance with Sb concentration for antimony-doped tin oxide films prepared by activated reactive evaporation. The reported optimum antimony concentration for minimum sheet resistance was 10 at% in Sn–Sb alloy. The optimum antimony concentration for lowest resistivity has been reported to range from 1–15 at% depending on the growth process [1, 79, 80].

Several workers [32, 81, 82] have studied the temperature dependence of ρ, μ_H and N of Sb-doped SnO_2 films. Figures 3.21 and 3.22 show typical variations of resistivity, mobility and carrier concentration with temperature for Sb-doped SnO_2 films. The resistivity is found to be independent of temperature for 1.4 mol% Sb doped films, while it decreases above room temperature for 10 mol% Sb doped tin oxide films. For 1.4 mol% Sb doped films, the carrier concentration is independent of temperature. This suggests that the impurity band formed at the shallow donor level has merged with the conduction band [77]. However, the carrier concentration increases with temperature above room temperature for 10 mol% Sb doped films, which is in agreement with other findings [81, 82].

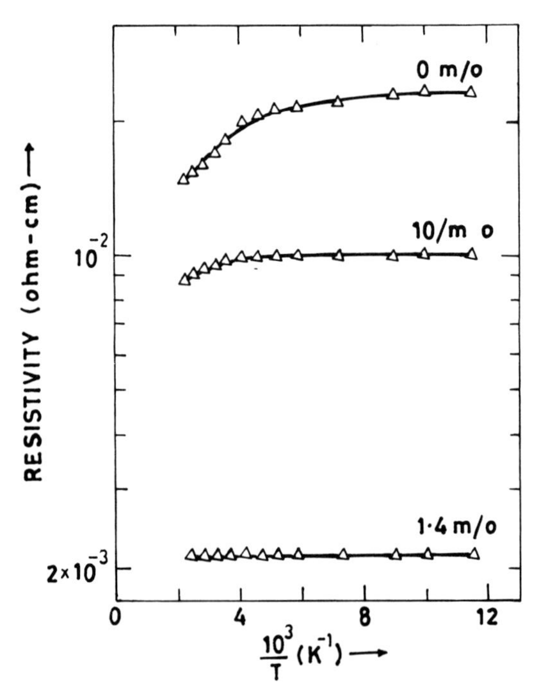

Figure 3.21 Variation of resistivity as a function of temperature for spray-deposited Sb-doped tin oxide films (from [32]).

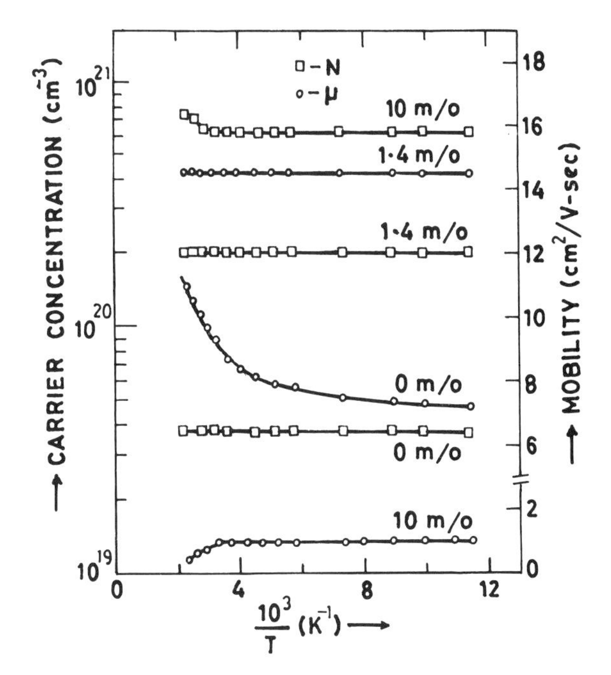

Figure 3.22 Variation of carrier concentration and mobility as a function of temperature for spray-deposited Sb-doped tin oxide films (from [32]).

(a) *Effect of deposition temperature*

Figure 3.23 shows the effect of substrate temperature on properties of Sb-doped SnO_2 films prepared by RF magnetron sputtering. The steady decrease in resistivity with increase [83] in substrate temperature is because of an increase in concentration of substitutional donor Sb^{5+} ions as a result of the more ordered crystal structure and larger grain size [84]. The mobility increases slowly with substrate temperature. This is because of the combined effect of the various scattering mechanisms involved, e.g. grain boundary scattering, lattice scattering and ionized impurity scattering.

Kaneko and Miyake [85] reported the dependence of resistivity, carrier concentration and Hall mobility on thickness of Sb-doped SnO_2 films prepared by a spray deposition method. Figures 3.24 and 3.25 show typical variations of Hall mobility, carrier concentration and resistivity with thickness of Sb-doped SnO_2 films deposited on fused quartz, borosilicate glass and soda-lime glass substrates at 600 °C. It may be observed that the Hall mobility and carrier concentration for films deposited on fused quartz and borosilicate glass are almost independent of film thickness. However, the carrier concentration and Hall mobility are dependent on film thickness for

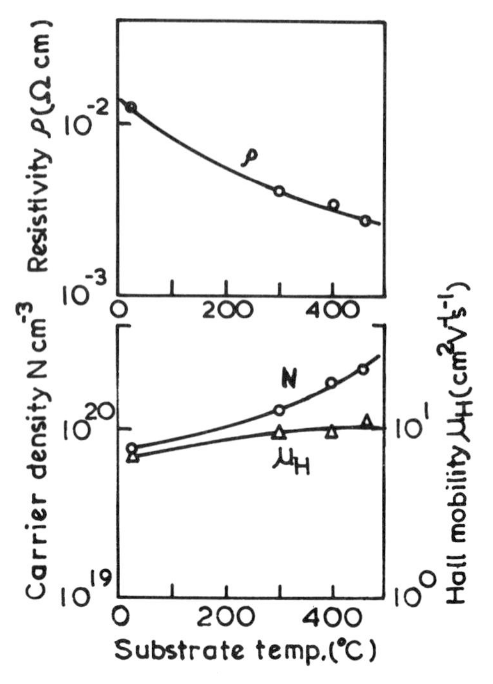

Figure 3.23 Variation of resistivity, carrier concentration and mobility as a function of substrate temperature for RF magnetron sputtered Sb-doped SnO_2 films (from [83]).

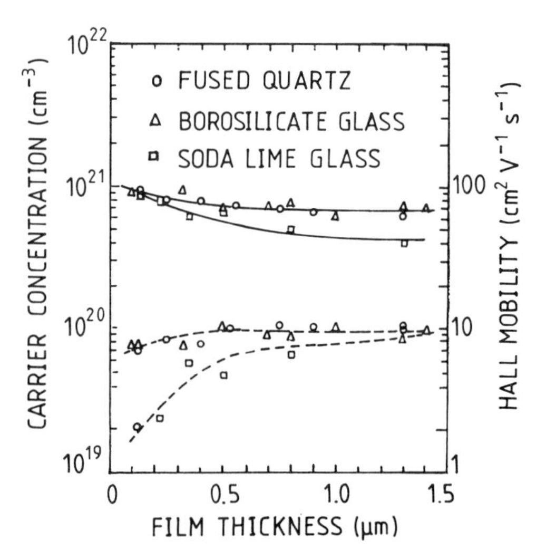

Figure 3.24 Variation of carrier concentration (———) and Hall mobility (– – – –) with thickness of spray-deposited Sb-doped SnO_2 films deposited on different substrates (from [85]).

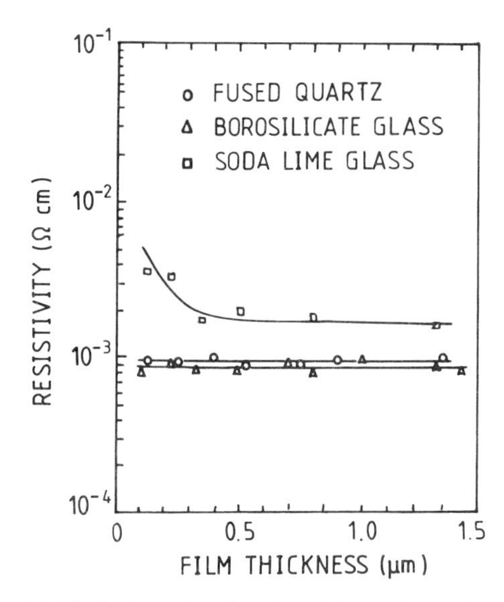

Figure 3.25 Variation of resistivity with thickness for spray-deposited Sb-doped SnO_2 films deposited on different substrates (from [85]).

films on soda-lime glass substrates. The resistivity of films on fused quartz and borosilicate glass is found to be independent of film thickness, and is about 9.5×10^{-4} and $8.6 \times 10^{-4}\ \Omega$ cm, respectively. The independent nature of resistivity with thickness is similar to that observed for undoped tin oxide films (figure 3.6). However, the resistivity of films on soda-lime glass substrates initially decreases with increasing film thickness (1000–4000 Å) and is almost constant for films thicker than 4000 Å. This decrease in resistivity cannot be attributed to size effects as the film thickness is far greater than the mean free path λ_f. The difference in electrical properties of films on soda-lime glass substrates from those of films deposited on fused quartz or borosilicate glass substrates is possibly due to the indiffusion of active ions such as sodium into the film from soda-lime glass substrates [86, 87]. This is possible as the films have been prepared at temperatures of 600 °C, which is near the softening point of the soda-lime glass.

(b) *Effect of post-deposition annealing*
Beensh-Marchwicka *et al* [58] investigated the effect of thermal annealing on the electrical resistance of antimony-doped tin oxide films prepared by DC reactive ion sputtering. The resistance changes in Sb-doped films are found to be three to four orders of magnitude lower than those observed in undoped SnO_2 films [52]. This has been attributed to the lower free energy of antimony-doped tin oxide films, resulting in increased oxidation during sputtering [58].

(c) *Effect of sputtering parameters*

Suzuki and Mizuhashi [83] investigated the dependence of electrical properties on sputtering conditions for antimony-doped tin oxide films prepared by RF magnetron sputtering. The gas composition was found to affect the electrical properties of the films as shown in figure 3.26. The change in carrier concentration with oxygen partial pressure is attributed to the change in the number of vacancies and excess metal ions due to non-stoichiometry. Though the presence of SnO phase is revealed by X-ray diffraction studies, the films formed in pure argon show a relatively higher carrier concentration. The films grown with 5% oxygen concentration have maximum carrier concentration and only SnO_2 phase is present, while films with higher oxygen concentration have lower carrier concentration due to a decrease in the number of oxygen vacancies. It can also be observed from figure 3.26 that the mobility is low for films deposited in pure argon. The presence of SnO phase is responsible for this low value of mobility. As the oxygen concentration increases, a decrease in the amount of SnO phase takes place thus resulting in an increase in mobility.

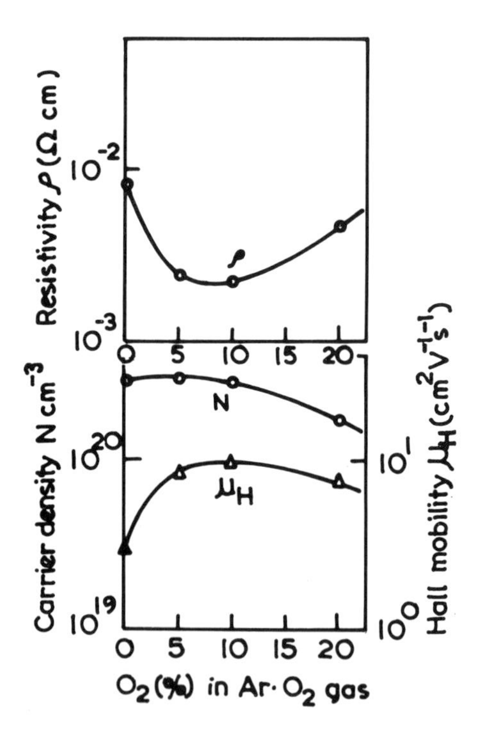

Figure 3.26 Variation of resistivity, carrier concentration and mobility as a function of oxygen concentration in sputtering gas (from [83]).

3.4.2 Fluorine-doped tin oxide films

Several workers [5, 48, 59, 69, 76, 89–91] have studied the effect of fluorine doping on the electrical characteristics of tin oxide films. Figure 3.27 illustrates the variation of sheet resistance R_s with the concentration of fluorine in SnO_2:F films prepared by spraying. The sheet resistance first decreases, reaches a minimum and then increases with fluorine concentration. Manifacier [5] observed a resistivity minimum in the solution for an atomic ratio of the order of (F)/(Sn) 0.2–0.3 for SnO_2:F films. This minimum corresponds to resistivity $\sim(4\text{–}6)\times10^{-4}\ \Omega$ cm. Pommier *et al* [89] also observed a similar type of behaviour. The initial decrease in sheet resistance can be attributed [5] to an increase in donor concentration and an increase in mobility. The subsequent increase in sheet resistance with fluorine concentration, after reaching a minimum, results from an increase in disorders. Shanthi *et al* [76] also studied the variation of electrical parameters as a function of fluorine concentration in spray-deposited F-doped tin oxide films. Figure 3.28 shows the variation of ρ, N and μ as a function of F concentration. The carrier concentration increases and the resistivity decreases with fluorine concentration up to 65 at% and saturates at higher concentrations. On the other hand, mobility initially increases up to 10 at% and saturates thereafter. Similar behaviour of ρ, N and μ with fluorine concentration has also been reported by Gottlieb *et al* [90] and Antonaia *et al* [69].

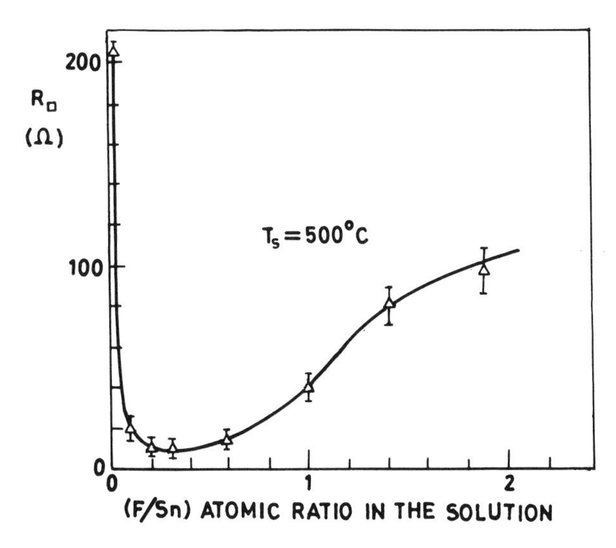

Figure 3.27 Variation of sheet resistance with atomic ratios in the solution (F/Sn) for spray-deposited fluorine-doped SnO_2 films (from [5]).

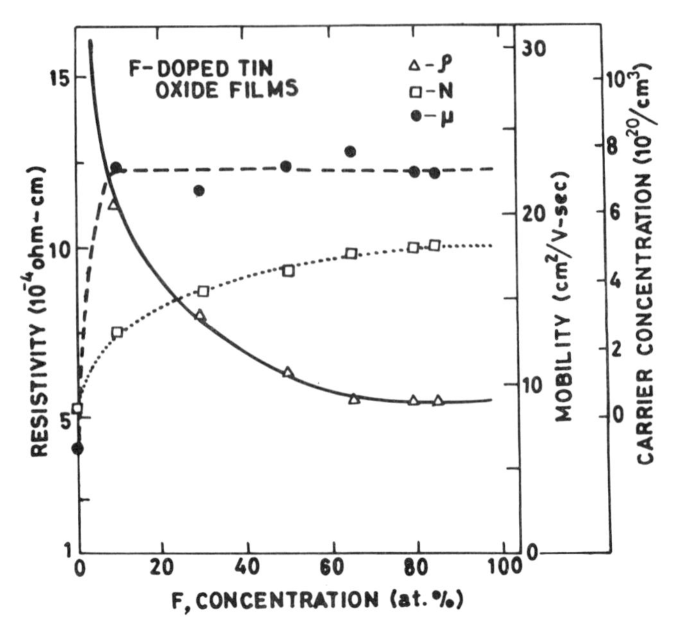

Figure 3.28 Variation of resistivity, carrier concentration and mobility with fluorine concentration for spray-deposited SnO_2 films (from [76]).

(a) *Effect of deposition temperature*

The substrate temperature T_s is found to have a significant effect on the electrical properties of SnO_2 films doped with fluorine. A minimum is observed in the relationship between resistivity and substrate temperature [5, 76, 89–91]. Figure 3.29 shows a typical variation of resistivity with substrate temperature for fluorine-doped tin oxide films with different fluorine contents. It can be observed that the shape of the curves is independent of the fluorine content and the minimum in resistivity occurs at ~375 °C in all cases. Figure 3.30 shows the typical variation of resistivity ρ, carrier concentration N and mobility μ_H of F-doped SnO_2 films as a function of substrate temperature. The resistivity minimum occurs at a substrate temperature $T_s \sim 400$ °C. The carrier concentration and the mobility show a maximum at the same temperature. The initial decrease in resistivity with an increase in T_s is due to an increase in the size of the crystallite which results in an increase of mobility. The subsequent increase in resistivity when T_s exceeds 400 °C can be partly attributed to the decreased number of oxygen vacancies and partly to film contamination by alkali ions [51, 67, 85, 87, 91, 92]. Such an influence of alkali doping in SnO_2:Sb films has also been reported [85, 87]. The effect of alkali ion concentration is observable only in low resistance films such as fluorine-doped SnO_2.

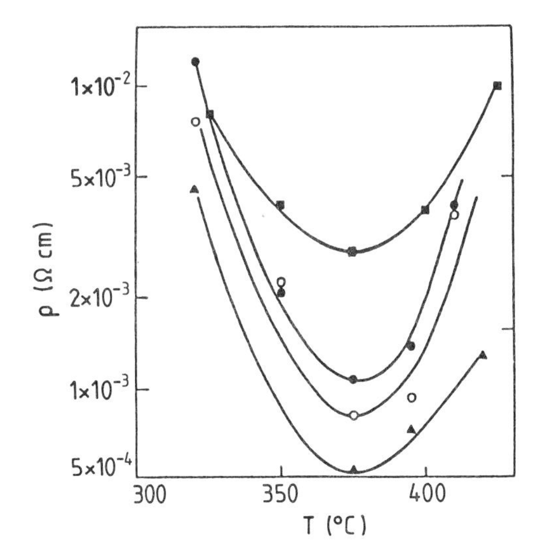

Figure 3.29 Resistivity ρ as a function of deposition temperature for fluorine-doped SnO_2 films with different fluorine concentrations: (■) undoped; (●) 1% F; (○) 5% F; and (▲) 10% F (from [90]).

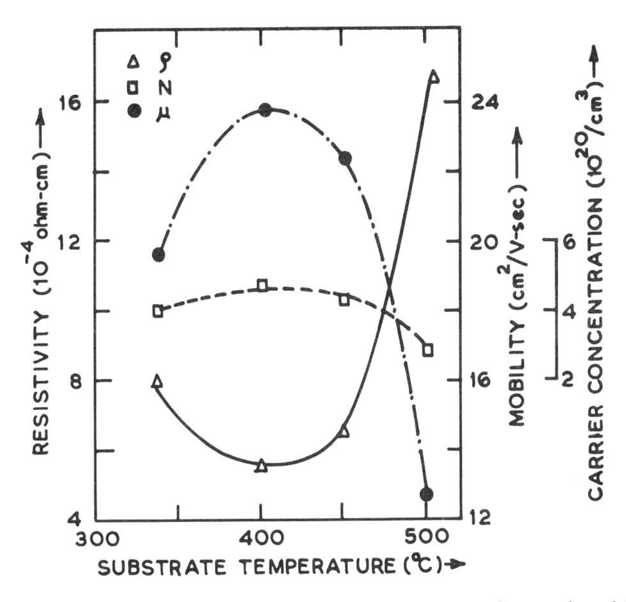

Figure 3.30 Variation of resistivity, carrier concentration and mobility of 65 at % F-doped SnO_2 films as a function of substrate temperature (from [76]).

(b) *Effect of film thickness*

Figure 3.31 shows the variation of sheet resistance R_S with film thickness for SnO_2:F film. It is observed that initially the decrease in sheet resistance with thickness is very rapid, whereas this decrease is very slow for thicknesses greater than ~0.5 μm. A similar variation in sheet resistivity with thickness has also been reported for undoped SnO_2 films (figure 3.7).

(c) *Effect of post-deposition annealing*

Jagadish *et al* [59] studied the effect of hydrogen annealing on the electrical properties of F-doped SnO_2 films. Figure 3.32 shows the variation of sheet resistance as a function of annealing temperature for both doped and undoped SnO_2 films. It can be observed from this figure that the value of R_S in as-grown undoped SnO_2 films is higher than that in as-grown SnO_2:F films. This decrease in sheet resistance with F-doping is due to the donor action of fluorine in the SnO_2 lattice. It may also be observed from figure 3.32 that the value of R_S in undoped films decreases with increase in annealing temperatures. This decrease in R_S is attributed to either (i) increase in grain size, and/or (ii) creation of oxygen vacancies. On the other hand, in the case of F-doped films the value of R_S increases with increase in annealing temperatures. This increase in R_S is due to the removal of fluorine from the SnO_2 lattice thereby decreasing the free carrier concentration in the films.

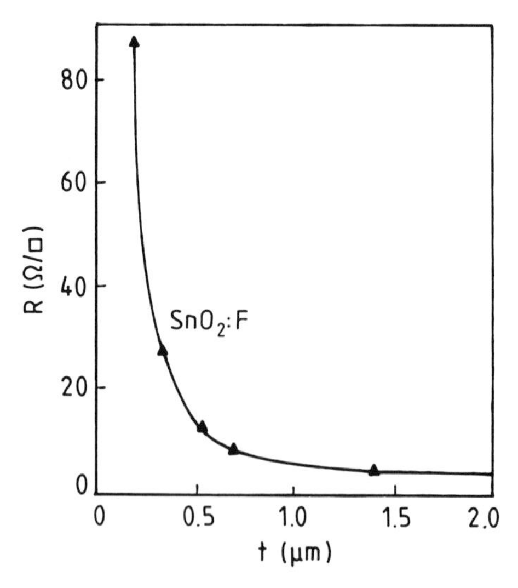

Figure 3.31 Variation of sheet resistance (R_s) with thickness for SnO_2:F films deposited by spray pyrolysis (from [89]).

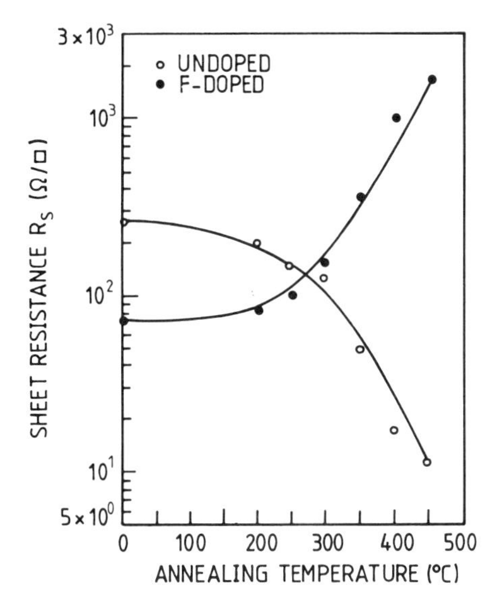

Figure 3.32 Variation of sheet resistance (R_s) with annealing temperature in undoped and F-doped SnO_2 films (from [59]).

(d) *Conduction mechanisms*

In general, SnO_2 films are polycrystalline in nature, and the conduction mechanism is governed by the carrier concentration in the films. For undoped and doped SnO_2 films, where carrier concentration is less than 10^{18} cm^{-3}, conduction is governed by the grain boundaries. When the carrier concentration is between 10^{18} and 10^{19} cm^{-3}, conduction is dominated by grain boundaries and bulk properties. However, in the case of films with carrier concentration greater than 10^{19} cm^{-3}, different conduction mechanisms have been reported. Shanthi *et al* [32] and Antonaia *et al* [69] suggested that grain boundary scattering dominates conduction even when carrier concentration is greater than 10^{19} cm^{-3}. Frank *et al* [93] have shown that in heavily doped tin oxide films, the free electron mobility is limited mainly by ionized impurity scattering. However, Bruneaux *et al* [48] reported that the effect of grain boundaries is entirely screened out by that of the bulk properties.

Bruneaux *et al* [48] analysed in detail their resistivity–carrier concentration data on the basis of various scattering mechanisms. Figure 3.33 depicts the variation of resistivity ρ versus carrier concentration N for undoped and fluorine-doped SnO_2 films. The dashed curve is the theoretical curve computed by taking into consideration the contributions of grain boundaries as well as that of the bulk. It can be observed that the resistivity changes by several orders of magnitude when the carrier density varies from

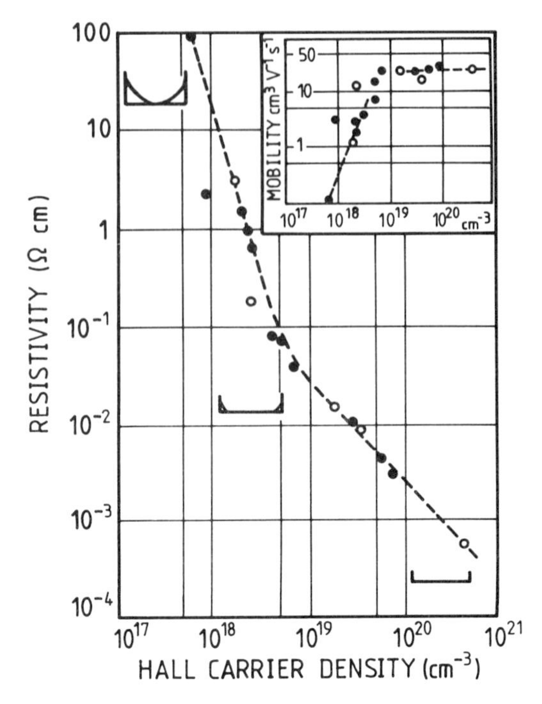

Figure 3.33 Resistivity versus Hall carrier density: (– – – –) calculated; (○) experimental data for fluorine-doped, and (●) for undoped SnO_2 samples (from [48]).

7×10^{17} cm^{-3} to 4.4×10^{20} cm^{-3}. Two distinct regions can be seen and the experimental data can be approximately represented by a power law with an exponent close to -1 for films with carrier density greater than 10^{19} cm^{-3}. However, for films with carrier density less than 10^{19} cm^{-3}, the variation is much steeper. The inset in figure 3.33 shows the experimental mobility data as well as the effective mobility calculated from the resistivity–carrier concentration curve. It can be seen that the mobility is independent of carrier concentration beyond $\sim 10^{19}$ cm^{-3}. The data have been analysed considering the contribution of grain boundaries in accordance with relation 3.29 or 3.42 and the computed results agree well with the experimental points for samples with carrier concentration less than 10^{19} cm^{-3}. However, these equations alone are unable to explain the data for samples with carrier concentration greater than 10^{19} cm^{-3}. At higher carrier densities, the influence of grain boundaries becomes negligible; the grain size is no longer an important factor in controlling film conductivity and resistivity is now governed by the properties of the bulk.

3.5 Electrical Properties of Indium Oxide Films

Indium oxide is widely used as a transparent conductor owing to its superior electrical and optical properties. This is largely due to higher mobility in indium oxide than in tin oxide. All the undoped In_2O_3 films are n-type conductive with complete degeneracy. The conduction is due to free electrons originating from oxygen vacancies and/or excess In atoms in the films, which form ionized impurity centres.

Many workers have optimized the growth conditions in order to prepare In_2O_3 films with high conductivity. Pan and Ma [99] reported the growth of highly conductive In_2O_3 films ($\rho \simeq 1.8 \times 10^{-4}\ \Omega$ cm) by a thermal evaporation technique in an oxygen ambient. This low value of resistivity is the best ever reported for In_2O_3 films and is comparable to the values reported for indium tin oxide and doped tin oxide films.

(a) *Effect of deposition temperature*

Several workers have studied the effect of various growth parameters on the electrical properties of In_2O_3 films. Figure 3.34 shows typical results of the effect of substrate temperature on the electrical properties of In_2O_3 [36] grown by vacuum evaporation. It can be observed from this figure that the resistivity decreases at first and then increases again as the substrate temperature is increased. On the other hand, the carrier concentration first remains practically constant with increase of substrate temperature and then starts decreasing with further increase of substrate temperature. The decrease in electron concentration at higher substrate temperatures confirms that the free carrier concentration is mainly because of oxygen vacancies, as suggested by Fan and Goodenough [92] and Bosnell and Waghorne [94].

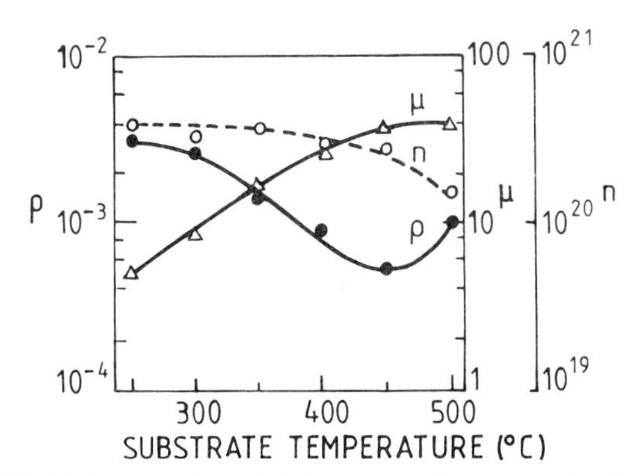

Figure 3.34 Variation of resistivity ρ, carrier concentration n and mobility μ of In_2O_3 films as a function of substrate temperature (from [36]).

The initial increase in mobility in the films with substrate temperature is because the crystallite size improves significantly, thus reducing the grain boundary scattering contribution. The observed initial increase in conductivity in figure 3.34 can also be understood using the reasoning that the crystallite size as well as the crystalline nature of the films improves with increase of substrate temperature, thus resulting in increased mobility and conductivity.

(b) *Film thickness effects*

Pan and Ma [99] deposited highly conductive indium oxide films by evaporating a mixture of 90% In_2O_3 + 10% In by weight in oxygen ambient onto glass at a substrate temperature of between 320 and 350 °C. Figure 3.35 shows the variation of sheet resistance as a function of film thickness. It can be seen that the sheet resistance decreases very significantly for very thin films, whereas the decrease is much slower for thicker films. Such behaviour is expected and has also been observed in SnO_2 films (figures 3.7 and 3.31). The sheet resistance attains a value 6.75 $\Omega\,\square^{-1}$, corresponding to a resistivity of $1.8 \times 10^{-4}\,\Omega$ cm at film thickness 2700 Å. These as-deposited undoped indium oxide films have a free carrier concentration of $4.6 \times 10^{20}\,cm^{-3}$ and Hall mobility $\sim 74\,cm^2\,V^{-1}\,s^{-1}$. This high value of mobility, which may be due to a lack of impurity scattering centres, is responsible for the high conductivity in the films.

(c) *Effect of oxygen partial pressure*

The oxygen partial pressure during the deposition process strongly influences the electrical properties of In_2O_3 films [100–102]. Figure 3.36 demonstrates the dependence of electrical conductivity, free carrier concentration

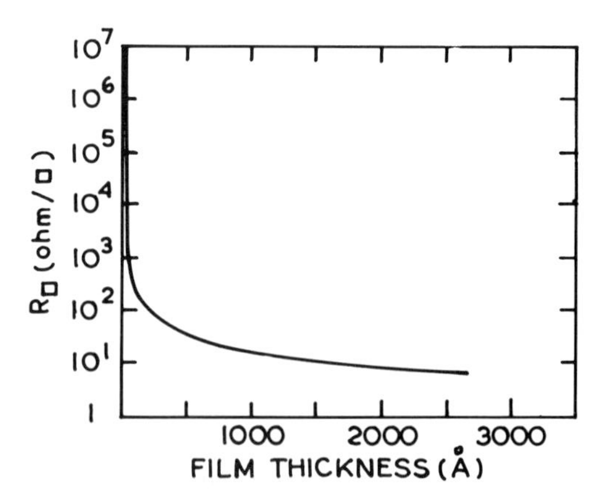

Figure 3.35 Variation of sheet resistance with thickness for reactively evaporated In_2O_3 films (from [99]).

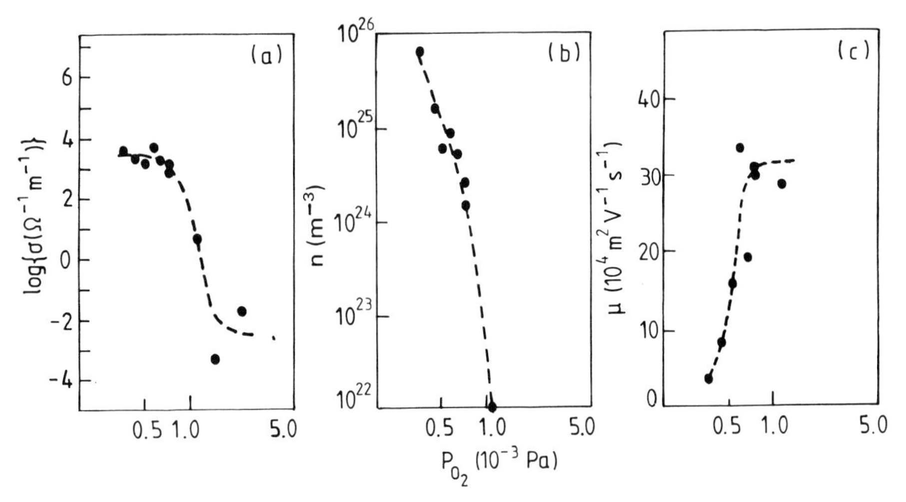

Figure 3.36 Variation of (a) conductivity, (b) carrier concentration and (c) mobility of RF sputtered In_2O_3 films with oxygen partial pressure (from [102]).

and Hall mobility on the oxygen partial pressure of the sputtering gas. It may be seen from this figure that the conductivity remains constant for low oxygen pressure and falls sharply at an oxygen partial pressure $\sim 8 \times 10^{-4}$ Pa (6×10^{-6} Torr). It may also be observed that the oxygen partial pressure of the sputter gas mainly affects the free carrier concentration in the films. This is associated with the non-stoichiometry of the films, resulting in oxygen vacancies. The sharp fall in conductivity is due to decreasing carrier concentration. The Hall mobility initially increases with oxygen partial pressure and attains a constant value 30 $cm^2\,V^{-1}\,s^{-1}$.

(d) *Conduction mechanisms*

Various workers have studied the temperature dependence of the electrical properties in In_2O_3 films in order to understand the conduction phenomenon in these films. Sundaram and Bhagavat [103] reported the electrical properties of indium oxide films obtained by oxidizing indium films at 300 and 400 °C. The variation of electrical conductivity σ with temperature is shown in figure 3.37. The conductivity variation indicates two distinct regions with activation energy 0.007 eV and 0.205 eV. The activation energy in region I corresponds to a shallow donor level near the bottom of the conduction band, while the activation energy in region II is indicative of another donor level. Typically the conductivity is found to be in the range 10–32 $\Omega^{-1}\,cm^{-1}$, mobility from 20 to 25 $cm^2\,V^{-1}\,s^{-1}$ and carrier concentration in the range 2.5×10^{18} to 1×10^{19} cm^{-3}.

Different types of scattering mechanism have been suggested for conduction in In_2O_3 films. Noguchi and Sakata [104] observed a weak dependence

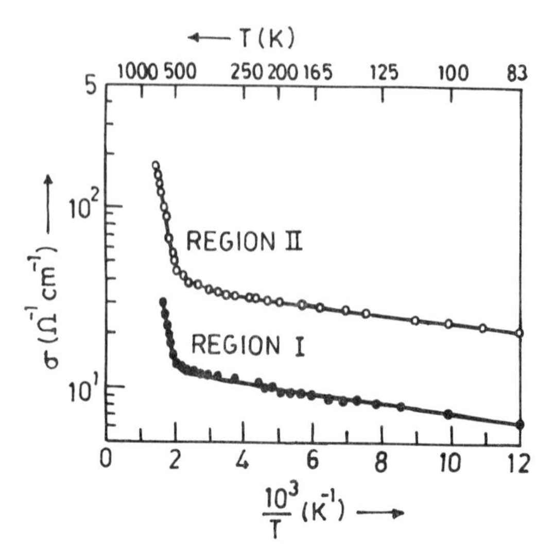

Figure 3.37 Variation of electrical conductivity (σ) as a function of temperature for In_2O_3 films prepared by oxidation at (○) 400 °C and (●) 300 °C (from [103]).

of mobility on temperature in In_2O_3 films grown by reactive evaporation. Scattering due to grain boundaries has been ruled out as the calculated value of mean free path 'l' (≈39 Å) of the charge carriers is found to be much less than the observed grain size of 1000 Å. The mean free path of the charge carriers has been estimated using the relation

$$l = (h/2e)(3N/\pi)^{1/3}\mu. \tag{3.46}$$

The contribution of neutral impurity scattering has not been considered at temperatures greater than 77 K. As there may be a high density of ionized impurity centres in the films due to oxygen vacancies and/or excess indium atoms, the data have been analysed in the light of ionized impurity scattering. Figure 3.38 shows the variation of Hall mobility with carrier concentration for In_2O_3 films. This figure also includes the theoretical curve deduced using the relationship due to Johnson and Lark–Horovitz [15] for completely degenerate samples:

$$\mu_H \simeq (4e/h)\,(\pi/3)^{1/3}N^{-2/3}. \tag{3.47}$$

The experimental mobility–carrier concentration relationship is found to be in good agreement with theory. This confirms that the dominant scattering mechanism in undoped In_2O_3 films is due to ionized impurity centres. Such a scattering mechanism has also been reported in In_2O_3 films made by spray pyrolysis [105] and sputtering [106]. Ovadyahu and Imry [107] studied

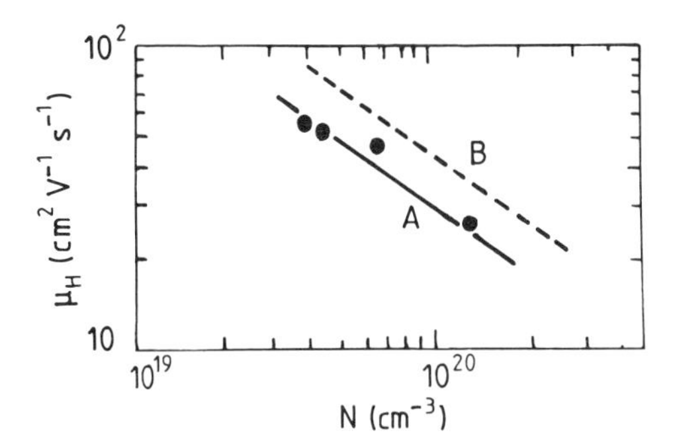

Figure 3.38 Variation of mobility versus carrier concentration (A) experimental (B) theoretical $\mu_H \propto N^{-2/3}$ (from [104]).

the dependence of conductivity on temperature for very thin 50 Å indium oxide films (effectively two dimensional). The sheet resistance of the films studied was in the range 30–250 kΩ $\square^{-1}$. It has been observed that a $\sigma \propto T$ law is followed for samples only in the narrow range of sheet resistance $R_S = 45$ to 67 kΩ $\square^{-1}$. However, for samples with $R_S \geqslant 90$ kΩ $\square^{-1}$, σ depends exponentially on temperature. Such behaviour is attributed to two-dimensional variable range hopping [108].

Graham *et al* [109] studied conductivity–temperature data in amorphous In_2O_3 films prepared by ion beam sputtering using an indium target. It was observed that in highly resistive samples, conduction was by variable range hopping in a coulomb gap located in states that were deep in the bandgap of indium oxide and strongly localized by structural disorders. In more conducting films, the Fermi level shifted towards the conduction band and hopping switched over to three-dimensional Mott variable range hopping.

3.6 Electrical Properties of Doped Indium Oxide Films

The incorporation of a dopant in In_2O_3 films changes the electrical properties significantly. Several investigators [5, 36, 89, 93, 106, 111–124, 130–173] have tried different dopants, e.g. tin, fluorine, titanium, antimony, lead and zirconium in indium oxide films to improve their electrical properties. However, tin is the most effective and most commonly used dopant in indium oxide films.

3.6.1 Tin-doped indium oxide films (ITO)

Tin-doped indium oxide films are designated ITO (indium tin oxide). In_2O_3 films are generally doped with tin because Sn^{4+} substitutes for the In^{3+}

cation creating a donor level in the energy band gap [110]. Tin doping of indium oxide films is reported [111–119] to decrease electrical resistivity significantly. All tin-doped indium oxide films have n-type conductivity with complete degeneracy.

The values of electrical parameters strongly depend on the dopant concentration. Figure 3.39 shows the typical variation of resistivity ρ, Hall mobility μ_H and free carrier concentration N with dopant level for tin-doped In_2O_3 films prepared by spray pyrolysis. The resistivity ρ of the films is found to initially decrease with increase of dopant level up to 5 wt% of tin, but increases at higher dopant levels. The minimum resistivity is $3 \times 10^{-4}\ \Omega$ cm corresponding to the optimum dopant level. Similar behaviour has also been reported for ITO films deposited by DC-sputtering [112], electron beam evaporation [113], and thermal evaporation [114, 115]. The optimum tin concentration for obtaining lowest resistivity has been found to vary in films prepared by different techniques [5, 89, 112, 114, 116, 117]. Noguchi and Sakata [114] obtained minimum resistivity $\sim 6 \times 10^{-4}\ \Omega$ cm at a doping ratio of Sn/In $\simeq 0.05$ for ITO films prepared by reactive evaporation. Manifacier [5] and Pommier *et al* [89] observed the minimum corresponding to a resistivity in the range $4\text{–}6 \times 10^{-4}\ \Omega$ cm for an atomic ratio in the solution of the order Sn/In $\simeq 0.2\text{–}0.3$ for In_2O_3:Sn films. Fraser and Cook [112] obtained a minimum resistivity of $1.77 \times 10^{-4}\ \Omega$ cm. However, the minimum resistivity in reactively evaporated film [114] is found at a doping level slightly lower than that for sputtered films, and the dependence of resistivity

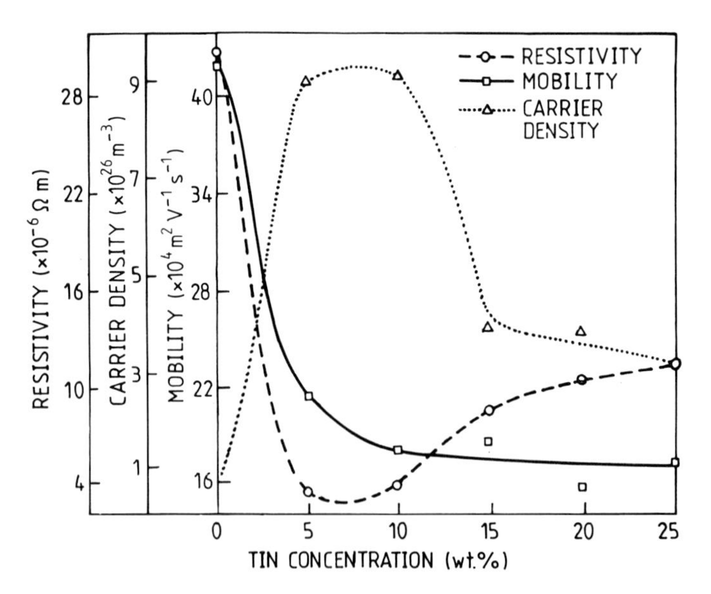

Figure 3.39 Variation of resistivity, Hall mobility and carrier concentration as a function of tin concentration for ITO films (from [111]).

on doping is greater than in the case of sputtered films [112]. It may also be observed from figure 3.39 that the free carrier concentration increases for levels of up to 5 wt% of tin. This initial increase in carrier concentration or decrease in resistivity may be due to the donor action of the tin impurity [5, 113]. Since the observed increase in carrier concentration is much higher than that expected due to dopant concentration alone, there is a significant contribution from oxygen vacancies. Moreover, Noguchi and Sakata [114] observed that only about 50% of the Sn dopant atoms are effective when Sn atoms are substituted in sites in the In_2O_3 films. The decrease in carrier concentration and increase in resistivity beyond 10 wt% of tin have been explained by Kostlin *et al* [110] and Mizuhashi [36]. According to Kostlin *et al* [110], for higher dopant levels, tin ions occupy the nearest neighbouring sites more frequently and thereby compensate their donor action. Mizuhashi [36] explained such results in terms of too much distortion of the crystal lattice. This is confirmed by loss of crystallinity in ITO films for higher dopant levels [113, 114]. Manifacier [5] also attributed the increase in resistivity to an increase in disorder, which decreases the mobility and free carrier concentration. It is further seen from figure 3.39 that Hall mobility decreases with increase of tin content. The decrease in mobility with increase of dopant level is due to enhancement of scattering mechanisms such as ionized impurity scattering.

(a) *Conduction mechanisms*

The mobility data for ITO films have been analysed by considering the important scattering mechanisms prevalent in polycrystalline semiconductors. Although Fistul and Vainshtein [120] postulated lattice scattering as the dominant scattering mechanism, most other workers [105, 106, 113, 114, 121–124] have interpreted their results in terms of ionized impurity scattering. Figure 3.40 shows a typical mobility and carrier concentration variation with temperature for In_2O_3 and SnO_2-doped In_2O_3 films. The grain boundary mechanism generally dominates in polycrystalline films with small crystallite size. However, it is found that the estimated mean free path is much smaller than the observed crystallite size in these films. This suggests that the contribution of grain boundary scattering mechanism is not significant in these films. Mobility data have usually been analysed by considering the contribution due to ionized impurity and lattice scattering. The contribution of ionized impurity scattering increases with increase of dopant level. Noguchi and Sakata [114] analysed mobility carrier concentration data to establish the scattering mechanisms occurring in ITO films. Figure 3.41 shows the variations of Hall mobility versus carrier concentration in indium tin oxide films obtained at different substrate temperatures and tin dopant levels. The continuous line shown in the figure has been calculated using the ionized impurity scattering theory of Johnson and Lark–Horovitz [15] as given by equation 3.24. It may be seen that most of the μ–N data, except for

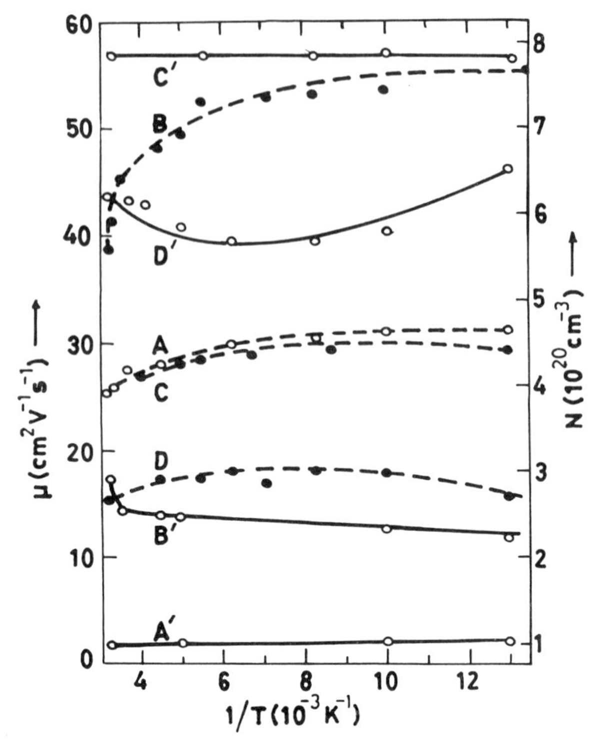

Figure 3.40 Variation of mobility (μ) (A, B, C, D) and carrier concentration (N) (A′, B′, C′, D′) as a function of temperature. (A, A′) undoped; (B, B′) 3 mol% SnO_2; (C, C′) 4 mol% SnO_2; and (D, D′) 7 mol% SnO_2 doped In_2O_3 films (from [113]).

some films doped above Sn/In = 0.1, fit quite well with the continuous line. This suggests that the dominant scattering mechanism in these ITO films doped below (Sn)/(In) = 0.10 is ionized impurity scattering, tin ions acting as ionized impurity scattering centres in addition to the oxygen vacancies and/or excess In atoms. This is in agreement with that in sputtered [106] and sprayed [93, 105, 121] ITO films. Noguchi and Sakata [114] suggested that for doping above (Sn)/(In) = 0.1, neutral impurities and other centres related to poor crystallinity may limit the conductivity.

Bellingham *et al* [122–124] observed that even in amorphous indium oxide and indium tin oxide films, the conductivity is mainly governed by scattering of the electrons due to ionized impurities. Figure 3.42 shows the room temperature resistivity–carrier concentration for amorphous In_2O_3 and ITO films prepared by ion beam sputtering. This figure also shows theoretical curves based on ionized impurity scattering. The dashed curve is based on

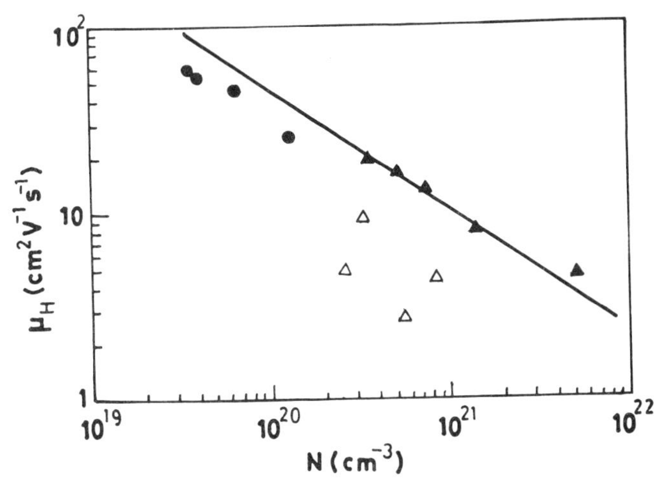

Figure 3.41 Mobility μ_H versus carrier concentration N for indium oxide films: ●, undoped; ▲, doped below Sn/In ≈ 0.10; △, doped above Sn/In ≈ 0.10 (from [114]).

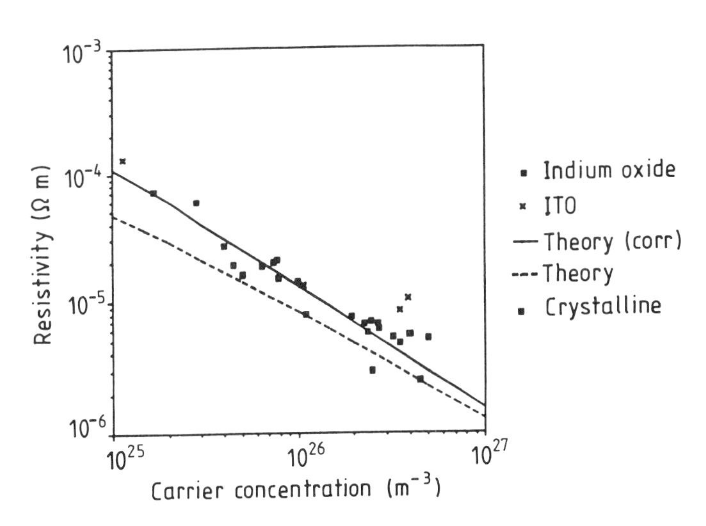

Figure 3.42 Resistivity versus carrier concentration for different indium oxide and indium tin oxide films (from [122]).

Dingle's model [125] for the ionized impurity contribution. According to Dingle [125], the resistivity is given by

$$\rho = [N_i Z^2 e^2 m^{*2}/24\pi^3(\varepsilon_o\varepsilon_s)^2\hbar^3 N^2]f(k_F) \quad (3.48)$$

where

$$f(k_F) = \ln(1+\beta^2) - \beta^2/(1+\beta^2) \quad (3.49)$$

and $\beta = 2k/k_{TF}$ where k_{TF} is the Thomos–Fermi wave vector and is given by

$$k_{TF}^2 = (3N/\pi^4)^{1/3} m^* e^2/\hbar^2\varepsilon_s\varepsilon_o. \quad (3.50)$$

The values of m^* and ε_s have been taken as $0.3m_o$ and 9, respectively.

The solid line in figure 3.42 is based on Dingle's calculations including Moore's correction [126] by considering higher order scattering terms. It can be observed from this figure that the experimental data fit very well with the model based on ionized impurity scattering. In order to study the effect of crystallinity on the scattering mechanism, data for crystalline In_2O_3 films [104, 127–130] are also shown in figure 3.42. The variation of crystalline resistivity with carrier concentration is also in agreement with the trend of the ionized impurity scattering calculations. Figure 3.42 confirms further that even in the case of amorphous structure, there is no need for invoking scattering by structural disorders.

(b) *Effect of deposition temperature*

Many workers [36, 89, 111, 113, 114, 116, 131–134] have studied the effect of deposition temperature on the electrical properties of ITO films prepared by different techniques. Most of the commonly used techniques for the growth of highly transparent and conductive ITO films require high substrate temperatures (>350 °C). Indium tin oxide films grown by RF sputtering [133] need temperatures greater than 450 °C and those prepared by spraying [114] or thermal evaporation [36] require a substrate temperature of about 400 °C. However, Agnihotry *et al* [113] prepared highly transparent and conducting films of indium tin oxide by an electron beam evaporation method at substrate temperatures as low as 200 °C. They obtained films with resistivity as low as $2.4 \times 10^{-4}\ \Omega$ cm with a carrier concentration of 8×10^{20} cm^{-3} and mobility of about 30 cm^2 V^{-1} s^{-1} at a doping concentration of 4 mol% SnO_2. Figure 3.43 illustrates the typical variation of electrical properties with substrate temperature for In_2O_3:Sn films prepared by a spray pyrolysis technique [111]. The substrate temperature is found to significantly affect the electrical properties. The resistivity initially decreases as the substrate temperature increases. This type of dependence of resistivity on substrate temperature may be due to the fact that the crystallinity of the films improves [111, 116] with increase in substrate temperature, thereby increasing the conductivity. At higher substrate temperatures (>733 K) the

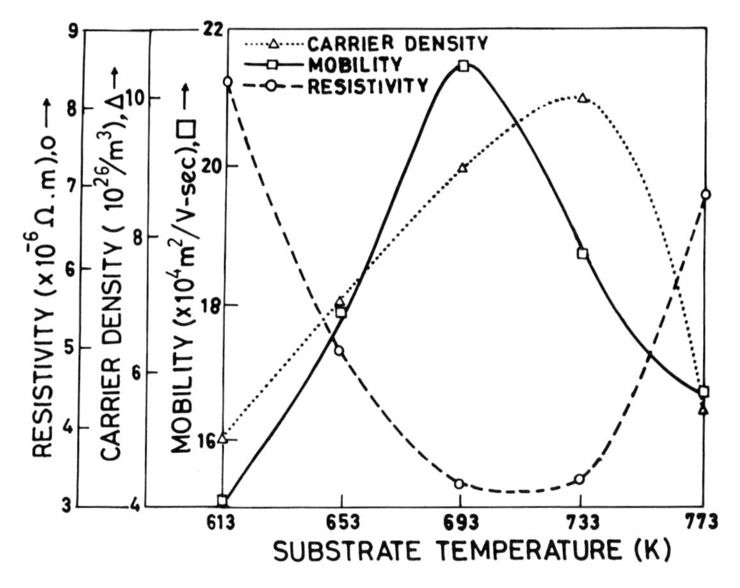

Figure 3.43 Variation of carrier density, mobility and resistivity in ITO films as a function of substrate temperature (from [111]).

resistivity increases again. This increase in resistivity may be attributed to the oxidation of InO_x:Sn films. This is further confirmed by the observed decrease in carrier concentration N in this temperature range. Similar variation in resistivity with substrate temperature has also been observed by Noguchi and Sakata [114] and Kulaszewicz *et al* [134] for ITO films prepared by reactive evaporation and a spray technique, respectively.

(c) *Film thickness effects*

The electrical parameters ρ, N and μ_H are shown [89, 121, 134–136] to depend on the thickness of the films. Figure 3.44 illustrates the typical variation of conductivity, carrier concentration and Hall mobility with thickness, for tin-doped In_2O_3 films deposited by RF sputtering. It may be seen that both the conductivity and Hall mobility increase with increasing film thickness. The low value of Hall mobility in very thin films can be explained by considering the scattering at intercrystalline boundaries. As thinner films contain more defects than thicker films, increased scattering of carriers takes place resulting in low mobility. The decrease in carrier concentration with the increase in film thickness may be due to a decrease in the defect contribution to the carrier concentration as a result of better growth characteristics. The increase in conductivity with film thickness is due to an improvement in crystallinity in ITO films [89, 134, 135].

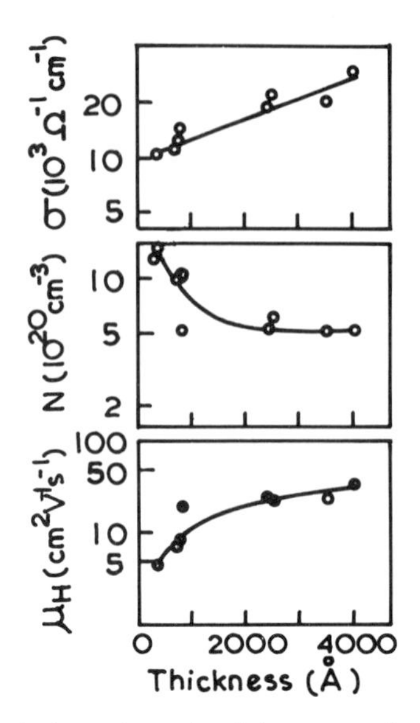

Figure 3.44 Variation of conductivity σ, carrier concentration N and mobility μ_H as a function of thickness for RF sputtered ITO films (from [135]).

(d) *Effect of sputtering parameters*

The electrical properties of sputtered ITO films have also been found [133, 137–141] to depend on other preparation parameters, e.g. sputtering power and oxygen partial pressure. In particular, it is essential to control the oxygen partial pressure during deposition to obtain high quality reproducible indium tin oxide films. The dependence of electrical resistivity on sputtering power is shown in figure 3.45. Resistivity decreases with increasing sputter power. For low values of sputter power, $P < 900$ W, the target surface is oxidized non-stoichiometrically. Sputtered particles can be further oxidized during their transport from target to substrate or on the substrate itself. However, for high sputter power, the number of sputtered atoms increases. As such, more oxygen atoms are used to oxidize these sputtered species, and the number of oxygen atoms available to oxidize the target decreases. The surface of the target becomes more metallic and so the deposited films have low resistivity.

Figure 3.46 illustrates the dependence of resistivity on partial oxygen pressure $p(O_2)$ for a typical ITO film prepared by DC magnetron sputtering using an alloy of indium and tin. It may be observed that the resistivity remains low $\sim 2 \times 10^{-5}$ Ω cm for $p(O_2) < 0.16$ Pa (1.2×10^{-3} Torr) and starts increasing thereafter. The low value of electrical resistivity for low $p(O_2)$ is due to non-stoichiometry, and films have a metallic nature. Low

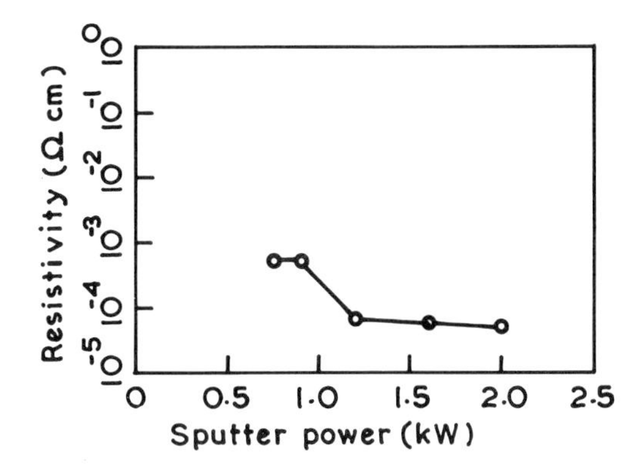

Figure 3.45 Variation of resistivity with sputtering power for DC magnetron sputtered ITO film (from [138]).

partial pressure of oxygen is not sufficient to oxidize the target and oxygen atoms are consumed in oxidizing the partially sputtered metal atoms. However, high partial pressure of oxygen [$p(O_2) > 0.23$ Pa (1.7×10^{-3} Torr)] is sufficient to continuously oxidize the target. Consequently, a stoichiometric oxide is sputtered, resulting in high resistivity. For intermediate oxygen partial pressure $0.23 \text{ Pa} > p(O_2) > 0.16 \text{ Pa}$ ($1.7 \times 10^{-3} \text{ Torr} > p(O_2) > 1.2 \times 10^{-3}$ Torr), both oxidation of the target and sputtering of metal atoms take place. Moreover, oxidation of the

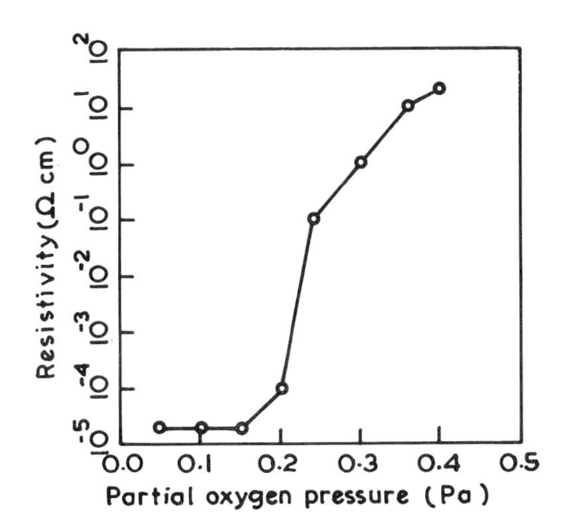

Figure 3.46 Resistivity versus partial oxygen pressure for DC sputtered ITO films. Power, 750 W; total pressure, 0.6 Pa (4.5×10^{-3} Torr) (from [138]).

sputtered species may take place during their transport or during growth of the films. This leads to an intermediate value of resistivity. Fan and co-workers [133, 137] obtained highly conductive and transparent ITO films only over a very narrow range of $p(O_2)$ from 3×10^{-5} to 4×10^{-5} Torr. Figures 3.47 and 3.48 show the variation of electrical resistivity ρ, carrier concentration N and Hall mobility μ_H as a function of oxygen partial pressure $p(O_2)$ for ITO films prepared by RF sputtering using two hot pressed targets of In_2O_3–SnO_2 of the same composition, namely In_2O_3–15 mol% SnO_2. It may be seen that the resistivity initially decreases with increasing oxygen partial pressure, reaches a minimum, and then increases sharply. The minimum value of resistivity is $\sim 3 \times 10^{-4}\ \Omega$ cm at $p(O_2) = 3$–4×10^{-5} Torr. The carrier concentration remains almost constant at ~ 5–7×10^{20} cm^{-3} up to $p(O_2) \sim 4 \times 10^{-5}$ Torr and then falls abruptly, resulting in a sharp increase in resistivity. The maximum value of Hall mobility is 35 cm^2 V^{-1} s^{-1} at $p(O_2) = 3$–4×10^{-5} Torr, where resistivity is minimum.

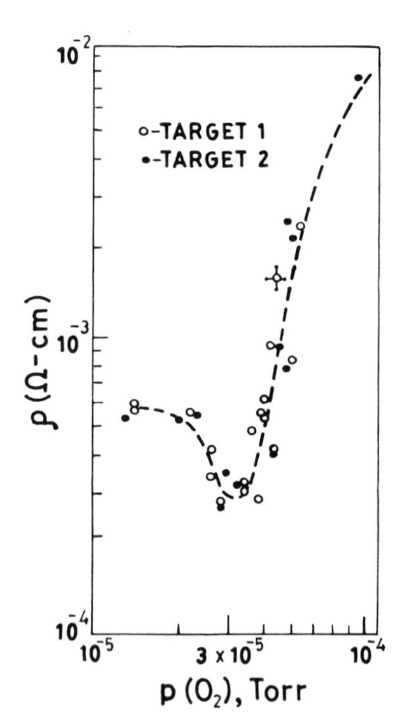

Figure 3.47 Variation of resistivity with oxygen partial pressure for RF sputtered ITO films (from [133]).

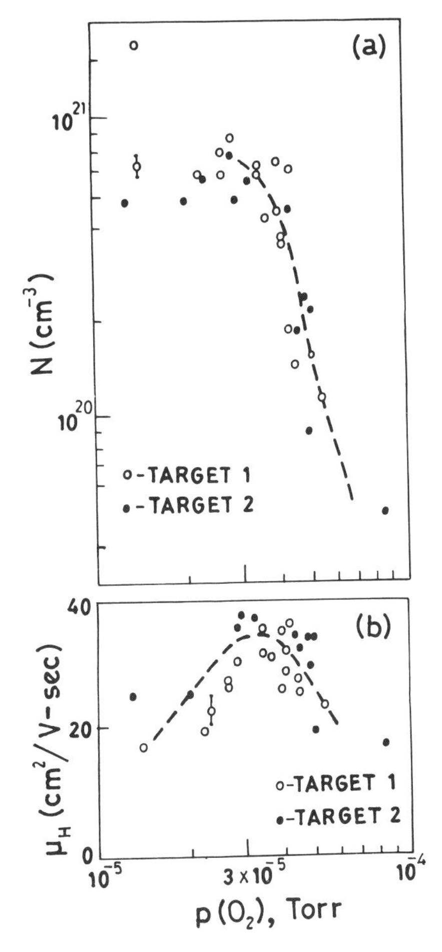

Figure 3.48 Variation of (a) carrier concentration and (b) Hall mobility as a function of oxygen partial pressure for RF sputtered ITO films (from [133]).

(e) *Post-deposition annealing effects*
Appropriate post-deposition annealing helps in producing highly conductive and transparent films of indium tin oxide. Many investigators [70, 93, 138, 142–147] have studied the annealing effects in ITO films. Too high an oxygen concentration in indium tin oxide films increases electrical resistivity, while too low an oxygen content makes the films more metallic. Thus it is necessary to optimize oxygen concentration to obtain highly conductive and transparent ITO films. This can be easily achieved by annealing either a highly resistive film in a reducing atmosphere or a low resistance (metallic)

film in an oxidizing atmosphere. Annealing a high resistance film in a reducing atmosphere lowers the oxygen content of the film, thereby making it more conductive. Figure 3.49 shows the influence of annealing on the resistivity of ITO films prepared by DC magnetron sputtering at different oxygen partial pressures. The films were annealed in forming gas (10% H_2–90% N_2) for 30 minutes at 485 °C. On comparing figure 3.49 with figure 3.46, it can be seen that annealing in a reducing atmosphere removes part of the oxygen from films which have too much oxygen incorporated ($p(O_2) > 0.23$ Pa (1.7×10^{-3} Torr)). However, the low resistivity of films which have less than the optimum amount of oxygen remains almost unchanged after annealing in the reducing atmosphere. Figure 3.50 shows

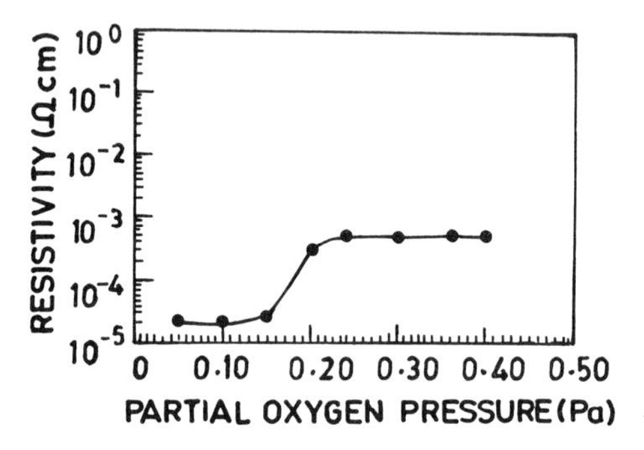

Figure 3.49 Resistivity of ITO films prepared at various partial oxygen pressures (figure 3.46) after subsequent annealing in forming gas at 485 °C for 30 min (from [138]).

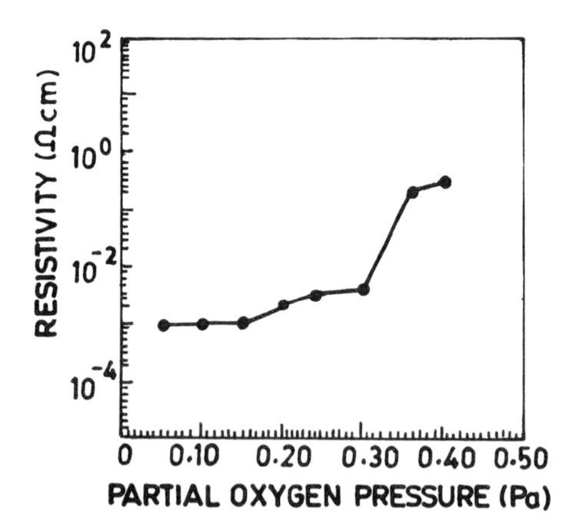

Figure 3.50 Resistivity of ITO films prepared at various partial pressures (figure 3.46) after subsequent annealing in oxygen at 600 °C for 30 min (from [138]).

the effect of annealing a low resistance film (metallic in nature) in an oxidizing atmosphere. The films were annealed in an oxygen atmosphere at 600 °C for 30 min. It is observed that the conductivity decreases for films deposited at low oxygen partial pressures after annealing in oxygen. However, there is no marked effect of annealing on films grown at high oxygen partial pressures. Enjouji *et al* [142] also observed similar results in ITO sputtered films.

3.6.2 Fluorine-doped indium oxide films

Fluorine has been used as a dopant in indium oxide films [151–154]. Avaritsiotis and Howson [151,152] used a reactive ion plating technique to prepare fluorine-doped indium oxide films in which CF_4 was used as a source of fluorine. The electrical properties of fluorine-doped indium oxide films depend strongly on fluorine content and fluorine partial pressure. Since the fluorine incorporated into the film depends on its partial pressure, control of fluorine partial pressure is important. Figure 3.51 shows the dependence of surface resistance (sheet resistance) on fluorine partial pressure. Typically, sheet resistance decreases from 12 $k\Omega\,\square^{-1}$ to 171 $\Omega\,\square^{-1}$ as fluorine partial pressure increases from 0.8 to 1.8 m Torr.

Maruyama and Fukui [153] studied the effect of reaction temperature (substrate temperature) on the resistivity of fluorine-doped indium oxide films prepared by a CVD technique. Figure 3.52 shows the electrical

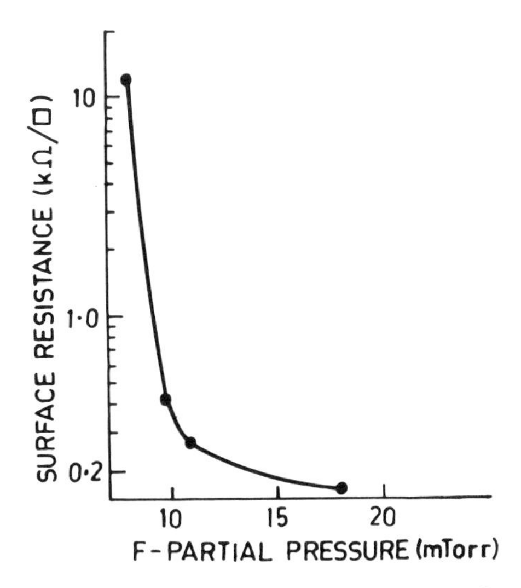

Figure 3.51 The effect of fluorine partial pressure on the sheet resistance of fluorine-doped In_2O_3 films prepared by reactive ion plasma (from [151]).

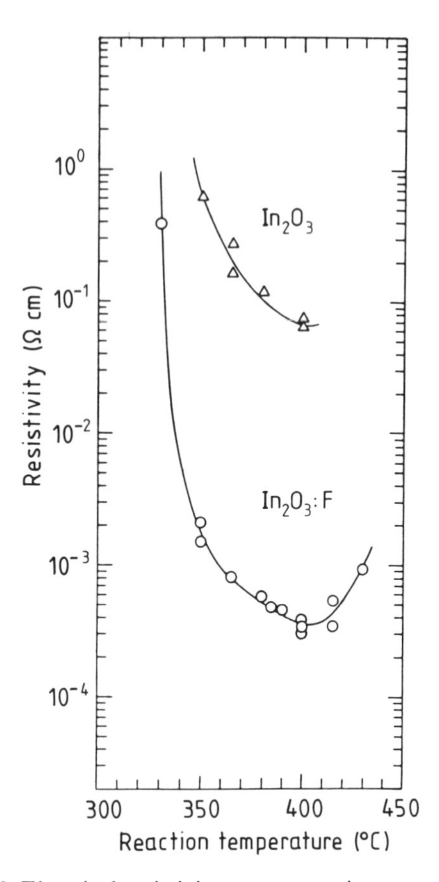

Figure 3.52 Electrical resistivity versus reaction temperature for In_2O_3 and fluorine-doped In_2O_3 films (from [153]).

resistivity of In_2O_3:F and In_2O_3 films as a function of reaction temperature. Though the nature of the curve is similar for both In_2O_3:F and In_2O_3 films, the values of resistivity in fluorine-doped films are almost two orders of magnitude lower than those of the undoped films. The lowest resistivity in these fluorine doped films was $2.89 \times 10^{-4}\ \Omega$ cm and the sheet resistance was $50\ \Omega\ \square^{-1}$ for 578 Å thick film.

Annealing in air also affects the electrical properties of fluorine-doped indium oxide films. Avaritsiotis and Howson [151] observed a significant improvement in the electrical properties of fluorine-doped In_2O_3 films. Typically, conductivity improved from 4.6×10^3 to $5.7 \times 10^3\ \Omega^{-1}\ cm^{-1}$, carrier concentration from 3.7×10^{20} to $3.2 \times 10^{21}\ cm^{-3}$ and mobility from 8 to 11 $cm^2\ V^{-1}\ s^{-1}$. These results have been explained on the basis of the assumption [151] that large numbers of unsaturated indium bonds acting as electron traps are present in fluorine-doped indium oxide films. On annealing, the unsaturated indium bonds are saturated with negative oxygen, thus improving the electrical properties.

3.7 Electrical Properties of Cadmium Stannate Films

Thin films of cadmium stannate or cadmium tin oxide (CTO) are n-type semiconductors with a wide band gap. There are two known phases of cadmium stannate, namely Cd_2SnO_4 and $CdSnO_3$. Thin films of Cd_2SnO_4 have extremely promising electrical properties such as high, metal-like electrical conductivity $\sim 10^3\ \Omega^{-1}\ cm^{-1}$, high carrier concentration $\sim 10^{20}\ cm^{-3}$ and sufficiently high mobility $\sim 45\ cm^2\ V^{-1}\ s^{-1}$. Many workers [174–190] have studied the electrical properties of Cd_2SnO_4 films grown by different techniques, namely DC reactive sputtering, RF sputtering, spray pyrolysis, etc. The electrical properties are reported to depend on deposition technique, deposition parameters and post-deposition annealing.

(a) *The effect of deposition temperature*
Substrate temperature (T_s) is an important deposition parameter in the growth of cadmium stannate films. Amorphous films [174] are formed if the substrate is maintained at or near room temperature. For the deposition of highly conductive polycrystalline films [175–178] high substrate temperatures are required. Ortiz [177] reported the deposition of transparent conducting films of Cd_2SnO_4, suitable for solar cells, using a spray pyrolysis technique. Figure 3.53 shows the effect of substrate temperature on the

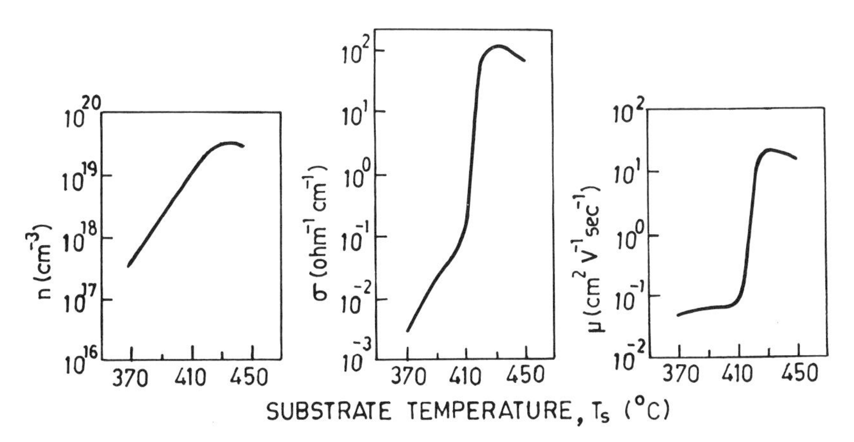

Figure 3.53 The effect of substrate temperature on the electrical properties of spray-deposited Cd_2SnO_4 films (from [177]).

electrical properties of Cd_2SnO_4 films. It can be observed from the figure that the electrical conductivity increases slowly with increase of substrate temperature up to 410 °C and then rises sharply as T_s goes from 410 to 430 °C. The initial increase in conductivity for substrate temperatures up to 410 °C is ascribed to the increase in carrier concentration. On the other hand, the carrier concentration value for films deposited above 410 °C suggests that the electrical conductivity is controlled mainly by mobility effects.

(b) *Effect of film thickness*

Film thickness is reported [179] to affect the resistivity of CTO films. Figure 3.54 shows the dependence of resistivity on the film thickness of CTO films deposited by RF sputtering. It is found that the resistivity decreases with film thickness up to 2000 Å, while films with thickness greater than 2000 Å have an almost constant value of resistivity $\sim 6.5 \times 10^{-4}\ \Omega$ cm. These results are consistent with those for other transparent conducting films.

(c) *Effect of oxygen partial pressure*

The electric properties of sputtered Cd_2SnO_4 films depend on the oxygen concentration in the sputtering gas mixture. Figure 3.55 illustrates the dependence of resistivity ρ, mobility μ and carrier concentration N on the pressure ratio of oxygen with respect to the total pressure for sputtered Cd_2SnO_4 films. It is seen that the resistivity passes through a minimum $\rho_{min} \sim 4.5 \times 10^{-4}\ \Omega$ cm. At the minimum of resistivity, carrier concentration and mobility have the values $\sim 4 \times 10^{20}$ cm^{-3} and 35 cm^2 V^{-1} s^{-1}, respectively. Similar qualitative results have also been obtained by Miyata *et al* [181] for Cd_2SnO_4 films prepared by DC magnetron–plasmatron sputtering.

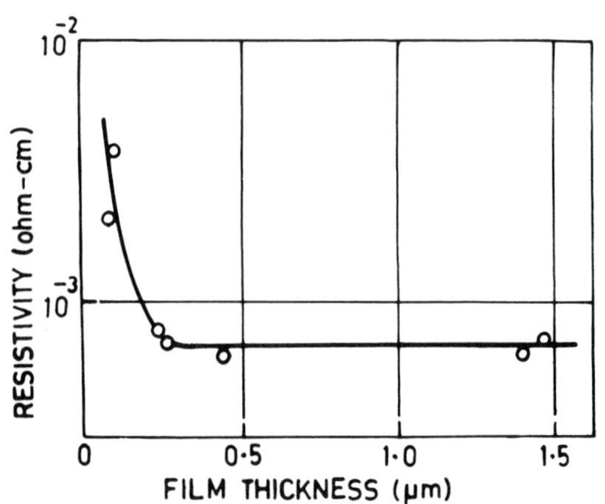

Figure 3.54 Effect of film thickness on the resistivity of RF sputtered CTO films (from [179]).

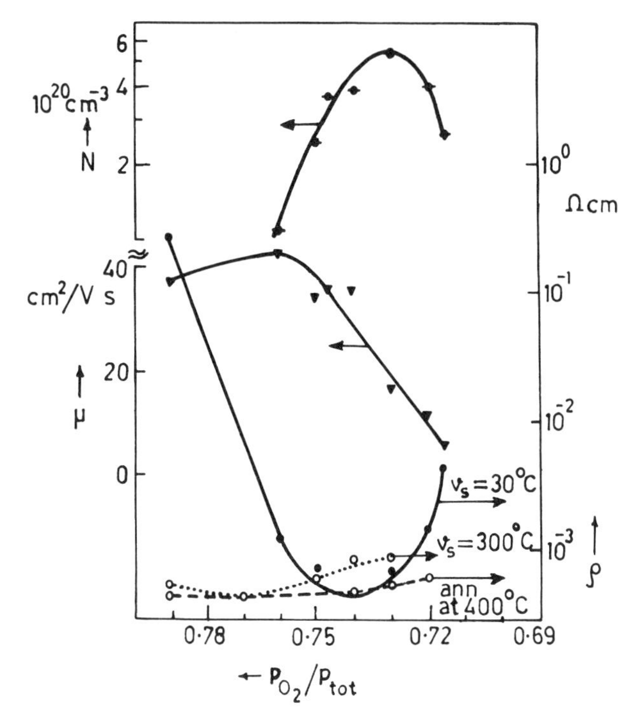

Figure 3.55 Dependence of resistivity ρ, mobility μ and carrier concentration N on pressure ratio of oxygen with respect to total pressure for DC reactively sputtered CTO films (from [180]).

(d) *Post-deposition annealing effects*

Post-deposition heat treatment of Cd_2SnO_4 films is found to affect their electric properties significantly. Many workers [175, 177, 179, 181–184] have made extensive investigations of the influence of post-deposition heat treatment in different ambients on the electrical properties of cadmium stannate films. Miyata and co-workers [181,184] observed a decrease in resistivity and an increase in carrier concentration after annealing of DC reactive sputtered Cd_2SnO_4 films either in vacuum, Ar or N_2 atmosphere at 300 °C. A typical variation of resistivity versus annealing time for CTO films annealed at 300 °C in a vacuum of 10^{-7} Torr is shown in figure 3.56. Pisarkiewicz *et al* [183] observed an increase in both carrier concentration and Hall mobility in Cd_2SnO_4 films prepared by DC reactive sputtering and annealed in vacuum at 10^{-4} Pa (7.5×10^{-7} Torr) at 670 K. The annealing effects may be explained by considering that the films deposited by sputtering are generally non-stoichiometric due to the formation of point defects that include oxygen deficiencies and metal interstitials as well as oxygen traps at the grain boundaries. Annealing in vacuum or inert gases causes

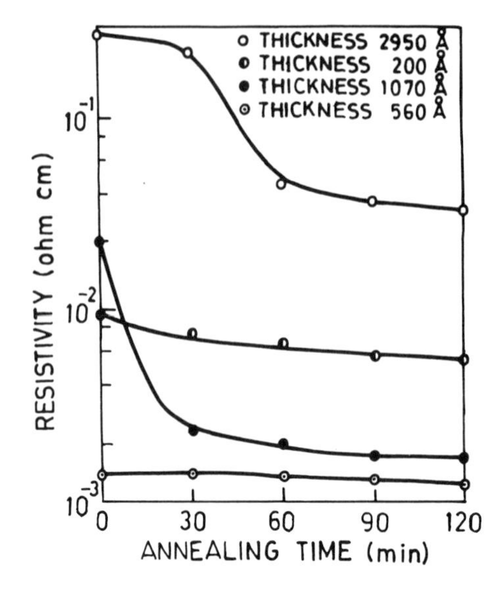

Figure 3.56 Resistivity of cadmium stannate films as a function of annealing time at 300 °C in vacuum (from [184]).

recrystallization of the films and generates oxygen deficiencies. The oxygen deficiencies provide donor states, thus affecting carrier concentration. Mobility increases due to the improvement in crystallinity of the films.

Pisarkiewicz *et al* [183] observed a further increase in carrier concentration but a slight decrease in mobility after subsequent reduction of the films in hydrogen at 670 K. The decrease in mobility has been ascribed to an increase in the effective mass of the charge carriers [183]. Ortiz [177] also reported an improvement of electrical properties after annealing the films in hydrogen at different temperatures. A typical variation of resistivity with annealing temperature for Cd_2SnO_4 films is shown in figure 3.57. A change of two orders of magnitude in resistivity is found in hydrogen-annealed films deposited at low substrate temperatures (370 °C; curve (a)). However, no appreciable change in resistivity is visible for films prepared at high substrate temperatures (430 °C; curve (b)).

(e) *Scattering mechanisms*

Figure 3.58 shows the typical variation of electrical conductivity as a function of I/T for CTO films prepared by an RF sputtering technique in an Ar atmosphere. The dependence of conductivity on temperature for films with different carrier concentrations reveals that films with $N = 8.2 \times 10^{17}\ \mathrm{cm^{-3}}$ have a donor ionization energy ~0.21 eV, whereas films with a carrier concentration of $7.2 \times 10^{18}\ \mathrm{cm^{-3}}$ have a donor activation

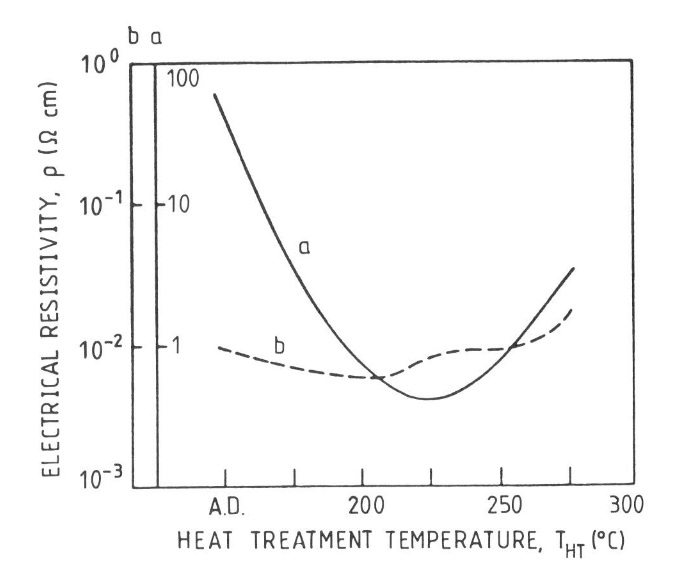

Figure 3.57 Variation of resistivity with annealing temperature in hydrogen atmosphere for spray-deposited CTO films grown at (a) 370 °C and (b) 430 °C (from [177]).

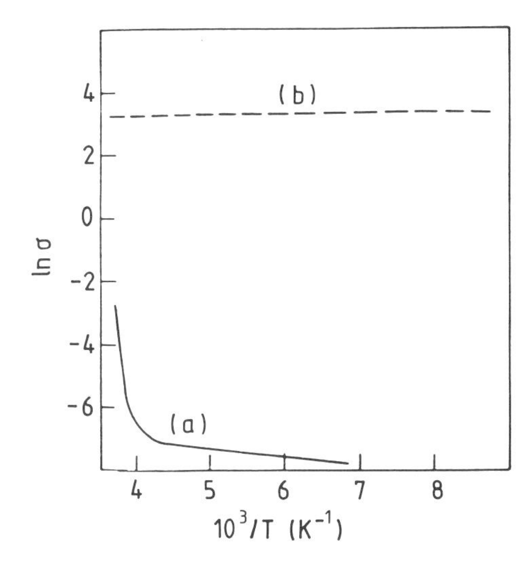

Figure 3.58 Electrical conductivity of spray-deposited cadmium stannate films as a function of temperature for films grown at (a) 370 °C and (b) 430 °C (from [177]).

energy ~0.014 eV [179]. Films with a carrier concentration greater than 4×10^{19} cm^{-3} are degenerate. Pisarkiewicz *et al* [183], Ortiz [177], and Leja *et al* [186] also observed that films with a carrier concentration more than 10^{19} cm^{-3} are degenerate. Figure 3.59 shows the temperature dependence of Hall mobility for cadmium stannate films grown using a sputtering technique. It can be observed that Hall mobility decreases with increasing temperature. Analysis of the data suggests that both acoustic phonon scattering and point defect scattering are the dominant scattering mechanisms in these films. The proposed point defects are oxygen vacancies, interestitial cadmium ions and neutral complexes $(CdO)^x$ [186]. Stapinski *et al* [182] reported that mobility increases with increasing carrier concentration for $N < 2.2 \times 10^{20}$ cm^{-3}. However, for samples with $N > 2.2 \times 10^{20}$ cm^{-3}, mobility decreases with increase of carrier concentration. This behaviour cannot be explained by considering ionized impurity scattering. Most likely, in Cd_2SnO_4 films, $(CdO)^x$ neutral complexes may exist, which decay with increase of carrier concentration. The most plausible scattering mechanism may be due to neutral defects. Hall mobility μ_n due to scattering by neutral defects depends only on the concentrations of neutral defects N_n [13]. The mobility μ_n is given by [13]

$$\mu_n \approx m^*/N_n \tag{3.51}$$

where m^* is the effective mass of the charge carriers. With increasing carrier concentration, N_n decreases resulting in an increase of Hall mobility. For films with $N > 2.2 \times 10^{20}$ cm^{-3}, the decrease in mobility with increase in carrier concentration is in agreement with the theory of electron scattering by ionized impurity centres for complete degeneracy [125].

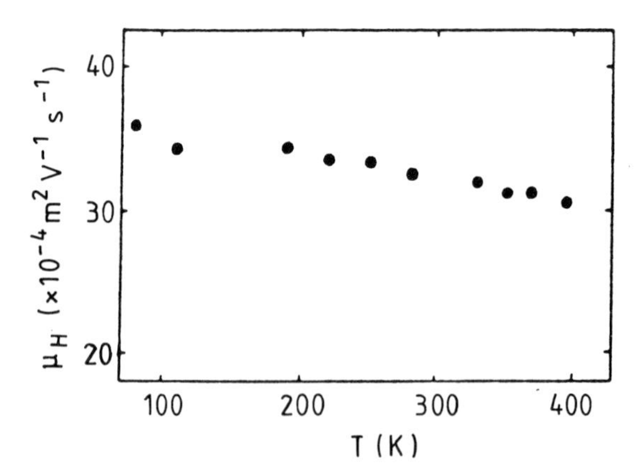

Figure 3.59 Temperature dependence of Hall mobility of cadmium stannate films (from [186]).

3.8 Electrical Properties of Zinc Oxide Films

Zinc oxide is a II–VI n-type wide band gap semiconductor; band gap $E_g = 3.2$–3.3 eV at room temperature. In undoped zinc oxide the n-type conductivity is due to deviations from stoichiometry. The free charge carriers result from shallow donor levels associated with oxygen vacancies and interstitial zinc [191–196], although interstitial oxygen and zinc deficiencies may also be present and produce acceptor states [197]. Regardless of the deposition method, all undoped ZnO conducting films have unstable electric properties in the long term. This stability is related to the change in surface conductance of ZnO films under oxygen chemisorption and desorption [197]. The electrical properties of ZnO films strongly depend on the deposition method, thermal treatment and oxygen chemisorption. Many workers [198–209, 213–240] have studied the electrical properties of undoped and doped ZnO films prepared by different methods.

(a) *Substrate temperature effects*
Figures 3.60(a) and 3.60(b) show the typical variations of room temperature conductivity, carrier concentration and Hall mobility as a function of substrate temperature for ZnO films deposited by the oxidation of diethyl zinc (DEZ). It may be seen that the conductivity varies from 10^{-2} to $50\ \Omega^{-1}\ \mathrm{cm}^{-1}$ as substrate temperature changes from 280 to 350 °C. In the case of substrate temperatures lower than 350 °C, the increase in electrical conductivity is due to an increase in carrier concentration, in the case of substrate temperatures above 350 °C, the decrease in electrical conductivity is consistent with the decrease in carrier concentration, probably due to more complete oxidation of organic products on the substrate surface yielding better stoichiometry. This variation of resistivity with substrate temperature is in agreement with the results of Webb *et al* [199], Tomar and Garcia [200] and Natsume and co-workers [201, 202] for undoped ZnO films. This behaviour is independent of the fabrication technique for these films.

Figure 3.60(b) shows that mobility increases with substrate temperature up to 350 °C. This increase in mobility has been attributed [198] to (i) better crystallization and (ii) a decrease in the grain boundary barrier potential of ZnO films prepared by a CVD technique.

(b) *Effect of sputtering parameters*
In addition to substrate temperature, oxygen partial pressure is another important parameter which affects the electrical properties of transparent conducting ZnO films prepared by a sputtering technique. Depending on oxygen concentration and substrate temperature, sputtered ZnO films can be catagorized into three groups. Figure 3.61 shows these three groups with

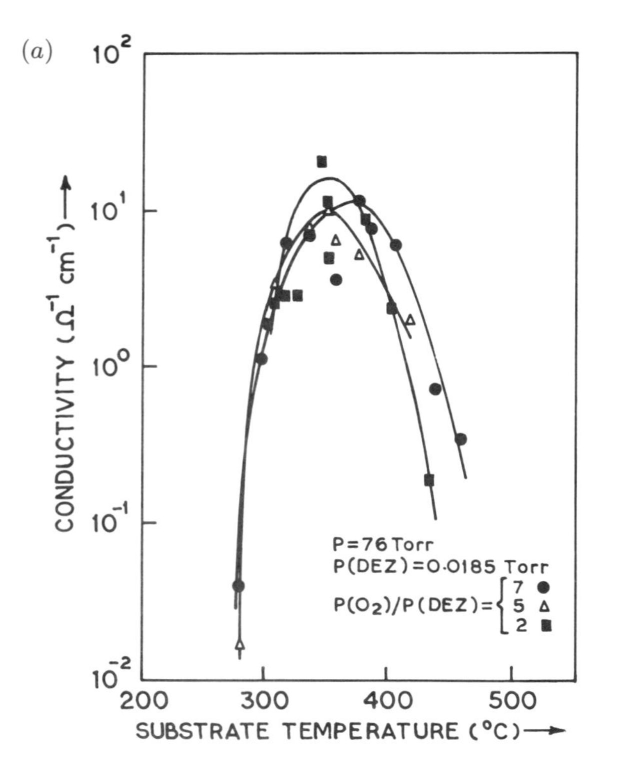

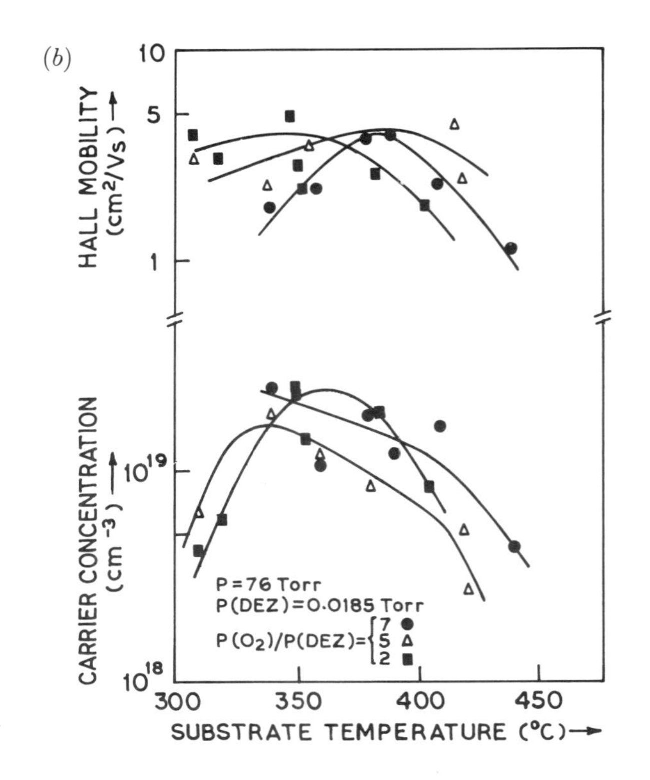

Figure 3.60 Variation of (*a*) conductivity and (*b*) Hall mobility and carrier concentration with substrate temperature of CVD grown ZnO films for different oxygen partial pressures (from [198]).

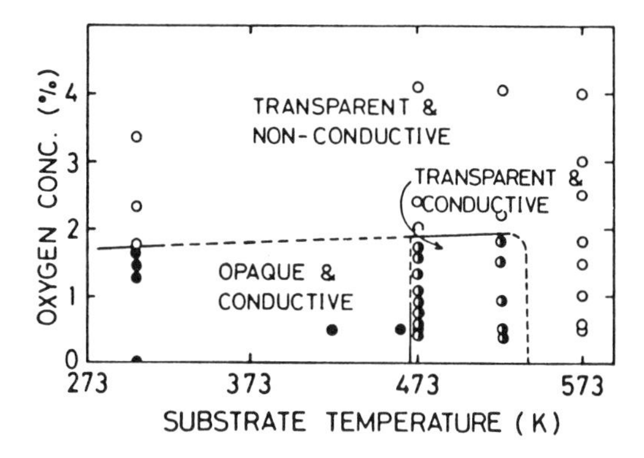

Figure 3.61 Effect of oxygen concentration and substrate temperature on conductivity and transparency of ZnO films: ●, opaque and conductive; ○, transparent and non-conductive; and ◑, transparent and conductive (from [203]).

respect to substrate temperature and oxygen contents. Films belonging to the first group, characterized as a mixture of metallic zinc and zinc oxide, are opaque and conductive. These are prepared at substrate temperatures less than 473 K with oxygen concentration less than 1.8%. Films of the third group have a composition close to the stoichiometry of ZnO and are transparent and non-conductive. These can be prepared either at substrate temperatures higher than 523 K or at low substrate temperature but with higher oxygen content (>1.8%). Since non-stoichiometry in the form of interstitial zinc atoms and/or deficient oxygen sites plays a crucial role in governing the electrical conductivity and optical transparency of ZnO films, the composition of films of the second group can be expressed as $Zn_{1+x}O$. These conductive transparent films can be prepared at substrate temperatures between 473 and 523 K with oxygen concentration less than 1.8%.

High conductivity of the order of $10^3\ \Omega^{-1}\ cm^{-1}$ has been reported in as-deposited ZnO films prepared by the sputtering technique [199, 204, 205]. However, the sputtered films require either sputtering in argon with hydrogen gas added [199, 204] or sputtering of a ZnO target with zinc dopant added [205]. Webb *et al* [199] achieved higher conductivity in sputtered zine oxide films by introducing hydrogen into the argon sputter gas to change the stoichiometry of the films. Figure 3.62 shows the variation of resistivity with added hydrogen. It may be observed that the resistivity initially decreases rapidly with increasing added hydrogen ($<10^{-5}$ Torr), reaches a minimum at an added hydrogen partial pressure of 1×10^{-5} Torr and then starts increasing. The initial decrease in resistivity is consistent with hydrogen removing oxygen and thereby increasing the zinc/oxygen ratio in the deposited films. This results in the formation of either oxygen vacancies or

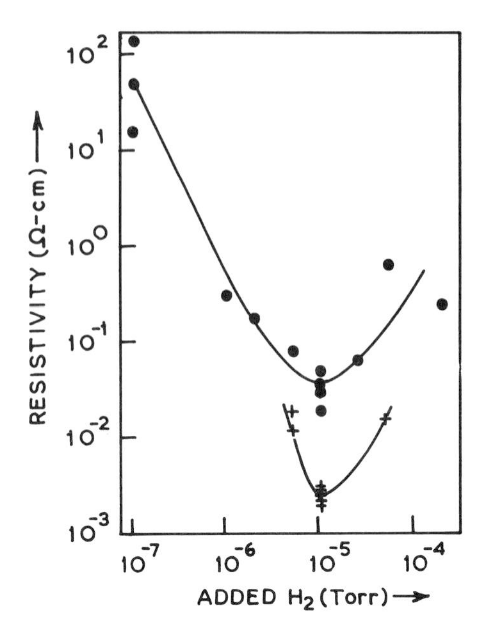

Figure 3.62 Variation of resistivity with added hydrogen for RF reactive magnetron sputtered ZnO films: ●, $T_S = 400$ K, deposit power = 200 W; +, $T_S = 315$ K and deposit power = 100 W (from [199]).

interstitial zinc giving rise to donor levels [197]. The carrier concentration is found to increase in this range. The increase in resistivity beyond 1×10^{-5} Torr added hydrogen is due to a decrease in carrier concentration. Films deposited at these pressures may show increased compensation from an increase in the density of acceptor levels resulting from changes in the growth characteristics [197, 206]. Nanto and co-workers [208, 209] used an external DC magnetic field to improve the conductivity of ZnO films prepared by an RF magnetron sputtering technique. They [208, 209] achieved a conductivity of the order $10^4\ \Omega^{-1}\ \text{cm}^{-1}$ in ZnO films prepared by RF magnetron sputtering in pure argon gas under an external DC magnetic field.

Caporaletti [204] reported a decrease in resistivity and an increase in carrier concentration in bias-sputtered ZnO films of several orders of magnitude by varying the bias voltage applied to the substrate, i.e. films deposited with larger bias voltage have lowest values of resistivity and highest value of carrier concentration. Typically, for bias voltage of -200 V, the values of resistivity and carrier concentration are $8 \times 10^{-3}\ \Omega$ cm and $8 \times 10^{19}\ \text{cm}^{-3}$, respectively.

(c) *Scattering mechanisms*

Figure 3.63 shows the variations of ρ, N and μ with temperature for ZnO films prepared by the oxidation of diethyl zinc. The interpretation of the mobility results is complicated by the possible simultaneous action of various scattering mechanisms. Roth and Williams [198] analysed their results by considering thermionic and thermal field emissions at the grain boundaries. In polycrystalline films, when the semiconductor is heavily doped or at low temperature, the dominating current is due to thermal field emission of carriers through the barrier [210, 211], while in lightly doped material or at high temperature, thermionic emission over the barrier dominates. Thermally activated mobility [19] can account for the thermionic emission over the grain boundary barriers in a lightly doped material:

$$\mu_{\mathrm{H}} \approx \mu_{\mathrm{g}} \approx \mu_{\mathrm{o}}' T^{-1/2} \exp\left(\frac{-e\phi_{\mathrm{b}}}{kT}\right). \tag{3.31}$$

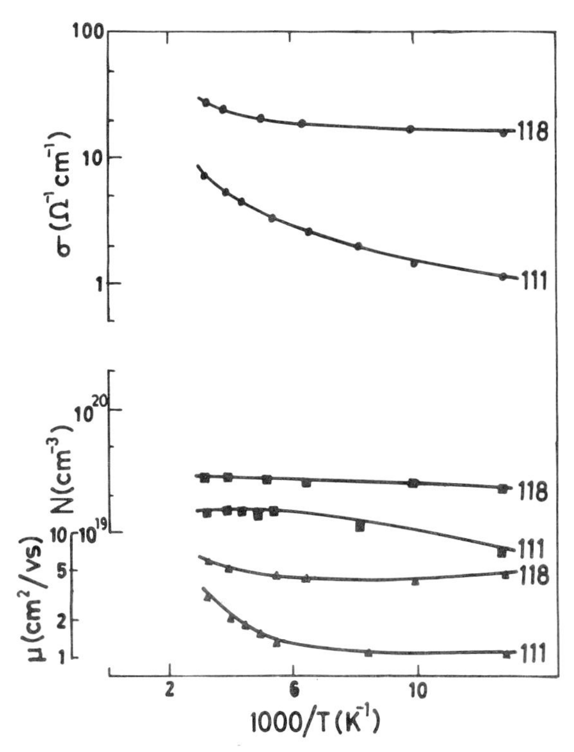

Figure 3.63 Temperature variation of σ, N and μ for CVD grown ZnO films. Sample (111) thickness 900 Å and sample (118) thickness 550 Å (from [198]).

On the other hand, for a current dominated by tunnelling in the case of a pure thermal field emission, the criterion is that $E_{oo} >> kT$ where E_{oo} is a parameter which determines the energy at which most of the tunnelling occurs.

The parameter E_{oo} is given by [211, 212]

$$E_{oo} = 18.5 \times 10^{-12}(N/m^*\varepsilon)^{1/2}\ \text{eV} \tag{3.52}$$

where ε is the static dielectric constant of the material. In ZnO films $m^* = 0.38$ and $\varepsilon = 8.5$ [213]. At room temperature, $E_{oo} \simeq kT$ for ZnO samples with $N = 6 \times 10^{18}\ \text{cm}^{-3}$. This means that the mobility in such samples will be limited by both thermionic and thermal field emission at the grain boundary. Figure 3.64 illustrates the variation of $\mu T^{1/2}$ versus $1/T$ for two samples of ZnO films. It may be seen that the mobility is dominated by tunnelling below 100 K in films with $N = 2.85 \times 10^{19}\ \text{cm}^{-3}$ (sample 118). Above 100 K, conduction is due to both tunnelling and thermionic emission. In the case of ZnO films with lower carrier concentration, $N = 1.5 \times 10^{19}\ \text{cm}^{-3}$ (sample 111), mobility is activated even at 77 K, indicating a mixed transport regime as expected with lower carrier concentration. It can be further observed that, in both samples, conduction is due to thermionic emission beyond 200 K. Blom *et al* [214] and Ogawa *et al* [202] also reported that thermionic emission is the dominating conduction mechanism across the grain barriers in ZnO films at the higher temperature range.

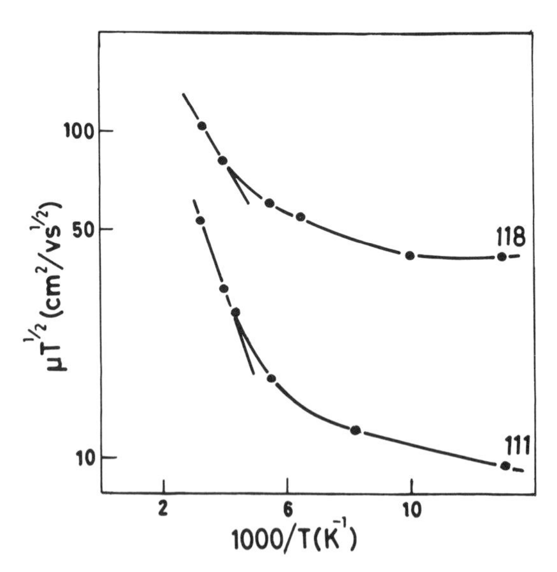

Figure 3.64 Variation of $\mu_H T^{1/2}$ versus $1/T$ for CVD grown ZnO films. Samples same as in figure 3.63 (from [198]).

Minami *et al* [215] investigated the stability of the electrical properties of ZnO films prepared by RF magnetron sputtering in various ambients. No significant change in electrical properties was reported for ZnO films exposed to air at room temperature for ten months. However, a change in electrical resistance was observed in ZnO films exposed to vacuum, inert gases and air ambients at high temperatures up to 400 °C. The resistance of all ZnO films was found to increase by one to three orders of magnitude after heat treatment in vacuum and inert gas ambients at 400 °C. Figure 3.65 shows a typical variation of resistance with temperature for sputtered ZnO films annealed in vacuum up to approximately 400 °C. It may be observed that on heat treatment, the increase in resistance is greater in sample A, which has a relatively low resistivity in the as-grown state, than in sample B. The increase in resistivity after heat treatment in vacuum and inert gases has been attributed mainly to a decrease in the carrier concentration. This decrease in carrier concentration is explained as being due to chemisorption into the ZnO films of oxygen which was trapped at defects such as grain boundaries [194, 216], because the chemisorption of oxygen on the surface

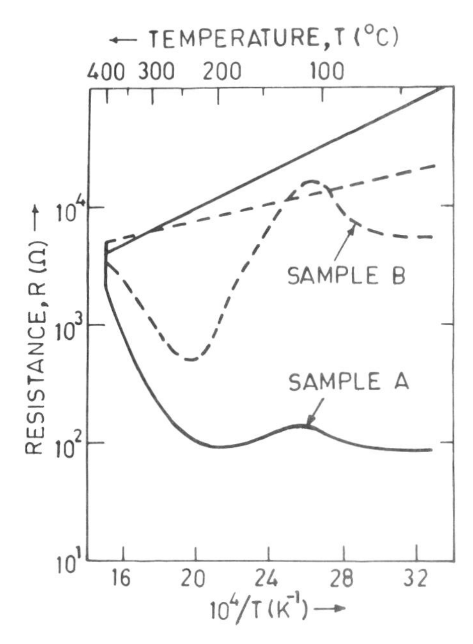

Figure 3.65 Resistance as a function of annealing temperature for sputtered ZnO films annealed in vacuum at a pressure of 1.3×10^{-4} Pa (9.8×10^{-7} Torr). Virgin state resistivity and Hall mobility of sample (A) 9.16×10^{-4} Ω cm, 13.6 $cm^2 V^{-1} s^{-1}$; (B) 3.54×10^{-2} Ω cm, 2.1 $cm^2 V^{-1} s^{-1}$ (from [215]).

of the films is negligible in vacuum and inert gases [217]. However, Major *et al* [218] observed a significant decrease in resistivity from 0.15 to 6×10^{-2} Ω cm after vacuum annealing at 625 K for 30 min an undoped ZnO film prepared by spray pyrolysis. Roth and Williams [198] also reported an increase in conductivity of ZnO films subjected to prolonged UV illumination under vacuum due to oxygen desorption. Minami *et al* [215] observed that the increase in resistivity after heat treatment in air is greater than that in vacuum and inert gases. The electrical resistance of the films is found to increase by three to ten orders of magnitude after heating the films in air at 400 °C. Caillaud *et al* [219] also observed a similar increase in resistivity as a result of annealing in air as well as in oxygen. The increase in resistivity after heat treatment in air is attributed to the chemisorption of oxygen on the surface of the films since the films are always exposed to air. In contrast to annealing in vacuum and in air, annealing in hydrogen at 400 °C is reported [215] to result in an increase of one order of magnitude only. This increase in resistivity after heat treatment in hydrogen is less than that for other ambients. This difference has been attributed to the formation of shallow donors as a result of doping by the diffusion of hydrogen impurities [188, 216, 220], and as a result of defects such as oxygen vacancies and interstitial zinc atoms due to oxygen deficiencies [191–196]. However, many workers [188, 204, 216, 220] have observed a reduction in resistivity of sputtered ZnO films after heat treatment in hydrogen ambient at a temperature around 400 °C. Call *et al* [221] observed a reduction in resistivity for aqueously deposited films of ZnO after annealing in forming gas (90% N_2–10% H_2) at 350 °C for 10 min.

3.9 Electrical Properties of Doped Zinc Oxide Films

Doping of ZnO films by aluminium, gallium or indium is reported [222–234] to have a significant effect on their electrical properties. Doping of ZnO films improves not only their electrical properties but also their stability. For example, ZnO films doped with 3 at% In are found [227] to exhibit thermal stability up to 650 K in vacuum and up to 450 K in oxygen ambients. Many workers [223–229] have reported highly conductive films of aluminium-doped ZnO prepared by different techniques. They have observed that aluminium-doped ZnO films have high carrier concentration and low mobility in comparison with those of undoped ZnO films. The high carrier concentration in aluminium-doped ZnO films is attributed to the contribution from Al^{3+} ions on substitutional sites of Zn^{2+} ions, and from interstitial aluminium in the ZnO lattice. Figures 3.66(a), (b) and (c) show the variation of conductivity, carrier concentration and mobility of aluminium-doped ZnO films grown by a CVD technique as a function of

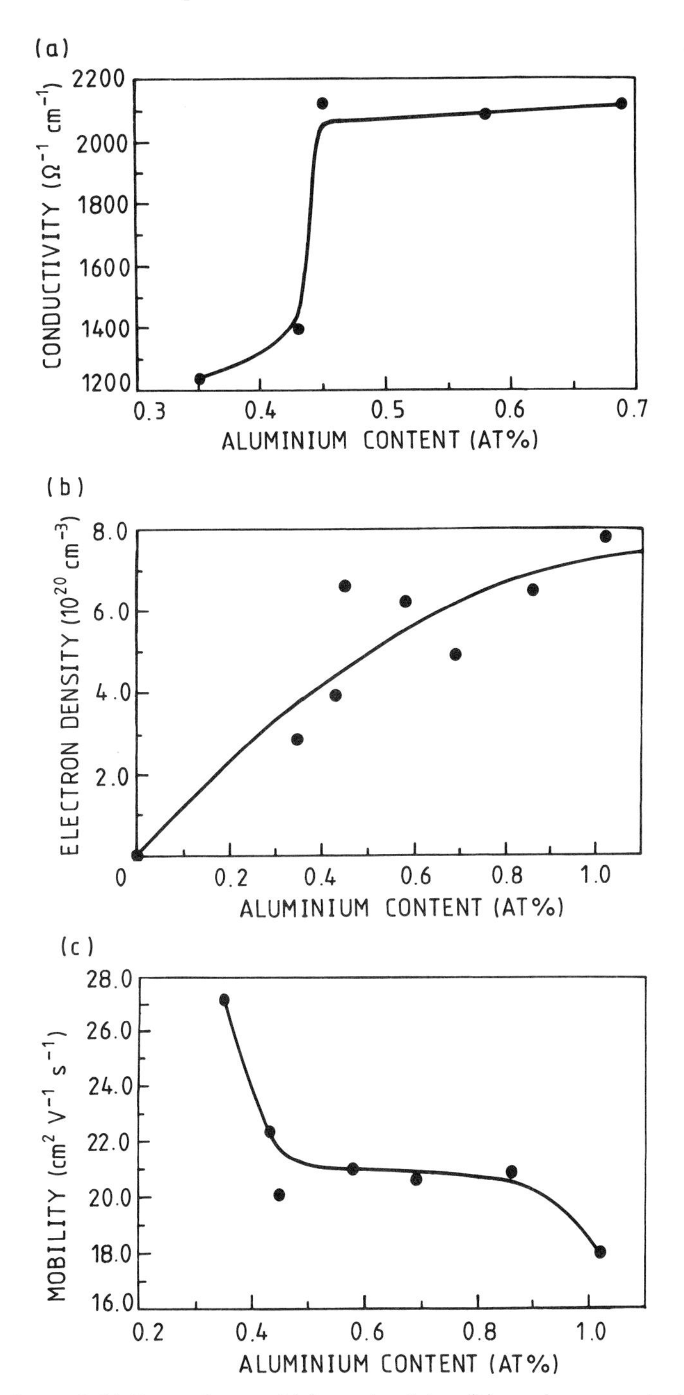

Figure 3.66 Dependence of (a) conductivity, (b) carrier concentration and (c) mobility on aluminium content in ZnO films (from [224]).

aluminium content. It can be observed from figure 3.66(a) that the conductivity increases rapidily with increase of aluminium concentration up to 0.45 at%; thereafter the conductivity almost stabilizes. The increase in conductivity is because the small amount of aluminium provides a large number of free electrons in the film. This can also be observed from figure 3.66(b). When the aluminium content is greater than 0.45 at%, non-conducting aluminium oxide phase due to extra aluminium atoms is formed. An equilibrium between the aluminium atoms contributing conduction electrons and those producing aluminium oxide is reached. For very high aluminium concentration, the conductivity would be expected to decrease again as a result of the large amount of non-conductive aluminium oxide in the films. Such a behaviour has been observed by Aktaruzzaman *et al* [223]. It can be observed from figure 3.66(c) that mobility decreases with increase in aluminium content. Aluminium atoms in the films produce not only conduction electrons but also ionized impurity scattering centres. It may also occupy interstitial positions and deform the crystal structure. Scattering by the ionized impurities and defects in the crystal result in a lower value of mobility. Similar effects have also been observed in gallium oxide doped ZnO films prepared by RF sputtering. The dependence of resistivity, carrier concentration and Hall mobility on gallium oxide content is shown in figure 3.67. It can be observed that resistivity decreases sharply with increase of

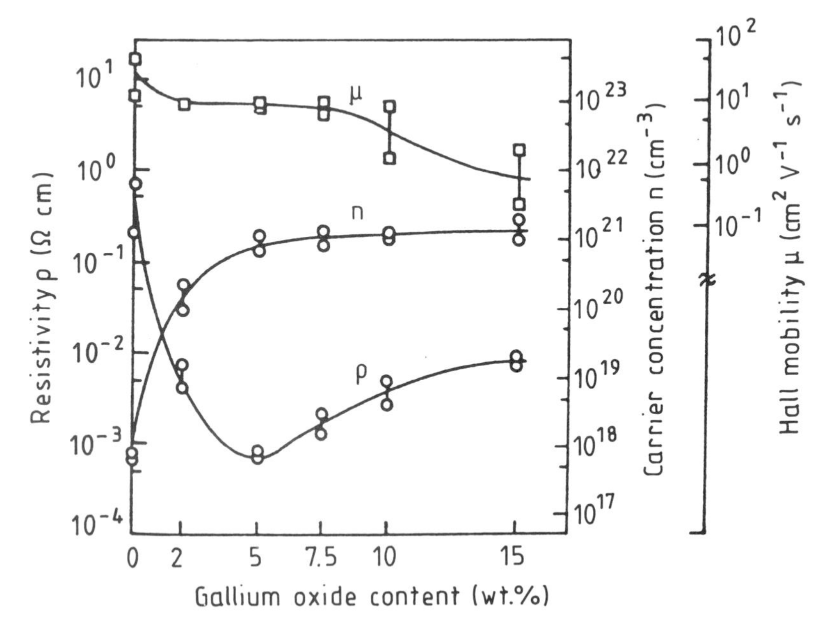

Figure 3.67 Dependence of conductivity, carrier concentration and mobility on Ga_2O_3 content in ZnO films (from [230]).

gallium oxide content up to 5 wt%, and then increases with further increase in Ga_2O_3 content. The carrier concentration first increases rapidly up to 5 wt% and then stabilizes for higher gallium oxide concentration. Up to 5 wt% of Ga_2O_3, it provides donors only, whereas above 5 wt% gallium segregation at the grain boundaries also takes place. It can be further observed from figure 3.67 that mobility decreases from ~30 $cm^2\ V^{-1}\ s^{-1}$ in undoped films to ~1 $cm^2\ V^{-1}\ s^{-1}$ in 15 wt% Ga_2O_3-doped ZnO films. The decrease in mobility is mainly due to ionized impurity scattering and additional scattering due to gallium segregation at grain boundaries.

(a) *Effect of substrate temperature and film thickness*

Doped ZnO films, like other transparent conducting films, show a dependence of electrical properties on substrate temperature and film thickness. Figure 3.68 shows the dependence of resistivity, carrier concentration and Hall mobility on substrate temperature for aluminium-doped ZnO films prepared by DC magnetron sputtering. The lowest resistivity, 3.5×10^{-4} cm, is observed to be independent of substrate temperature between 250 and 350 °C. The gradual decrease in the value of resistivity with increase in substrate temperature up to 250 °C is mainly related to the

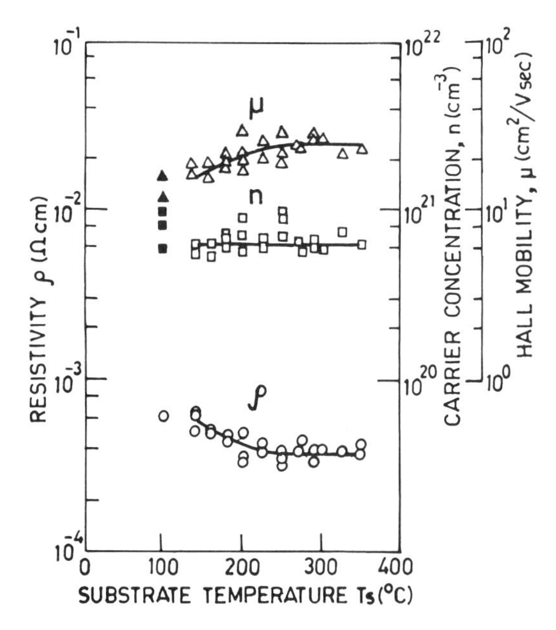

Figure 3.68 Variation of resistivity, carrier concentration and mobility with substrate temperature for A1:ZnO films. Solid points indicate the data for films grown on unheated substrates (from [225]).

increase in mobility. The increase in mobility with increase in T_s up to 250 °C is due to an improvement in crystallinity in these films. This improvement in crystallinity has been confirmed [225] using XRD studies. Aktaruzzaman *et al* [223] observed a weak dependence of electrical resistivity on substrate temperature in the range 300–500 °C in spray-deposited aluminium-doped ZnO films. Figure 3.69 shows the dependence of resistivity on film thickness for gallium-doped ZnO films prepared by RF sputtering. It can be observed that the resistivity decreases with increase of film thickness when the thickness is less than 2500 Å. However, after that it becomes almost independent of film thickness. The variation of resistivity in thinner films is due to surface scattering as discussed earlier. A similar behaviour has been observed by Minami *et al* [209, 226] and Major *et al* [222] for aluminium- and indium-doped ZnO films, respectively.

(b) *Thermal stability*

As discussed earlier, for practical applications, undoped ZnO films are not suitable because of their thermal instability. Doped ZnO films, however, have stable electrical and optical properties. Figure 3.70 shows the change in resistivity of undoped and doped (3 at% indium) ZnO films, when heated in oxygen and in vacuum. It may be observed that there is a marked variation in the resistivity in the case of undoped ZnO films as a result of heating both

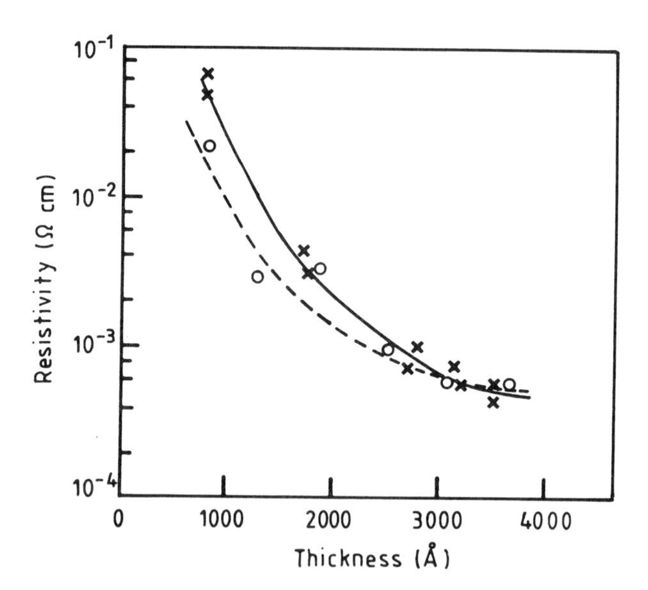

Figure 3.69 Variation of resistivity as a function of film thickness for Ga_2O_3-doped ZnO films sputtered with power densities of: (O) 0.42 W/cm^2; and (×) 0.84 W/cm^2 (from [230]).

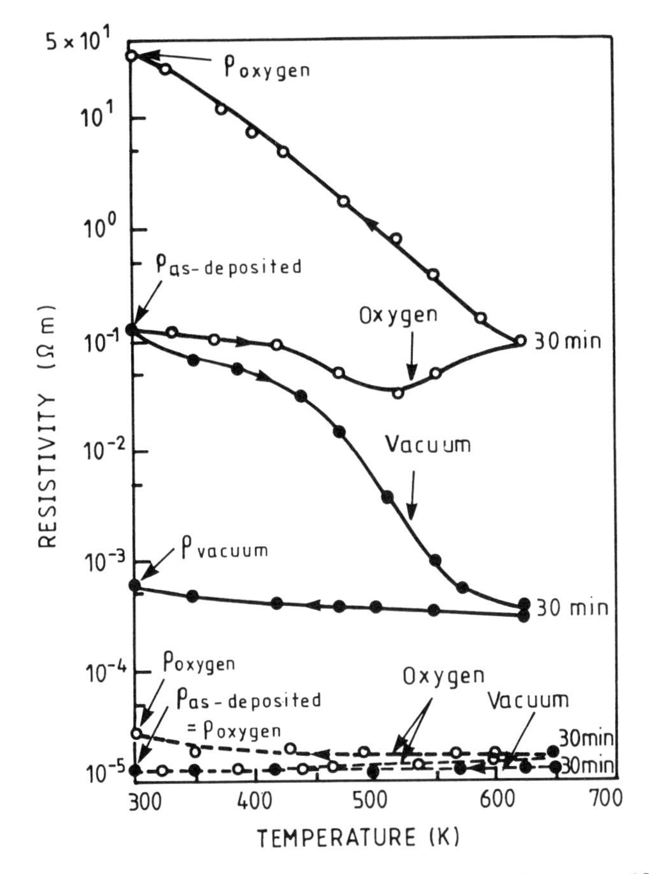

Figure 3.70 Variation of resistivity with temperature in vacuum (●) and oxygen (○) ambients for undoped ZnO (———) and 3 at% indium-doped ZnO (– – – –) films (from [222])

in vacuum and in oxygen. On the other hand, ZnO films doped with 3 at% indium show high thermal stability up to 650 K both in vacuum and in oxygen.

3.10 Conclusions

In general, the electrical properties of transparent conducting oxide films depend on the growth techniques and the parameters involved. Table 3.1 summarizes the important results on the electrical properties of various semiconducting transparent oxide films grown under optimum conditions.

Table 3.1 Electrical properties of important transparent conducting oxide films.

Material	Deposition technique	Sheet resistivity R_s ($\Omega\,\square^{-1}$)	Electrical conductivity ($\Omega^{-1}\,cm^{-1}$)	Carrier concentration N (cm^{-3})	Hall mobility ($cm^2\,V^{-1}\,s^{-1}$)	Reference
SnO_2	Spray	—	—	4.0×10^{19}	53	69
SnO_2	Spray	—	2.5×10^2	1.3×10^{20}	11	50
SnO_2	Spray	—	40	10^{20}	8	32
SnO_2	Spray	6.3	—	1.53×10^{20}	8.5	34
SnO_2	Spray	—	3×10^2	1.5×10^{20}	10	90
SnO_2	Sputtering	—	1.56	2.68×10^{18}	3.64	71
SnO_2	Sputtering	—	10^2	1.2×10^{20}	—	44
SnO_2	Sputtering	—	1.6×10^2	1.3×10^{20}	7.72	42
SnO_2	Sputtering	600	10^2	—	—	241
SnO_2	Sputtering	600	1.1×10^4	—	—	242
SnO_2	Sputtering	100	3×10^2	—	—	43
SnO_2	CVD	420	7.0×10^2	—	—	41
SnO_2	CVD	23	2.6×10^2	1.27×10^{20}	12.8	243
SnO_2	Laser evaporation	—	3.3×10^2	—	—	244
SnO_2:Sb	Spray	10	10^3	7×10^{20}	10	85
SnO_2:Sb	Spray	43	10^3	4×10^{20}	17	165
SnO_2:Sb	Spray	—	5×10^2	2×10^{20}	15	32
SnO_2:Sb	Sputtering	—	5×10^2	3×10^{20}	10	83
SnO_2:F	Spray	—	—	3.5×10^{20}	46	245
SnO_2:F	Spray	9.2	2×10^3	4.6×10^{20}	27.7	76
SnO_2:F	Spray	4	10^5	10^{21}	20	246
SnO_2:F	Spray	14.5	—	5×10^{20}	50	247
SnO_2:F	Spray	—	2×10^3	4.4×10^{20}	30	48
SnO_2:F	Spray	—	2×10^3	1.5×10^{21}	17	90

SnO_2:F	Spray	—	2×10^3	1.2×10^{21}	10	259
SnO_2:As	CVD	15	6×10^3	9×10^{20}	45	248
SnO_2:Mo	Reactive evaporation	3×10^3	—	8×10^{20}	—	68
In_2O_3	Spray	46	6×10^3	4.2×10^{19}	89.7	165
In_2O_3	Sputtering	—	3×10^3	3.6×10^{20}	—	123
In_2O_3	Thermal evaporation	—	5×10^3	4×10^{20}	70	99
In_2O_3	Reactive evaporation	—	5×10^2	3.5×10^{19}	25–60	104
In_2O_3	Reactive evaporation	—	3×10^3	10^{20}	—	249
In_2O_3	Ion plating	—	6.7×10^3	7.0×10^{20}	70	260
In_2O_3:F	CVD	50	3.5×10^3	—	—	153
In_2O_3:F	Ion plating	53	5.7×10^3	3.2×10^{21}	11	151
ITO	Spray	4	—	10^{21}	70	148
ITO	Spray	—	10^4	2×10^{20}	33	88
ITO	Spray	10	2×10^3	5×10^{20}	30	149
ITO	Spray	—	4×10^3	6×10^{20}	36	121
ITO	Spray	5.6	5.5×10^3	10^{21}	30	165
ITO	Sputtering	8.5	—	10^{20}	12	143
ITO	Sputtering	—	2.5×10^3	10^{21}	35	135
ITO	Sputtering	140	1.5×10^3	8×10^{20}	12	250
ITO	Sputtering	—	8×10^3	10^{21}	43	147
ITO	Sputtering	—	0.5	6.2×10^{20}	48.6	251
ITO	Sputtering	—	1.4×10^3	7×10^{20}	16	132
ITO	Sputtering	—	2.7×10^3	7×10^{20}	24.5	145
ITO	CVD	1.9	6×10^3	8.8×10^{20}	43	252
ITO	CVD	23	5×10^3	10^{21}	10	119
ITO	e-beam evaporation	—	4.1×10^3	8×10^{20}	30	113
ITO	e-beam evaporation	21.6	2×10^3	5.8×10^{20}	16.8	253
ITO	e-beam evaporation	5	5×10^3	7×10^{20}	24	254
ITO	e-beam evaporation	—	2.3×10^4	1.4×10^{21}	103	261

Continued

Table 3.1 *Continued.*

Material	Deposition technique	Sheet resistivity R_s ($\Omega\,\square^{-1}$)	Electrical conductivity ($\Omega^{-1}\,cm^{-1}$)	Carrier concentration N (cm^{-3})	Hall mobility ($cm^2\,V^{-1}\,s^{-1}$)	Reference
ITO	Reactive evaporation	—	1.6×10^3	7×10^{20}	15	114
ITO	Ion plating	18.7	5×10^3	9.1×10^{20}	53.6	166
ITO	Ion plating	21	10^3	—	—	161
ITO	Post-oxidized	—	5×10^2	10^{20}	20	115
ITO	Sol–gel	10^4	30	—	—	72
ITO:F	Sputtering	31	1.5×10^3	6×10^{20}	16	255
Cd_2SnO_4	Sputtering	400	1.1×10^2	2×10^{19}	—	177
Cd_2SnO_4	Sputtering	—	2×10^3	5×10^{20}	40	180
Cd_2SnO_4	Sputtering	—	1.5×10^3	—	20	256
Cd_2SnO_4	Sputtering	1	6.5×10^3	10^{21}	—	175, 189
Cd_2SnO_4	Sputtering	14	2×10^3	5×10^{20}	22	185
Cd_2SnO_4	Sputtering	—	—	5×10^{19}	40	183
ZnO	Sputtering	85	5×10^2	5×10^{19}	8	199
ZnO	CVD	—	10	10^{21}	14	258
ZnO	Reactive evaporation	—	10^3	10^{20}	10	257
ZnO	Reactive evaporation	—	2.9×10^2	$4–12 \times 10^{19}$	10–40	262
ZnO:In	Spray	—	10^3	2.2×10^{20}	24	222
ZnO:In	Sputtering	—	3×10^2	10^{20}	12.6	228
ZnO:In	Sputtering	—	50	7×10^{19}	1.9	234
ZnO:Al	Sputtering	—	4×10^3	8×10^{20}	20	225
ZnO:Al	Sputtering	—	4×10^3	10^{21}	30–40	226
ZnO:Al	Sputtering	—	7×10^3	10^{21}	25	227
ZnO:Al	Sputtering	—	10^2	4.68×10^{20}	1.47	228
ZnO:Al	CVD	4000	3×10^3	8×10^{20}	35	224

The principle common features of their electrical properties are:

(i) All transparent conducting oxide films are known to be n-type.

(ii) Substrate temperature has a significant effect on the electrical properties of all oxide films. Increasing substrate temperature creates more oxygen vacancies and hence higher conductivity in these oxide films.

(iii) The electrical properties are influenced by the thickness variation of these oxide films. This may be due to increased grain size, improved crystallinity and the presence of oxygen on surface layer.

(iv) Doping of transparent conducting oxide films with suitable impurities improves the electrical properties of these oxide films markedly. In general, it results in an increase in carrier concentration as well as mobility. However, the upper limit of electron density is determined by the solubility of the dopants. In the case that excess impurities are added, these form clusters in the lattice and distort it, in addition to producing additional scattering centres. Although mobility depends on the dopant used, in general, it increases initially, reaches a maximum value, and then decreases at higher dopant concentrations.

(v) In films having carrier concentration greater than 10^{20} cm^{-3}, both carrier concentration and mobility are almost independent of temperature, indicating a degenerate semiconductor behaviour. The conduction mechanism is governed by carrier concentration in these films. In general, when carrier concentration is less than 10^{18} cm^{-3}, conduction is always limited by grain boundary scattering. On the other hand, when carrier concentration is greater than 10^{20} cm^{-3}, ionized impurity scattering is the dominant scattering mechanism.

(vi) A number of different conduction mechanisms have been proposed for films having carrier concentration between 10^{18} and 10^{20} cm^{-3}. Much work is still needed to understand fully the scattering mechanisms involved in the conduction process in these oxide films.

(vii) The properties of the films are influenced by the process of diffusion of oxygen either into the film or out of it. Post-deposition annealing in different ambients such as H_2, NH_4 etc. may further improve their electrical properties.

(viii) Of all the transparent conducting oxide coatings, fluorine-doped tin oxide and indium tin oxide have the best electrical properties. It is worth mentioning that recently aluminium-doped zinc oxide films have also emerged as an alternative to F–SnO_2 and ITO, due to their comparable electrical properties. In addition, these coatings are cost-effective and can be fabricated with ease.

References

[1] Vossen JL 1977 *Phys. Thin Films* **9** 1
[2] Jarzebski ZM 1982 *Phys. Status Solidi* a **71** 13
[3] Chopra KL, Major S and Pandya DK 1983 *Thin Solid Films* **102** 1
[4] Dawar AL and Joshi JC 1984 *J. Mater. Sci.* **19** 1
[5] Manifacier JC 1982 *Thin Solid Films* **90** 297
[6] Valdes LB 1954 *Proc. IRE* **42** 420
[7] Fuchs K 1938 *Proc. Cambridge Phil. Soc.* **34** 100
[8] Bardeen J and Shockley W 1950 *Phys. Rev.* **80** 72
[9] Huston AR 1961 *J. Appl. Phys.* **32** 2287
[10] Seeger K 1973 *Semiconductor Physics* (New York: Springer)
[11] Ehrenreich H 1961 *J. Appl. Phys.* **32** 2155
[12] Putley EH 1968 *The Hall Effect and Semiconductor Physics* (New York: Dover)
[13] Erginsoy C 1950 *Phys. Rev.* **79** 1013
[14] Brooks H 1955 *Advances in Electronics and Electron Physics* Vol 78, ed L Martin (New York: Academic) p 85
[15] Johnson VA and Lark-Horovitz K 1947 *Phys. Rev.* **71** 374
[16] Appel J 1961 *Phys. Rev.* **122** 1760
[17] Bate RT, Baxter RD, Reid FJ and Beer AC 1965 *J. Phys. Chem. Solids* **26** 1205
[18] Petritz RL 1956 *Phys. Rev.* **104** 1508
[19] Seto JYW 1975 *J. Appl. Phys.* **46** 5247
[20] Orton JW, Goldsmith BJ, Powell MJ and Chapman AJ 1980 *Appl. Phys. Lett.* **37** 557
[21] Berger H 1961 *Phys. Status Solidi* **1** 739
[22] Mankarious RG 1964 *Solid State Electron.* **7** 702
[23] Kazmerski LL, Ayyagari MS and Sanborn GA 1975 *J. Appl. Phys.* **46** 4685
[24] Kazmerski LL, Ayyagari MS, White FR and Sanborn GA 1976 *J. Vac. Sci. Technol.* **13** 139
[25] Kazmerski LL and Juang YJ 1977 *J. Vac. Sci. Technol.* **14** 769
[26] Kazmerski LL, White FR, Ayyagari MS, Juang YJ and Patterson RP 1977 *J. Vac. Sci. Technol.* **14** 65
[27] Anderson JC 1970 *Adv. Phys.* **19** 311
[28] Efror AL and Shklovskii BI 1975 *J. Phys. C: Solid State Phys.* **8** L 49
[29] Mott NF 1969 *Phil. Mag.* **19** 835
[30] Lou JC, Lin MS, Chyl JI and Shieh JH 1983 *Thin Solid Films* **106** 163
[31] Croitoru N and Bannett E 1981 *Thin Solid Films* **82** 235
[32] Shanthi E, Dutta V, Banerjee A and Chopra KL 1980 *J. Appl. Phys.* **51** 6243
[33] Muranoi T and Furukoshi M 1978 *Thin Solid Films* **48** 309
[34] Vasu V and Subrahmanyam A 1990 *Thin Solid Films* **193/194** 973
[35] Vasu V and Subrahmanyam A 1991 *Thin Solid Films* **202** 283
[36] Mizuhashi M 1980 *Thin Solid Films* **70** 91
[37] Dewall H and Simonis F 1981 *Thin Solid Films* **77** 253
[38] Murty NS, Bhagavat GK and Jawalekar SR 1982 *Thin Solid Films* **92** 347
[39] Kuznetsov AYa 1960 *Sov. Phys.–Solid State* **2** 30
[40] Omar OA, Ragaie HF and Fikry WF 1990 *J. Mater. Sci. Mater. Electron.* **3** 79

[41] Srinivasa Murthy N and Jawalekar SR 1983 *Thin Solid Films* **108** 277
[42] De A and Ray S 1991 *J. Phys. D: Appl. Phys.* **24** 719
[43] Stjerna B and Granqvist CG 1990 *Solar Energy Mater.* **20** 225
[44] Stjerna B and Granqvist CG 1990 *Thin Solid Films* **193/194** 704
[45] Croitoru N, Seidman A and Yassin K 1985 *J. Appl. Phys.* **57** 102
[46] Croitoru N, Seidman A and Yassin K 1984 *Thin Solid Films* **116** 327
[47] Tarng ML 1978 *J. Appl. Phys.* **49** 4069
[48] Bruneaux J, Cachet H, Froment M and Messad A 1991 *Thin Solid Films* **197** 129
[49] Belanger D, Dodelet JP, Lombos BA and Dickson JI 1985 *J. Electrochem. Soc.* **132** 1398
[50] Islam MN and Hakim MO 1986 *J. Phys. D: Appl. Phys.* **19** 615
[51] Manifacier JC, Szepessy L, Bresse JF, Perotin M and Stuck R 1979 *Mater. Res. Bull.* **14** 109
[52] Jarzebski SM and Marton JP 1976 *J. Electrochem. Soc.* **123** 299C
[53] Sabnis AG and Feisel CD 1976 *IEEE Trans. Parts. Hybrids Packag.* **12** 357
[54] Sabnis AG 1978 *J. Vac. Sci. Technol.* **15** 1565
[55] Leja E, Korecki J, Krop K and Toll K 1979 *Thin Solid Films* **59** 147.
[56] Platteuw JC and Meyer G 1956 *Trans. Faraday Soc.* **52** 1066
[57] Shanthi E, Banerjee A, Dutta V and Chopra KL 1980 *Thin Solid Films* **71** 237
[58] Beensh-Marchwicka G, Krol-Stepniewska L and Misiuk A 1984 *Thin Solid Films* **113** 215
[59] Jagadish C, Dawar AL, Sharma S, Shishodia PK, Tripathi KN and Mathur PC 1988 *Mater. Lett.* **6** 149
[60] Baliga BJ and Ghandhi SK 1976 *J. Electrochem. Soc.* **123** 941
[61] Hsu YS and Ghandhi SK 1979 *J. Electrochem. Soc.* **126** 1434
[62] Reddy MHM, Jawalekar SR and Chandorkar AN 1990 *Thin Solid Films* **187** 171
[63] Viscrian I and Georgescu V 1979 *Thin Solid Films* **3** R17
[64] Tang ML 1978 *Appl. Phys.* **49** 4069
[65] Islam MN and Hakim MO 1985 *J. Phys. D: Appl. Phys.* **18** 71
[66] Melsheimer J and Ziegler D 1983 *Thin Solid Films* **109** 71
[67] Kane J and Schweizer HP 1975 *J. Electrochem. Soc.* **122** 1144
[68] Casey V and Stephenson MI 1990 *J. Phys. D: Appl. Phys.* **23** 1212
[69] Antonaia A, Menna P, Addonizio ML and Crocchiolo M 1992 *Solar Energy Mater. and Solar Cells* **28** 167
[70] Hamdi AH, Laugal RCO, Catalan AB, Micheli AL and Schubring NW 1991 *Thin Solid Films* **198** 9
[71] Sanjines R, Damarne V and Levy F 1990 *Thin Solid Films* **193/194** 935
[72] Mattox DM 1991 *Thin Solid Films* **204** 25
[73] Maddalena A, Dalmaschio R, Dire S and Raccanelli A 1990 *J. Non-Cryst. Solids* **121** 365
[74] Vincent CA 1972 *J. Electrochem. Soc.* **119** 515
[75] Fonstad CG and Redicker RH 1971 *J. Appl. Phys.* **42** 2911
[76] Shanthi E, Banerjee A, Dutta V and Chopra KL 1982 *J. Appl. Phys.* **53** 1615
[77] Rohatgi A, Viverit TR and Slack LH 1974 *J. Amer. Ceram. Soc.* **57** 278
[78] Randhawa HS, Matthews MD and Bunshah RF 1981 *Thin Solid Films* **83** 267
[79] Haacke G 1977 *Ann. Rev. Mater. Sci.* **7** 73

[80] Feist WM, Steele SR and Ready DW 1975 *Phys. Thin Films* **5** 237
[81] Imai I 1960 *J Phys. Soc. Japan* **15** 937
[82] Inagaki T, Nishimura Y and Saaki H 1969 *Japan. J. Appl. Phys.* **8** 625
[83] Suzuki K and Mizuhashi M 1982 *Thin Solid Films* **97** 119
[84] Carroll AF and Slack LH 1976 *J. Electrochem. Soc.* **123** 1889
[85] Kaneko H and Miyake K 1982 *J. Appl. Phys.* **53** 3629
[86] Mizuhashi M 1980 *J. Non-Cryst. Solids* **38/39** 329
[87] Kane J, Schweizer HP and Kern W 1976 *J. Electrochem. Soc.* **123** 270
[88] Kulaszewicz S, Jarmoc W, Lasocka I, Lasocki Z and Turowska K 1984 *Thin Solid Films* **117** 157
[89] Pommier R, Gril C and Marucchi J 1981 *Thin Solid Films* **77** 91
[90] Gottlieb B, Koropecki R, Arce R, Crisalle R and Ferron J 1991 *Thin Solid Films* **199** 13
[91] Manifacier JC, Szepessy L, Bresse JF, Perotin M and Stuck R 1979 *Mater. Res. Bull.* **14** 163
[92] Fan JCC and Goodenough JB 1977 *J. Appl. Phys.* **48** 3524
[93] Frank G, Kauer E, Kostlin H and Schmitte FJ 1983 *Solar Energy Mater.* **8** 387
[94] Bosnell JR and Waghorne R 1973 *Thin Solid Films* **15** 141
[95] Lida H, Mishuku T, Ito A, Kata K, Yamanaka M and Hayashi Y 1988 *Solar Energy Mater.* **17** 407
[96] Abass AK, Bakr H, Jassin SA and Fahad TA 1988 *Solar Energy Mater.* **17** 425
[97] Haitjema H, Elich JJPh and Hoogendoorn CJ 1989 *Solar Energy Mater.* **18** 283
[98] Unaogu AL and Okeke CE 1990 *Solar Energy Mater.* **20** 29
[99] Pan CA and Ma TP 1981 *J. Electron. Mater.* **10** 43
[100] Bellingham JR, Mackenzie AP and Phillips WA 1991 *Appl. Phys. Lett.* **58** 2506
[101] Howson RP, Ridge MI and Suzuki K 1983 *Proc. SPIE* **428** 14
[102] Szczyrbowski J, Dietrich A and Hoffmann H 1982 *Phys. Status Solidi* a **69** 217
[103] Sundaram KB and Bhagavat GK 1981 *Phys. Status Solidi* a **63** K15
[104] Noguchi S and Sakata H 1980 *J. Phys. D: Appl. Phys.* **13** 1129
[105] Clanget R 1973 *Appl. Phys.* **2** 247
[106] Hoffmann H, Pickl J and Schmidt M 1978 *Appl. Phys.* **16** 239
[107] Ovadyahu Z and Imry Y 1985 *J. Phys. C: Solid State Phys.* **18** L19
[108] Mott NF and Davis EA 1979 *Electronic Processes in Non-crystalline Materials* 2nd edn, (Oxford: Clarendon)
[109] Graham MR, Bellingham JR and Adkins CJ 1992 *Phil. Mag.* B **65** 669
[110] Kostlin H, Jost R and Lems W 1975 *Phys. Status Solidi* a **29** 87
[111] Vasu V and Subrahmanyam A 1990 *Thin Solid Films* **193/194** 696
[112] Fraser DB and Cook HD 1972 *J. Electrochem. Soc.* **119** 1368
[113] Agnihotry SA, Saini KK, Saxena TK, Nagpal KC and Chandra S 1985 *J. Phys. D: Appl. Phys.* **18** 2087
[114] Noguchi S and Sakata H 1981 *J. Phys. D: Appl. Phys.* **114** 1523
[115] Mizuhashi M 1981 *Thin Solid Films* **76** 97
[116] Nath P, Bunshah RF, Basol BM and Staffud M 1980 *Thin Solid Films* **72** 463
[117] Blandenet G, Court M and Lagarde Y 1981 *Thin Solid Films* **77** 81
[118] Maruyama T and Fukui K 1991 *Thin Solid Films* **203** 297
[119] Maruyama T and Fukui K 1991 *J. Appl. Phys.* **70** 3848
[120] Fistul VI and Vainshtein VM 1967 *Sov. Phys.-Solid State* **8** 2769

[121] Saxena AK, Singh SP, Thangaraj R and Agnihotri OP 1984 *Thin Solid Films* **117** 95
[122] Bellingham JR, Phillips WA and Adkins CJ 1990 *J. Phys.: Condens. Matter* **2** 6207
[123] Bellingham JR, Phillips WA and Adkins CJ 1991 *Thin Solid Films* **195** 23
[124] Bellingham JR, Phillips WA and Adkins CJ 1992 *J. Mater. Sci. Lett.* **11** 263
[125] Dingle RB 1955 *Phil. Mag.* **46** 831
[126] Moore EJ 1967 *Phys. Rev.* **160** 618
[127] Pan CA and Ma TP 1980 *Appl. Phys. Lett.* **37** 163
[128] Chen T, Ma TP and Barker RC 1983 *Appl. Phys. Lett.* **43** 901
[129] Ovadyahu Z, Ovryn B and Kraner HW 1983 *J. Electrochem. Soc.* **130** 917
[130] Hamberg I and Granqvist CG 1986 *J. Appl. Phys.* **60** R123
[131] Latz R, Michael K and Scherer M 1991 *Japan. J. Appl. Phys.* **30** L149
[132] Ishibashi S, Higuchi Y, Ota Y and Nakamura K 1990 *J. Vac. Sci. Technol.* A **8** 1399,1403
[133] Fan JCC, Bachner FJ and Foley GH 1977 *Appl.Phys. Lett.* **31** 773
[134] Kulaszewicz S, Jarmoc W and Turowska K 1984 *Thin Solid Films* **112** 313
[135] Buchanan M, Webb JB and Williams DF 1981 *Thin Solid Films* **80** 373
[136] Yao JL, Hao S and Wilkinson JS 1990 *Thin Solid Films* **189** 27
[137] Fan JCC 1981 *Thin Solid Films* **80** 125
[138] Theuwissen AJP and Declerck GJ 1984 *Thin Solid Films* **121** 109
[139] Bawa SS, Sharma SS, Agnihotry SA, Biradar AM and Chandra S 1983 *Proc. SPIE* **428** 22
[140] Kawada A 1990 *Thin Solid Films* **191** 297
[141] Hoheisel M, Heller S, Mrotzek C and Mitwalsky A 1990 *Solid State Commun.* **76** 1
[142] Enjouji K, Murata K and Nishikawa S 1983 *Thin Solid Films* **108** 1
[143] Sreenivas K, Sudersena Rao T and Mansingh A 1985 *J. Appl. Phys.* **57** 384
[144] Harding GL and Window B 1990 *Solar Energy Mater.* **20** 367
[145] Weijtens CHL 1991 *J. Electrochem Soc.* **138** 3432
[146] Krokoszinski HJ and Oesterlein R 1990 *Thin Solid Films* **187** 179
[147] Shigesato Y, Takaki S and Haranon T 1992 *J. Appl. Phys.* **71** 3356
[148] Frank G, Kauer E and Kostlin H 1981 *Thin Solid Films* **77** 107
[149] Manifacier JC, Fillard JP and Bind JM 1981 *Thin Solid Films* **77** 67
[150] Frank G and Kostlin H 1982 *Appl. Phys.* A **27** 197
[151] Avaritsiotis JN and Howson RP 1981 *Thin Solid Films* **77** 351
[152] Avaritsiotis JN and Howson RP 1981 *Thin Solid Films* **80** 63
[153] Maruyama T and Fukui K 1990 *Japan. J. Appl. Phys.* **29** L1075
[154] Singh SP, Tiwari LM and Agnihotri OP 1986 *Thin Solid Films* **139** 1
[155] Groth R 1966 *Phys. Status Solidi* a **14** 69
[156] Maruyama T and Tago T 1994 *Appl. Phys. Lett.* **64** 1395
[157] Lee CH and Wang CS 1994 *Mater. Sci. Eng.* **B 22** 233
[158] Tueta R and Braguier M 1981 *Thin Solid Films* **80** 143
[159] Buchanan M, Webb JB and Williams DF 1980 *Appl. Phys. Lett.* **37** 213
[160] Habermeier HU 1981 *Thin Solid Films* **80** 157
[161] Machet J, Guille J, Saulnier P and Robert S 1981 *Thin Solid Films* **80** 149
[162] Budzynska K, Leja E and Skrzypek S 1985 *Solar Energy Mater.* **12** 57
[163] Coutts TJ and Naseem S 1985 *Appl. Phys. Lett.* **46** 164

[164] Jain VK and Kulshreshtha AP 1981 *Solar Energy Mater.* **4** 151
[165] Siefert W 1984 *Thin Solid Films* **121** 275
[166] Yuanri C, Xinghao Xu, Zhaoting J, Chuancai P and Shuyun X 1984 *Thin Solid Films* **115** 195
[167] Balasubramanian N and Subrahmanyam A 1989 *J. Phys. D: Appl. Phys.* **22** 206
[168] Higuchi M, Uekusa S, Nakano R and Yokogawa K 1993 *J. Appl. Phys.* **74** 6710
[169] Davis L 1993 *Thin Solid Films* **236** 1
[170] Lee SB, Pincenti JC, Cocco A and Naylor DL 1993 *J. Vac. Sci. Technol.* A **11** 2742
[171] Howson RP, Avaritsiotis JN, Ridge MI and Bishop CA 1979 *Appl. Phys. Lett.* **35** 161
[172] Ichihara K, Inoue N, Okubo M and Yasuda N 1994 *Thin Solid Films* **245** 152
[173] Fan JCC 1979 *Appl. Phys. Lett.* **34** 515
[174] Nojik AJ 1972 *Phys. Rev.* B **6** 453
[175] Haacke G 1976 *Appl. Phys. Lett.* **28** 622
[176] Haacke G, Mealmaker WE and Siegel LA 1978 *Thin Solid Films* **55** 67
[177] Ortiz RA 1982 *J. Vac. Sci. Technol.* **20** 7
[178] Enoki H, Satoh T and Ecnigoya J 1991 *Phys. Status Solidi* a **126** 163
[179] Miyata N, Miyake K, Fukushima T and Koga K 1979 *Appl. Phys. Lett.* **35** 542
[180] Schiller S, Beister G, Buedke E, Becker HJ and Schicht H 1982 *Thin Solid Films* **96** 113
[181] Miyata N, Miyake K and Nao S 1979 *Thin Solid Films* **58** 385
[182] Stapinski T, Leja E and Pisarkiewicz T 1984 *J. Phys. D: Appl. Phys.* **17** 407
[183] Pisarkiewicz T, Zakrzewska K and Leja E 1987 *Thin Solid Films* **153** 479
[184] Miyata N and Miyake K 1979 *Surf. Sci.* **86** 384
[185] Miyata N, Miyake K, Koga K and Fukushima T 1980 *J. Electrochem. Soc.* **127** 918
[186] Leja E, Stapinski T and Marszalek K 1985 *Thin Solid Films* **125** 119
[187] Ma YY 1977 *J. Electrochem Soc.* **124** 1430
[188] Aranovich J, Armando OR and Bube RH 1979 *J. Vac. Sci. Technol.* **16** 994
[189] Haacke G 1977 *Appl. Phys. Lett.* **30** 380
[190] Nakazawa T and Ito K 1988 *Japan. J. Appl. Phys.* **27** 1630
[191] Neumann G 1981 *Phys. Status Solidi* b **105** 605
[192] Kroger FA 1964 *The Chemistry of Imperfect Crystals*, (Amsterdam: North-Holland) p 691
[193] Kenigsberg NL and Chernets AN 1969 *Sov. Phys. Solid State* **10** 2235
[194] Schoenes J, Kanazawa K and Kay E 1977 *J. Appl. Phys.* **48** 2537
[195] Ziegler E, Heinrich A, Oppermann H and Stover G 1981 *Phys. Status Solidi* a **66** 635
[196] Hagemark KI and Chacka LC 1975 *J. Solid State Chem.* **15** 261
[197] Gopel W and Lampe U 1980 *Phys. Rev.* B **22** 6447
[198] Roth AP and Williams DF 1981 *J. Appl. Phys.* **52** 6685
[199] Webb JB, Williams DF and Buchanan M 1981 *Appl. Phys. Lett.* **39** 640
[200] Tomar MS and Garcia FJ 1982 *Thin Solid Films* **90** 419
[201] Natsume Y, Sakata H, Hirayama T and Yangida H 1991 *J. Mater. Sci. Lett.* **10** 810

[202] Ogawa MF, Natsume Y, Hirayama T and Sakata H 1990 *J. Mater. Sci. Lett.* **9** 1351
[203] Tsuji N, Komiyama H and Tanaka K 1990 *Japan. J. Appl. Phys.* **29** 835
[204] Caporaletti O 1982 *Solar Energy Mater.* **7** 65
[205] Nayar PS and Catalano A 1981 *Appl. Phys. Lett.* **39** 105
[206] Mackrodt WC and Stewart RF 1980 *J. Physique* C **41** 6–64
[207] Hu J and Gordon RG 1991 *Solar Cells* **30** 437
[208] Nanto H, Minami T, Shooji S and Takata S 1984 *J. Appl. Phys.* **55** 1029
[209] Minami T, Nanto H and Takata S 1984 *Japan. J. Appl. Phys.* **23** L280
[210] Padovani FA and Stratton R 1966 *Solid State Electron.* **9** 695
[211] Crowell CR and Rideout VL 1969 *Solid State Electron.* **12** 89
[212] Tansley TL, Neeley DF and Foley CP 1984 *Thin Solid Films* **117** 19
[213] Reyonds DC, Litton CW and Collin TC 1965 *Phys. Status Solidi* **12** 3
[214] Blom FR, Van De Pol FCM, Bauhuis G and Popma ThJA 1991 *Thin Solid Films* **204** 365
[215] Minami T, Nanto H, Shooji S and Takata S 1984 *Thin Solid Films* **111** 167
[216] Caporaletti O 1981 *Solid State Commun.* **42** 109
[217] Yen JC 1975 *J. Vac. Sci. Technol.* **12** 47
[218] Major S, Banerjee A and Chopra KL 1988 *Solar Energy Mater.* **17** 319
[219] Caillaud F, Smith A and Baumard JF 1991 *J. Europ. Ceram. Soc.* **7** 379
[220] Bube RH 1960 *Photoconductivity of Solids* (New York: Wiley) p 160
[221] Call RL, Jaber NK, Seshan K and Whyte Jr JR 1980 *Solar Energy Mater.* **2** 373
[222] Major S, Banerjee A and Chopra KL 1985 *Thin Solid Films* **125** 179
[223] Aktaruzzaman AF, Sharma GL and Malhotra LK 1991 *Thin Solid Films* **198** 67
[224] Hu J and Gordon R 1992 *J. Appl. Phys.* **71** 880
[225] Minami T, Oohashi K, Takata S, Mouri T and Ogawa N 1990 *Thin Solid Films* **193/194** 721
[226] Minami T, Nanto H and Takata S 1985 *Thin Solid Films* **124** 43
[227] Igasaki Y and Saito H 1991 *J. Appl. Phys.* **70** 3613
[228] Ghosh S, Sarkar A, Bhattacharya S, Chaudhury S and Pal AK 1991 *J. Cryst. Growth* **108** 534
[229] Minami T, Sato H, Sonoda T, Nanto H and Takata S 1989 *Thin Solid Films* **171** 307
[230] Choi BH, Im HB, Song JS and Yoon KH 1990 *Thin Solid Films* **193/194** 712
[231] Major S, Banerjee A and Chopra KL 1983 *Thin Solid Films* **108** 333
[232] Aoki M, Tada K and Murai T 1981 *Thin Solid Films* **83** 283
[233] Smith FTJ 1983 *Appl. Phys. Lett.* **43** 1108
[234] Sarkar A, Ghosh S, Chaudhury S and Pal AK 1991 *Thin Solid Films* **204** 255
[235] Ruth M, Tuttle J, Goral J and Noufi R 1989 *J. Cryst. Growth* **96** 363
[236] Kuroyanagi A 1989 *J. Appl. Phys.* **66** 5492
[237] Wu P, Gao YM, Baglio J, Kershaw R, Dwight K and Wold A 1989 *Mater Res. Bull.* **24** 905
[238] Igasaki Y, Ishikawa M and Shimaoka G 1988 *Appl. Surf. Sci.* **33–34** 926
[239] Minami T, Nanto H, Sato H and Takata S 1988 *Thin Solid Films* **164** 275
[240] Schropp R and Madan A 1989 *J. Appl. Phys.* **66** 2027
[241] Howson RP, Barankova H and Spencer AG 1990 *Proc. SPIE* **1275** 75
[242] Howson RP, Barankova H and Spencer AG 1991 *Thin Solid Films* **196** 315
[243] Sanon G, Rup R and Mansingh A 1990 *Thin Solid Films* **190** 287

[244] Dai CM, Su CS and Chuu DS 1990 *Appl. Phys. Lett.* **57** 1879
[245] Simonis F, Leij MV and Hoogendoorn CJ 1979 *Solar Energy Mater.* **1** 221
[246] Agashe C, Takwale MG, Marathe BR and Bhide VG 1988 *Solar Energy Mater.* **17** 99
[247] Mavrodiev G, Gajdardziska M and Novkovski N 1984 *Thin Solid Films* **113** 93
[248] Vishwakarma SR, Upadhyay JP and Prasad HO 1989 *Thin Solid Films* **176** 99
[249] Golan A, Bragman J and Shapira Y 1990 *Appl. Phys. Lett.* **57** 2205
[250] Jachimowski M, Brudnik A and Czternastek H 1985 *J. Phys. D: Appl. Phys.* **18** L145
[251] Shigesato Y, Hayashi Y, Masui A and Haranou T 1991 *Japan. J. Appl. Phys.* **30** 814
[252] Ryabova LA, Salun VS and Serbinov IA 1982 *Thin Solid Films* **92** 327
[253] Banerjee A, Das D, Ray S, Batabyal AK and Barua AK 1986 *Solar Energy Mater.* **13** 11
[254] Robosto PF and Braunstein R 1990 *Phys. Status Solidi* a **119** 155
[255] Geoffroy C, Campet G, Partier J, Salardenne J, Couturier G, Baourrel M, Chabagno JM, Ferry D and Quet C 1991 *Thin Solid Films* **202** 77
[256] Leja E, Budzynska K, Pisarkiewicz T and Stapinski T 1983 *Thin Solid Films* **100** 203
[257] Swamy HG and Reddy PJ 1990 *Semicond. Sci. Technol.* **5** 980
[258] Shimizu M, Horu T, Shiosaki T and Kawabata A 1982 *Thin Solid Films* **96** 149
[259] Dutta J, Roubeau P, Emeraud T, Laurent JM, Smith A, Lebelanc F and Perrin J 1994 *Thin Solid Films* **239** 150
[260] Leong JI, Moon JH, Hong JH, Kang JS and Lee YP 1994 *Appl. Phys. Lett.* **64** 1215
[261] Rauf IA 1993 *Mater. Lett.* **18** 123
[262] Ma J and Li SY 1994 *Thin Solid Films* **237** 16

4. Optical Properties

4.1 Introduction

The optical properties of thin conducting transparent films depend strongly on the deposition parameters, microstructure, level of impurities and growth technique. Literature in this area is fairly scattered. Brief reviews [1–5] have been given by various workers, but the main emphasis has been on the electrical properties. Hamberg and Granqvist [6] have, however, reviewed the optical properties of indium oxide and indium tin oxide. This chapter briefly discusses various optical constants, and then, more comprehensively, the optical properties of films of a number of transparent semiconducter materials.

4.2 Determination of Optical Constants

4.2.1 Determination of refractive index and extinction coefficient

(a) *Absorbing films*
If the thickness of a film is t, and its reflectance R and transmittance T measured at normal incidence are known, it is possible in principle to derive the optical constants, namely refractive index n and extinction coefficient k of the complex refractive index $n^* = n - \mathrm{i}k$. Basic equations have been derived by Heavens [7]. For a layer of index $n^* = n - \mathrm{i}k$ and thickness t, and substrate of index $n_1^* = n_1 - \mathrm{i}k_1$, the reflectance and transmittance can be written as:

$$R = \frac{(g_1^2 + h_1^2)\,\mathrm{e}^{2\alpha_1} + (g_2^2 + h_2^2)\,\mathrm{e}^{-2\alpha_1} + A\cos 2\gamma_1 + B\sin 2\gamma_1}{\mathrm{e}^{2\alpha_1} + (g_1^2 + h_1^2)(g_2^2 + h_2^2)\,\mathrm{e}^{-2\alpha_1} + C\cos 2\gamma_1 + D\sin 2\gamma_1} \qquad (4.1)$$

and

$$T = \left(\frac{n_1}{n_0}\right) \frac{[(1+g_1)^2 + h_1^2][(1+g_2)^2 + h_2^2]}{e^{2\alpha_1} + (g_1^2 + h_1^2)(g_2^2 + h_2^2)\,e^{-2\alpha_1} + C\cos 2\gamma_1 + D\sin 2\gamma_1} \tag{4.2}$$

where

$$g_1 = \frac{n_0^2 - n^2 - k^2}{(n_0 + n)^2 + k^2} \qquad g_2 = \frac{n^2 - n_1^2 + k^2 - k_1^2}{(n - n_1)^2 + (k + k_1)^2}$$

$$h_1 = \frac{2n_0 k}{(n_0 + n)^2 + k^2} \qquad h_2 = \frac{2(nk_1 - n_1 k)}{(n + n_1)^2 + (k + k_1)^2}$$

$$\alpha_1 = \frac{2\pi k t}{\lambda} \qquad \gamma_1 = \frac{2\pi n t}{\lambda}$$

$$A = 2(g_1 g_2 + h_1 h_2) \qquad B = 2(g_1 h_2 - g_2 h_1)$$

$$C = 2(g_1 g_2 - h_1 h_2) \qquad D = 2(g_1 h_2 + g_2 h_1).$$

and where n_0 and λ are the refractive index of air and the wavelength of incident light, respectively. Equations (4.1) and (4.2) can be further simplified for the case of a transparent film on a transparent substrate, i.e. $h_1 = h_2 = 0$, $\alpha_1 = 0$, $A = C = 2g_1g_2$ and $B = D = 0$. The expressions for R and T reduce to

$$R = \frac{g_1^2 + g_2^2 + 2g_1 g_2 \cos 2\gamma_1}{1 + g_1^2 g_2^2 + 2g_1 g_2 \cos 2\gamma_1} \tag{4.3}$$

$$T = \left(\frac{n_1}{n_0}\right) \frac{(1+g_1)^2(1+g_2)^2}{1 + g_1^2 g_2^2 + 2g_1 g_2 \cos 2\gamma_1}. \tag{4.4}$$

Equations (4.1) and (4.2) relating R and T explicitly in terms of the optical constants of the film and substrate are complicated and have multiple solutions, as shown by Bennett and Booty [8], Nilsson [9], Campbell [10] and Tomlin [11]. According to Tomlin [11], for a single film on a substrate, the reflectance and transmittance measured at normal incidence are related to the film and substrate parameters by the formulae:

$$\begin{aligned}\frac{1+R}{T} = \frac{1}{4n_0 n_1 (n+k)} & [(n_0^2 + n^2 + k^2)\{(n^2 + n_1^2 + k^2 + k_1^2)\cosh 2\alpha_1 \\ & + 2(nn_1 + kk_1)\sinh 2\alpha_1\} + (n_0^2 - n^2 - k^2) \\ & \times \{(n^2 - n_1^2 + k^2 - k_1^2)\cos 2\gamma_1 + 2(nk_1 - n_1 k)\sin 2\gamma_1\}]\end{aligned} \tag{4.5}$$

$$\begin{aligned}\frac{1-R}{T} = \frac{1}{2n_1(n^2 + k^2)} & [n\{(n^2 + n_1^2 + k^2 + k_1^2)\sinh 2\alpha_1 \\ & + 2(nn_1 + kk_1)\cosh 2\alpha_1 + k\{(n^2 - n_1^2 + k^2 - k_1^2)\sin 2\gamma_1 \\ & - 2(nk_1 - n_1 k)\cos 2\gamma_1\end{aligned} \tag{4.6}$$

where

$$\gamma_1 = \frac{2\pi nt}{\lambda} \qquad \alpha_1 = \frac{2\pi kt}{\lambda}$$

The expressions (4.5) and (4.6) not only give $(1 + R)/T$ explicitly in terms of optical parameters but are relatively simpler. A computer solution of these equations may be carried out by rearranging them in the form $f_1(n, k) = 0$ and $f_2(n, k) = 0$, respectively. One may readily solve $f_2(n, k) = 0$, finding k for a given value of n by successive approximation using Newton's method. Each value of n and the corresponding value of k are substituted in turn into $f_1(n, k) = 0$.

(b) *Weakly absorbing films*

For very weakly absorbing films, the measurement of transmission of light through a film in the region of transparency is sufficient to determine the real and imaginary parts of the complex refractive index $n^* = n - ik$. Hall and Ferguson [12] and Lyashenko and Milosolavskii [13] developed a method using successive approximations and interpolations to calculate these constants. Manifacier *et al* [14] developed a method which is much simpler and the computation is far easier. If the incident light has unit amplitude then the amplitude of the transmitted wave would be

$$A = \frac{t_1 t_2 \exp(-2\pi i n^* t/\lambda)}{1 + r_1 r_2 \exp(-4\pi i n^* t/\lambda)} \tag{4.7}$$

where, t_1, t_2, r_1 and r_2 are the transmission and reflection coefficients of the front and rear surfaces and are given by

$$t_1 = \frac{2n_0}{n_0 + n} \qquad t_2 = \frac{2n}{n + n_1}$$

$$r_1 = \frac{n_0 - n}{n_0 + n} \qquad r_2 = \frac{n - n_1}{n + n_1}.$$

The transmission of the layer is given by

$$T = \frac{n_1}{n_0}|A|^2. \tag{4.8}$$

In the case of weak absorption, $k^2 \ll (n - n_0)^2$ and $k^2 \ll (n - n_1)^2$ so that T is given as

$$T = \frac{16 n_0 n_1 n^2 \alpha}{C_1^2 + C_2^2\alpha^2 + 2C_1C_2\alpha\cos(4\pi nt/\lambda)} \tag{4.9}$$

where $C_1 = (n + n_0)(n_1 + n)$, $C_2 = (n - n_0)(n_1 - n)$, and

$$\alpha = \exp\frac{-4\pi kt}{\lambda}. \tag{4.10}$$

Generally outside the region of fundamental absorption, the dispersion of n and k is not large. The maxima and minima of T in equation (4.9) occurs for

$$\frac{4\pi nt}{\lambda} = m\lambda \tag{4.11}$$

where m is the order number. In most cases $n > n_1$ and C_2 is negative, and the extreme values of transmission are given by the formulae

$$T_{\max} = 16n_0n_1n^2\alpha/(C_1 + C_2\alpha)^2 \tag{4.12}$$

$$T_{\min} = 16n_0n_1n^2\alpha/(C_1 - C_2\alpha)^2. \tag{4.13}$$

$T_{\min}$ and $T_{\max}$ can be considered as a continuous function of λ through $n(\lambda)$ and $\alpha(\lambda)$. These functions are the envelopes of the maxima $T_{\max}(\lambda)$ and minima $T_{\min}(\lambda)$ in the transmission spectrum. The ratio $T_{\max}/T_{\min}$ can be used to find the value of α thus:

$$\alpha = \frac{C_1[1 - (T_{\max}/T_{\min})^{1/2}]}{C_2[1 + (T_{\max}/T_{\min})^{1/2}]} \tag{4.14}$$

From equations (4.12) and (4.13)

$$n = [N + (N^2 - n_0^2n_1^2)^{1/2}]^{1/2} \tag{4.15}$$

where

$$N = \frac{n_0^2 + n_1^2}{2} + 2n_0n_1\frac{T_{\max} - T_{\min}}{T_{\max}T_{\min}}.$$

Equation (4.15) shows that n is explicitly determined from $T_{\max}$, $T_{\min}$, n_1 and n_0, measured at the same wavelength. Knowing n, one can determine α from equation (4.14). Using this method, one can also find the thickness of the film, which can be calculated from the two maxima or minima using the relation (4.11)

$$t = \frac{M\lambda_1\lambda_2}{2[n(\lambda_1)\lambda_2 - n(\lambda_2)\lambda_1]} \tag{4.16}$$

where M is the number of oscillations between two extrema. Knowing t and α, one can calculate k using equation (4.10).

(c) *Lossless films*

If one is interested only in the refractive index in the transparency region, then there are much simpler formulae which can be used for this purpose. Rewriting equation (4.3) once again for zero absorption:

$$R = \frac{g_1^2 + g_2^2 + 2g_1g_2\cos 2\gamma_1}{1 + g_1^2g_2^2 + 2g_1g_2\cos 2\gamma_1} \tag{4.3}$$

where

$$g_1 = \frac{n_0 - n}{n_0 + n} \qquad g_2 = \frac{n - n_1}{n + n_1}$$

and

$$\gamma_1 = \frac{2\pi nt}{\lambda}.$$

From equation (4.3) it can be seen that maxima and minima of the reflectance curve occur at values of nt given by

$$\begin{aligned} nt &= (2m + 1)\lambda/4 \quad \text{(maxima)} \\ nt &= (2m + 2)\lambda/4 \quad \text{(minima).} \end{aligned} \tag{4.17}$$

In terms of transmission, these values correspond to minima and maxima, respectively. Knowing the order of the minima or maxima and the thickness, one can find the refractive index. It should be noted that optical parameters are usually determined only for normal incidence and at room temperature.

For transparent films, other techniques for the determination of n involve the measurement of Brewster's angle θ_B, given by

$$\tan\theta_B = n. \tag{4.18}$$

At an angle θ_B, the parallel component of light reflected at the air–film interface vanishes.

Other methods which can be used for the measurement of these optical constants include ellipsometry [15], wideband spectrophotometry [16–18], a modified Valeer [19] turning point method, the Nestell and Christy [20] method, and an algebraic inversion method [21]. Critical evaluation and comparison of these methods has been conducted by the Optical Materials and Thin Films Technical Group of the Optical Society of America and the results are available in the literature [22].

4.2.2 Determination of band-gap

Transparent semiconducting materials are, in general, electrically conductive, and optically they act as a selective transmitting layer, being transparent in the visible and near-infrared range and reflective to thermal infrared radiation. Figure (4.1) shows the typical spectral dependence of these materials. For large wavelengths, high reflection due to free electrons is observed, and for very low wavelengths, absorption due to the fundamental band-gap dominates. In the low-wavelength region, one can estimate the value of band-gap using standard relations [24].

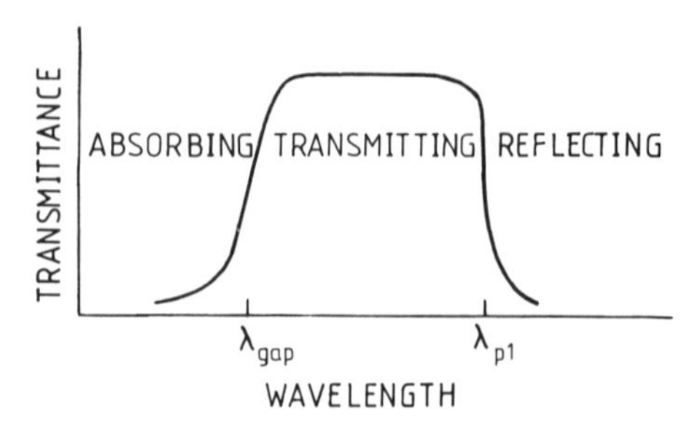

Figure 4.1 Spectral dependence of a semiconducting transparent material: λ_{gap} and λ_{pl} are the wavelengths at which the band-gap absorption and free electron plasma absorption take place (from [23]).

The variation of the imaginary part of the dielectric constant ε_2 ($\varepsilon_2 = 2nk$) with photon energy for direct allowed band-to-band transition is of the form

$$\begin{aligned} (h\nu)^2\varepsilon_2 &= \varepsilon_0(h\nu - \Delta E)^{1/2} \qquad h\nu > \Delta E \\ \varepsilon_2 &= 0 \qquad h\nu < \Delta E \end{aligned} \tag{4.19}$$

where $h\nu$ is the photon energy, ε_0 is approximately constant independent of photon energy and ΔE is the band-gap. The curve $(h^2\nu_2^2\varepsilon_2)^2$ versus $h\nu$ extrapolated to zero gives the value of ΔE.

For allowed indirect transitions, i.e. non-vertical transitions, for a single phonon process, the imaginary part of the dielectric constant ε_2, is given by

$$\begin{aligned} (h\nu)^2\varepsilon_2^a &= \frac{A'}{1 - e^{E_p/kT}}(h\nu - E_g + E_p)^2 \qquad h\nu > (E_g - E_p) \\ \varepsilon_2^a &= 0 \qquad h\nu < (E_g - E_p) \end{aligned} \tag{4.20}$$

where A' is a constant independent of photon energy, E_p is the phonon energy and E_g is the indirect band-gap. The corresponding form of phonon emission is given by

$$\begin{aligned} (h\nu)^2\varepsilon_2^{(e)} &= \frac{A'}{e^{E_p/kT} - 1}(h\nu - E_g - E_p)^2 \qquad h\nu > (E_g + E_p) \\ \varepsilon_2^{(e)} &= 0 \qquad h\nu < (E_g + E_p). \end{aligned} \tag{4.21}$$

If $h\nu > (E_g + E_p)$ the total value of ε_2 is given by

$$\varepsilon_2 = \varepsilon_2^a + \varepsilon_2^e \tag{4.22}$$

A plot of $h\nu\varepsilon_2^{1/2}$ versus $h\nu$ when extrapolated to low energies (corresponding to $h\nu\varepsilon_2^{1/2} = 0$) gives the value of $(E_g - E_p)$. A similar plot of $h\nu(\varepsilon_2 - \varepsilon_2^a)^{1/2}$, corresponding to the phonon emission term of equation (4.21), gives a photon energy intercept at $(E_g + E_p)$. From these two intercept values of $(E_g - E_p)$ and $(E_g + E_p)$, one can obtain values of E_g and E_p.

4.3 Correlation of Optical and Electrical Properties

In the infrared range, the optical phenomenon can be understood on the basis of Drude's theory for free electrons in metals. The interaction of free electrons with an electromagnetic field may lead to polarization of the field within a material and thereby influence the relative permittivity ε. According to Drude's theory, for free electrons in metals, ε can be expressed as

$$\varepsilon = (n - \mathrm{i}k)^2$$

$$\varepsilon' = n^2 - k^2 = \varepsilon_\infty\left(1 - \frac{\omega_\mathrm{p}^2}{\omega^2 + \gamma^2}\right) \tag{4.23}$$

and

$$\varepsilon'' = 2nk = \frac{\omega_\mathrm{p}^2 \gamma \varepsilon_\infty}{\omega(\omega^2 + \gamma^2)}. \tag{4.24}$$

The plasma resonance frequency ω_p is given by

$$\omega_\mathrm{p} = (4\pi N e^2/\varepsilon_0 \varepsilon_\infty m_e^*)^{1/2} \tag{4.25}$$

where ε_∞ and ε_0 represent the dielectric constants of the medium and free space, respectively. m_e^* is the effective mass of the charge carriers and N is the carrier concentration. γ is equal to $1/\tau$ where τ is the relaxation time, which is assumed to be independent of frequency and is related to mobility as

$$\gamma = \frac{1}{\tau} = -\frac{e}{m_\mathrm{e}^* \mu}. \tag{4.26}$$

The values of n and k determine the reflectance and absorptance of the surface.

Drude's theory discussed above correlates the optical constants n and k with the electrical quantities N and μ. Drude's theory is based on the classical description of the interaction between radiation and free electrons under the assumption that the relaxation time τ $(= 1/\gamma)$ is frequency independent.

Three different frequency regions can be distinguished for the free carriers.

(a) $0 < \omega\tau < 1$ (*absorbing region*)
In this region, the imaginary part ε'' (equation (4.24)) is much larger than the real part ε' (equation (4.23)), so that transparent conducting films are strongly reflecting.

In this case equations (4.23) and (4.24) become

$$\varepsilon' = \varepsilon_\infty(1 - \omega_\mathrm{p}^2 \tau^2) \tag{4.27}$$

$$\varepsilon'' = \varepsilon_\infty \frac{\omega_\mathrm{p}^2 \tau}{\omega} \gg 1. \tag{4.28}$$

This leads to the Hagen–Ruben relation [25]

$$n^2 = k^2 \frac{\varepsilon_\infty \omega_p^2 \tau}{2\omega}. \tag{4.29}$$

When the thickness of the layer is greater than the skin depth δ ($= \lambda/4\pi k$), the reflectivity of the layer is given by

$$R = \frac{(n-1)^2 + k^2}{(n+1)^2 + k^2}. \tag{4.30}$$

Using equation (4.29) in equation (4.30), one obtains

$$R = 1 - 2\left(\frac{2\omega}{\varepsilon_\infty \omega_p^2 \tau}\right)^{1/2}. \tag{4.31}$$

(b) $1/\tau < \omega < \omega_p$ (*reflecting region*)
This is the relaxation region in which $\omega^2\tau^2 > 1$ and the absorption coefficient falls rapidly. In this region, the real part ε' (equation (4.23)) is negative and we have almost total reflection. In this case equations (4.23) and (4.24) take the form

$$\varepsilon' = \varepsilon_\infty \left[1 - \left(\frac{\omega_p}{\omega}\right)^2\right] < 0 \tag{4.32}$$

$$\varepsilon'' = \varepsilon_\infty \frac{\omega_p^2}{\omega^3 \tau} > 1. \tag{4.33}$$

This results in

$$n = \frac{\omega_p \varepsilon_\infty^{1/2}}{2\omega^2 \tau}$$

and

$$k = \sqrt{\left(\frac{\omega_p}{\omega}\right)^2 - 1} \simeq \frac{\omega_p}{\omega} \varepsilon_\infty^{1/2}. \tag{4.34}$$

When the thickness of the layer is greater than the skin depth δ, the reflectivity is given by

$$R = 1 - \frac{2}{\omega_p \tau \varepsilon_\infty^{1/2}}. \tag{4.35}$$

(c) $\omega > \omega_p$ (*transparent region*)
In this region, the real part ε' (equation (4.23)) becomes positive and the reflective power becomes minimum; the films become transparent. In this case equations (4.23) and (4.24) become

$$\varepsilon' = \varepsilon_\infty \left(1 - \left(\frac{\omega_p}{\omega}\right)^2\right) \tag{4.36}$$

$$\varepsilon'' = \varepsilon_\infty \frac{\omega_p^2}{\omega^3 \tau} \ll 1. \tag{4.37}$$

This leads to

$$n \simeq \varepsilon_\infty^{1/2} \sqrt{1 - \left(\frac{\omega_p}{\omega}\right)^2} \simeq \varepsilon_\infty^{1/2} \tag{4.38}$$

and

$$k \simeq \varepsilon_\infty^{1/2} \frac{\omega_p^2}{2\omega^3 \tau} \simeq 0. \tag{4.39}$$

Figure 4.2 shows the variation of n and k with wavelength for a typical transparent semiconductor film. For $\lambda < \lambda_{pl}$, the plasma resonance wavelength, the value of n is constant ($\simeq \varepsilon_\infty^{1/2}$). Near λ_{pl}, the value of n approaches unity, which implies a very low reflectance. For $\lambda > \lambda_{pl}$, both n and k increase rapidly, which leads to high reflectivity.

The effect of carrier concentration N is two-fold. First, N determines the plasma wavelength in accordance with relation (4.25); the higher the carrier concentration, the lower will be the plasma wavelength. Second, N governs the maximum achievable reflectivity in the infrared region. Figure 4.3 shows a theoretical curve of reflectivity versus carrier concentration for a sample with mobility = 15 cm^2/V s and $m^* = 0.25m_0$. It can be observed that for high values of carrier concentration, reflectivity approaches unity.

The effect of mobility on the infrared reflectivity of transparent conducting films is similar. However, plasma wavelength is not directly influenced by the change in value of mobility.

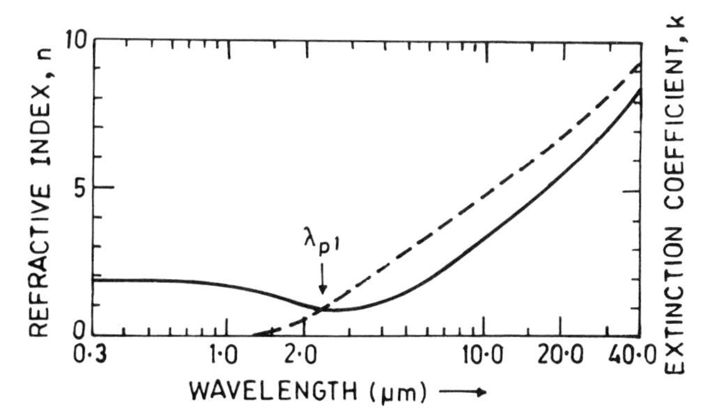

Figure 4.2 Typical curves for n and k for SnO_2 films (———) n and (- - - -) k (from [23]).

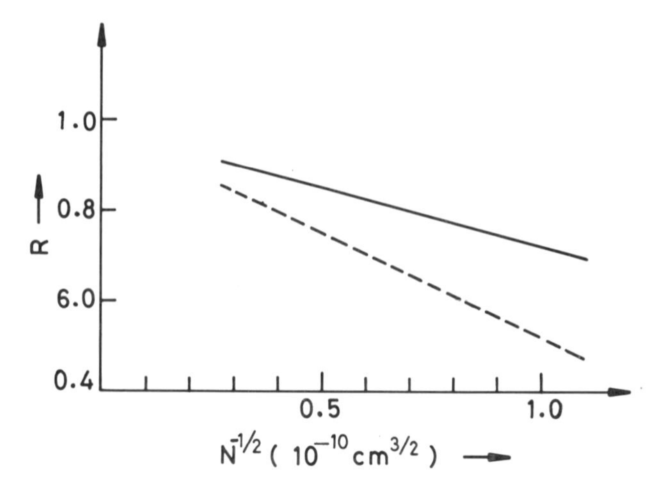

Figure 4.3 Theoretical curves of IR reflectance versus carrier concentration for two wavelengths: - - - -, 10 μm; ———, 40 μm.

Knowledge of the values of n and k in the plasma resonance region from optical data, and N and μ from electrical data helps in estimating the values of m_e^* with the use of equations (4.23)–(4.26). In addition, Drude's approach is useful for computing the dielectric function in transparent conducting films in order to estimate the electron density and mobility from optical data.

The validity of Drude's theory has been tested by various workers [6, 23, 26–32]. Figures 4.4, 4.5 and 4.6 compare experimental results with the theoretical model based on Drude's theory, for tin oxide, fluorinated indium tin oxide and indium tin oxide, respectively. It can be observed that in

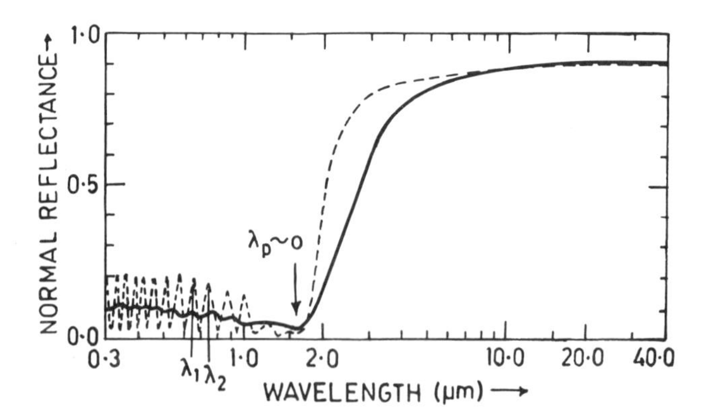

Figure 4.4 Normal reflectance, calculated according to Drude's theory (- - - -) and experiment (———) for fluorine-doped SnO_2 films with $N = 3.5 \times 10^{20}$ cm^{-3} and $\mu = 46$ cm^2/V s (from [23]).

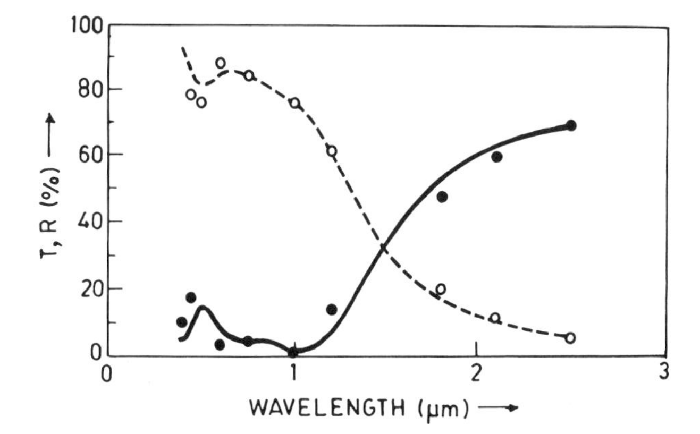

Figure 4.5 Optical transmittance and reflectance versus wavelength for fluorinated indium tin oxide films. Experimental (○, ●) and calculated (– – – –, ———) (from [26]).

general the experimental results for transmittance and reflectance in these films fit well with Drude's theory. Although (figure 4.4) Drude's theory is in fairly good agreement with the experimental results for higher wavelengths, this theory does not fully explain the experimental results near the plasma wavelength. Hu and Gordon [28] compared the values of electron density N and mobility μ for ZnO films determined from Hall studies with those from an optical analysis based on Drude's model (table 4.1). It can be seen that the agreement is quite good for most samples.

From the above results it can be inferred that Drude's theory is able to explain the results fairly well only when applied to a limited spectral region.

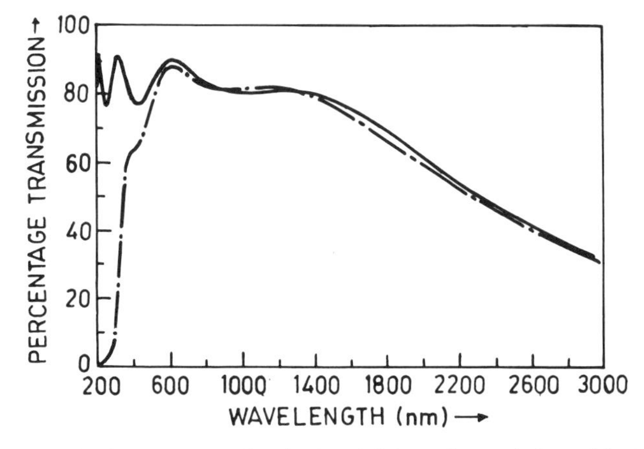

Figure 4.6 Optical transmission in the visible and near-infrared for ITO films. Experimental (–·–·–) and Drude's fit (———) (from [27]).

Table 4.1 Comparison of electron density N and mobility μ determined from Hall measurements and from optical analysis for Al-doped ZnO films [28].

	Electrical		Optical	
Sample	N (10^{20} cm^{-3})	μ (cm^2 V^{-1} s^{-1})	N (10^{20} cm^{-3})	μ (cm^2 V^{-1} s^{-1})
A	6.74	11.2	8.21	9.2
B	5.91	16.8	5.82	17.0
C	6.73	17.6	5.27	22.4
D	7.78	18.1	4.83	29.1
E	6.45	20.9	4.34	31.0

This may be due to an inherent limitation of Drude's theory because it assumes a constant relaxation time over the whole spectral region, whereas the relaxation time is reported to be frequency dependent [6, 28]. Various workers [6, 33–35] have tried to modify Drude's theory in order to explain their results over a much wider spectral region. Hamberg and Granqvist [6], Frank *et al* [34] and Grosse *et al* [35] suggested that the relaxation time is frequency dependent because there are many sources of electron scattering which may influence the electrical properties. These include electron–defect scattering, electron–lattice scattering and electron–electron scattering. Using this technique Frank *et al* [34] and Grosse *et al* [35] explained their results on the optical properties of doped SnO_2 films within the visible and near-infrared wavelength ranges. Hamberg and Granqvist [6] analysed their results on ITO films over a wider spectral range (0.25–50 μm).

4.4 Optical Properties of Tin Oxide Films

4.4.1 Undoped tin oxide films

Figure 4.7 shows typical transmission and reflection characteristics of

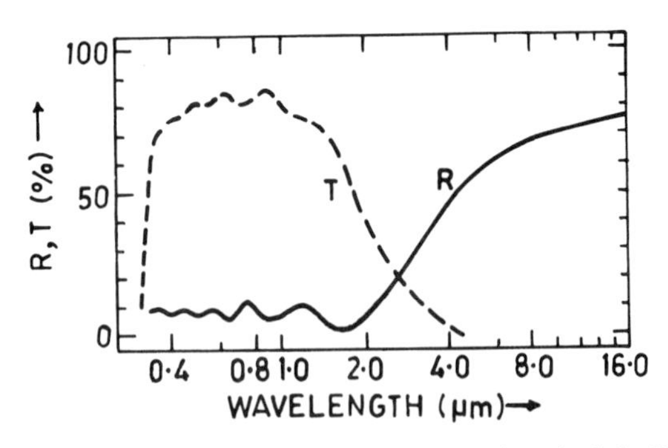

Figure 4.7 Transmittance and reflectance of typical SnO_2 film (from [36]).

undoped SnO_2 films deposited on glass. In general the films are transparent in the range 0.4–2 μm. This range of transparency and the exact values of transmission and reflection depend on various parameters, such as growth conditions, thickness, doping concentration, etc. Several workers [23, 36–56, 58–66] have studied the dependence of optical properties on various growth parameters. We discuss below some of the representative results concerning the effect of growth parameters on the optical properties of undoped tin oxide films.

(a) *Band-gap*
Figures 4.8 and 4.9 show the typical dependence of $\alpha^{1/2}$ and α^2, respectively, on photon energy for spray-deposited SnO_2 films. The α^2 versus photon

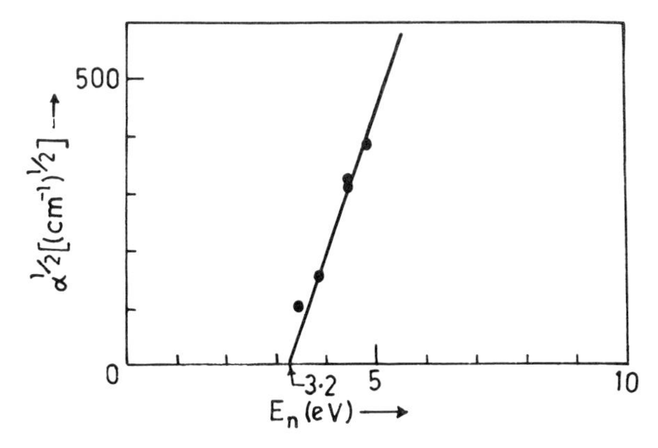

Figure 4.8 Photon energy dependence of $\alpha^{1/2}$ for sprayed SnO_2 films (from [37]).

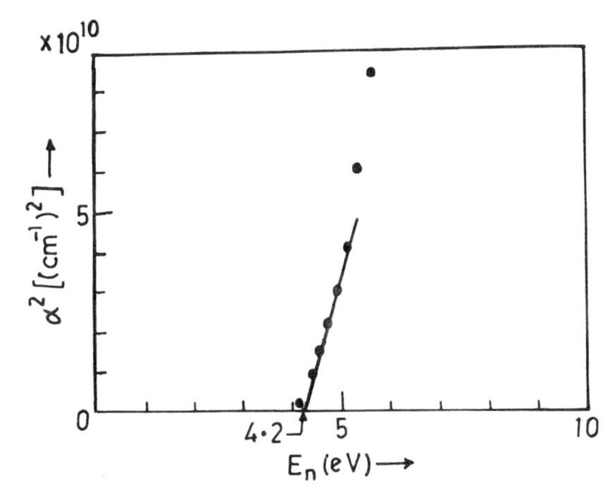

Figure 4.9 Photon energy dependence of α^2 for sprayed SnO_2 films (from [37]).

energy curve has been used for the estimation of direct band-gap in accordance with relation (4.19), and the curve $\alpha^{1/2}$ versus photon energy has been used for estimation of the indirect band-gap. Instead of using $(h^2\nu^2\varepsilon_2)^2$ and $(h^2\nu^2\varepsilon_2)^{1/2}$ the authors have used α^2 and $\alpha^{1/2}$, thus assuming that the value of n is constant in this energy range. The values for direct and indirect band-gaps for sprayed films have been estimated to be 4.2 eV and 3.2 eV, respectively. Shanthi *et al* [38] used similar plots for the estimation of direct band-gap for SnO_2 films sprayed from $SnCl_4$:propanol and HCl mixture. The estimated value of the direct band-gap was ≈4.1 eV. Recently Afify *et al* [39] used plots of $(\alpha\Delta t)^2$ versus $h\nu$ (figure 4.10) to evaluate the value of direct band-gap for sprayed SnO_2 films of different thicknesses. The value obtained from these plots was ≈4.11 eV, irrespective of film thickness. Casey and Stephenson [40] also obtained a value of ≈4.1 eV for direct band-gap, for thin films of tin oxide prepared by a reactive evaporation technique. Although the above studies indicate that the value of direct band-gap in undoped SnO_2 films is ≈4.1–4.2 eV in most cases, some workers have reported different values. Demiryont *et al* [41] observed that in SnO_2 films prepared by spraying a solution of monobutyl tin chloride and methyl alcohol, α obeys a square power law, giving two straight lines with different slopes in the energy range E greater than 2.5 eV. (This value of E corresponds to extrapolated values at $\alpha = 0$.) For photon energy $E > 3.5$ eV, $(\alpha E)^{1/2}$ increases at a faster rate with photon energy suggesting the existence of a second absorption process at a transition energy of $E_g \simeq 3.65$ eV. This value of direct band-gap is significantly lower than the value of 4.1–4.2 eV found by other researchers. These results suggest that SnO_2 films grown by spraying a solution of monobutyl tin chloride might not be of single-phase SnO_2: the films may contain many other phases, i.e. SnO or $SnO_{(2-X)}$. Recently De and Ray [42] have shown the dependence of the direct band-gap of SnO_2 films on the oxygen content in the film. The results are shown in figure 4.11. For low values of oxygen content, where the major phase is SnO,

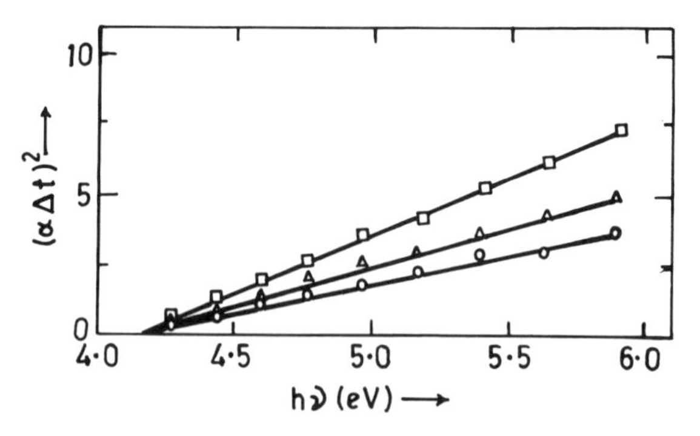

Figure 4.10 Variation of $(\alpha\Delta t)^2$ versus $h\nu$ for different thicknesses (t) of SnO_2 thin films: ○, 25 nm; △, 40 nm; □, 85 nm (from [39]).

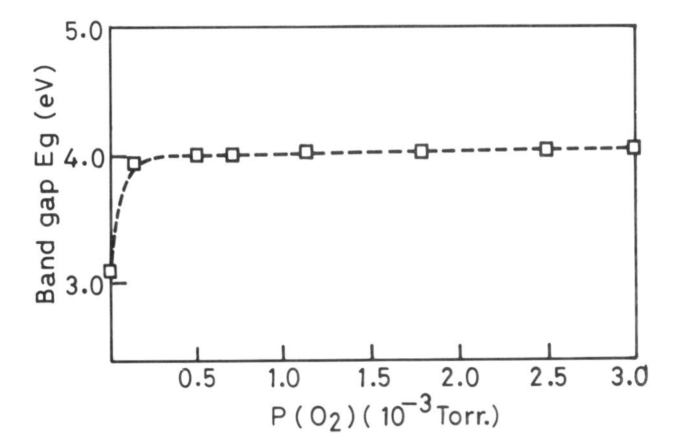

Figure 4.11 Variation of optical band-gap with oxygen partial pressure (from [42]).

the value of direct band-gap is ≈3.1 eV. With an increase of oxygen content, the predominant phase is SnO_2 and the value of band-gap approaches 4.13 eV.

The effect of growth temperature on the optical absorption properties of CVD-grown SnO_2 films is shown in figures 4.12 and 4.13. These figures show α^2 versus photon energy and $\alpha^{1/2}$ versus photon energy for estimation of the direct and indirect band-gap, respectively, for films grown at different

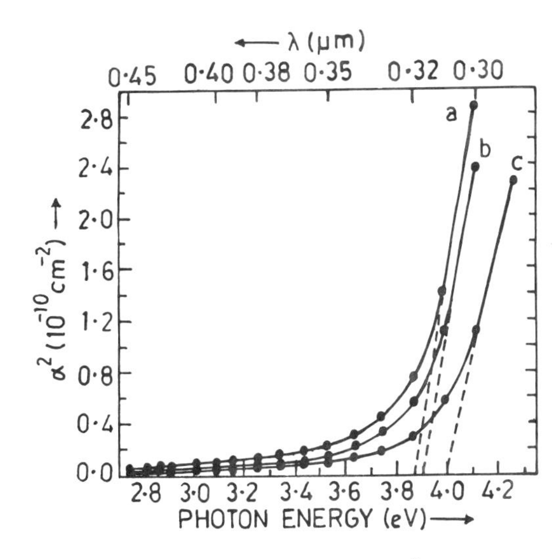

Figure 4.12 Photon energy dependence on α^2 for CVD-grown SnO_2 films grown at different temperatures (a) 400 °C; (b) 450 °C; (c) 500 °C (from [43]).

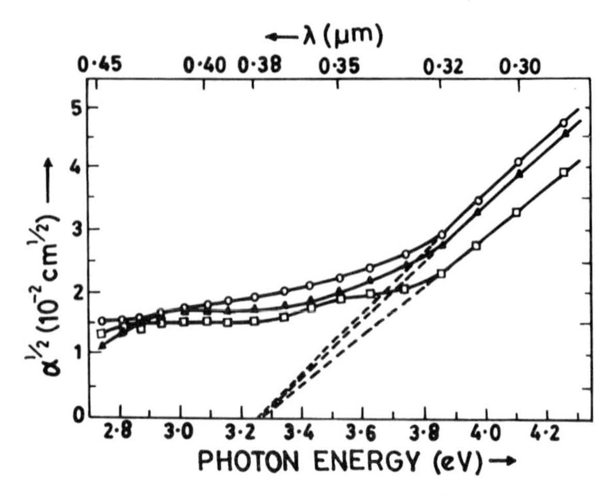

Figure 4.13 Photon energy dependence of $\alpha^{1/2}$ for CVD-grown SnO_2 films grown at different temperatures ○, 400 °C; ▲, 450 °C; □, 500 °C (from [43]).

substrate temperatures, namely 400 °C, 450 °C and 500 °C. The values for direct transition vary from 3.87 eV for films grown at 400 °C to 3.99 eV for films grown at 500 °C. These values are slightly lower than the 4.1–4.2 eV observed by other workers. On the other hand, the values observed for indirect band-gap (3.26–3.28 eV) in CVD-grown films are consistent with the 3.20 eV observed by Demichelis *et al* [37] for sprayed films. However, these values differ significantly from 2.25–3.05 eV, as reported by Spence [44] for sprayed SnO_2 films. De and Ray [42] also studied the effect of substrate temperature on the band-gap of magnetron sputtered tin oxide films. At a substrate temperature of 150 °C, the tin oxide films had a band-gap of 3.83 eV, which increased to 4.13 eV when the substrate temperature was increased to 450 °C. The low value of band-gap at lower growth temperature is due to the presence of SnO phase, in addition to the poor crystalline nature of these SnO_2 films. A higher growth temperature enhances the growth of SnO_2 phase in the films and improves the crystallinity significantly, thereby increasing the band-gap.

(b) *Refractive index and extinction coefficient*

The optical constants n and k depend significantly on the process as well as the substrate material used. On the basis of studies of the variation of n and k in the narrow wavelength range 0.7–3 μm, Shanthi *et al* [38] reported that initially the value of refractive index remains almost constant (2.0) for sprayed films, but in the near-infrared (in the plasma region) the value decreases monotonically as shown in figure 4.14. The value of k, which is almost zero in the visible region, increases appreciably in the infrared

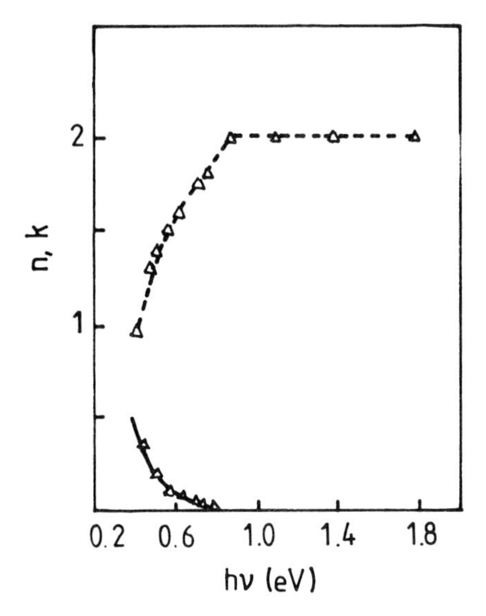

Figure 4.14 Variation of n and k as a function of photon energy for sprayed SnO_2 films (from [38]).

region. Demiryont *et al* [41], on the other hand, studied the variation of n and k in the range 0.4–3.0 μm. Their results for sprayed films using mobobutyl tin trichloride are shown in figure 4.15. These films were found to be non-dispersive at $\lambda > 0.8\,\mu$m with a refractive index value of 1.95. The abrupt increase in n and k at shorter wavelengths ($\lambda < 0.5\,\mu$m) is associated with the fundamental band-gap absorption in the films. Between the

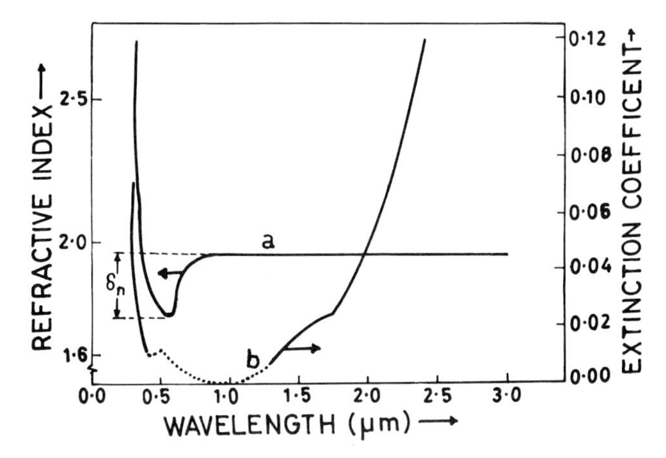

Figure 4.15 Variation of n (curve a) and k (curve b) as a function of wavelength λ for sprayed SnO_2 films (from [41]).

dispersive and non-dispersive region, a pronounced minimum at 0.5 μm is observed in the n versus λ plot. The dip in refractive index δn, as shown in figure 4.15, depends on film thickness; the thicker the film the larger is the value of δn. A similar dip in the refractive index near the ultraviolet region has also been reported by Afify *et al* [39]. This anomaly observed in refractive index spectra has been explained on the basis of the existence of multioriented tin oxide micro-crystallites; each orientation having a different refractive index spectrum. The increase in the value of k in the infrared region, as shown in figure 4.15, is expected on the basis of Drude's theory, as discussed earlier. However, the constant nature of the n versus λ plot observed by Demiryont *et al* [41] cannot be explained on the basis of Drude's theory. The results suggest that the sprayed films studied by Demiryont *et al* [41] are not of single-phase SnO_2, but contain many other phases, such as SnO and $SnO_{(2-x)}$. Melsheimer and Zifgler [45] studied the dependence of n on wavelength in the range 300–800 nm, and also found dispersion as observed by Demiryont *et al* [41] and Afify *et al* [39]. This type of dispersion is expected near the band edge. However, no abrupt change in n has been observed by them [45], indicating that their films had single-phase SnO_2 only. Roos and Hedenqvist [46] studied the variation of n and k with wavelength in SnO_2 films grown on aluminium instead of glass. Their results are shown in figure 4.16. Comparing figures 4.14 and 4.15 with 4.16 it may be seen that the value of extinction coefficient k is more significant throughout the visible region for SnO_2 films grown on aluminium than for films grown on glass. This suggests that the substrate affects the optical properties of tin oxide films.

(c) *Reflection, transmission and absorption studies*

Transmission, reflection and absorption in SnO_2 films depend on many factors, such as the process itself, growth temperature, flow rates of $SnCl_4$

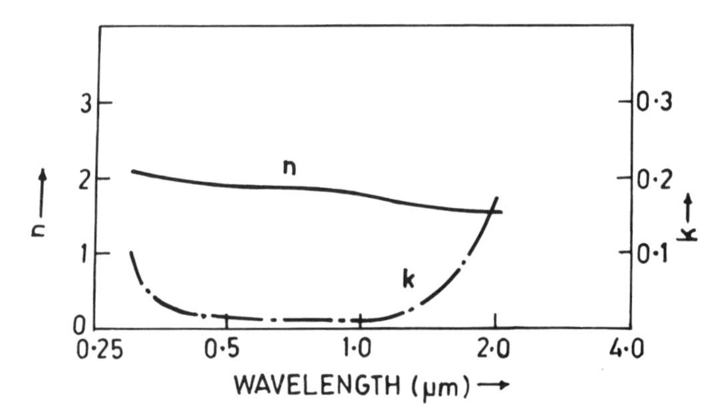

Figure 4.16 Variation of n and k of SnO_2 films deposited on aluminium (from [46]).

vapour, O_2 flow rates, thickness and annealing conditions, etc. Figures 4.7 and 4.17 show the typical transmission T, reflection R and absorption $A = (1 - R - T)$ curves for spray-deposited SnO_2 films. The transmission of these films is ≈80% over a large wavelength range from 0.4–2.0 μm, after which it decreases. The reflection curve indicates that plasma resonance occurs at ≈3.0 μm. The absorption curve (figure 4.17) shows the presence of a peak near 3 μm which is due to the free electrons in the films.

Figure 4.18 shows the effect of oxygen flow rate and growth temperature on the absorption coefficient and transmission for SnO_2 films grown by oxidation of $SnCl_2$. From this figure it can be seen that the absorption coefficient increases with increasing deposition temperature. This increase is due to an increase in free carriers with rise in T_S, since the conductivity observed in the films grown at higher temperature (typically 500 °C) was about five times that of films grown at 400 °C. As conductivity increases, carrier concentration and hence absorption by the carriers also increases. It can be further observed from figure 4.18 that for films deposited at the same temperature T_S, but with different oxygen flow rates, namely 1.35, 1.8 and 2.5 l min^{-1}, the value of absorption coefficient is a maximum for an oxygen flow rate of 1.8 l min^{-1}. The variation of transmission with oxygen flow rate is further elaborated in figure 4.19, wherein it is seen that the transmission is also minimum for oxygen flow rates 1.8 l min^{-1}. These results are consistent with the data on electrical properties (figure 3.8). The electrical data shows that, for a given deposition temperature, the resistivity passes through a minimum at an oxygen flow rate of 1.8 l min^{-1}. At very low flow rate (<1.8 l min^{-1}) not all the tin atoms are oxidized to SnO_2 phase, which results in SnO phase, which is an insulator. On the other hand, at faster flow

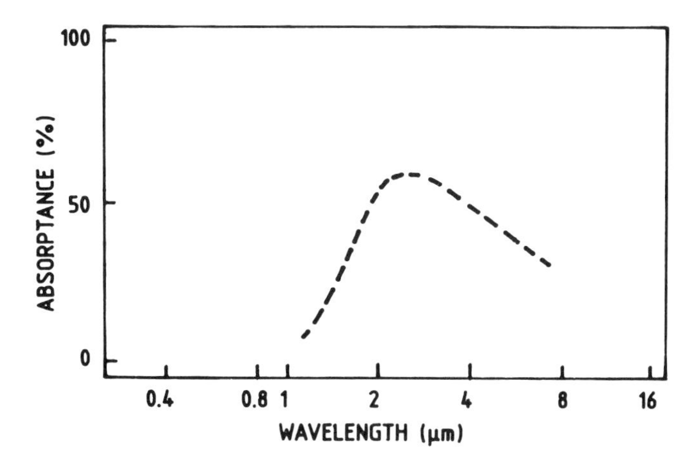

Figure 4.17 Spectral dependence of absorption A for sprayed SnO_2 films (from [36]).

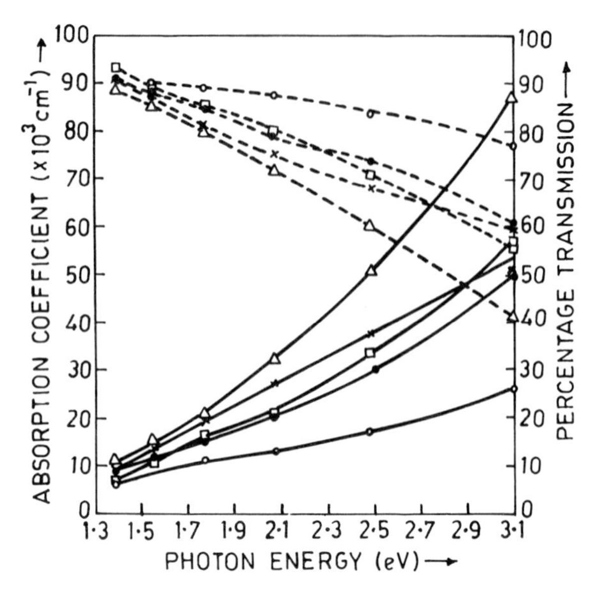

Figure 4.18 Dependence of absorption coefficient (———) and percentage transmission (– – – –) for CVD-grown SnO_2 films with O_2 flow rate 1.8 l min^{-1} and deposition temperatures (○, 400 °C; □, 450 °C; △, 500 °C) and at constant $T = 500$ °C for flow rates: ×, 1.35 l min^{-1}; ●, 2.5 l min^{-1} (from [50]).

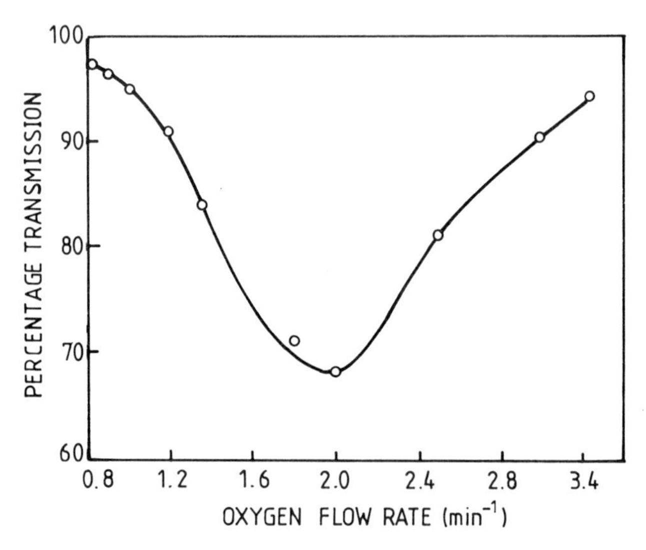

Figure 4.19 Dependence of transmission (at 0.6 μm) on oxygen flow rate in CVD-grown SnO_2 films deposited at 500 °C (from [51]).

rates (>1.8 l min^{-1}), the oxygen content increases thereby filling the oxygen vacancies in the SnO_2 phase and hence the oxygen-deficient SnO_2 becomes nearly stoichiometric. The effect of oxygen partial pressure on the transmission of SnO_2 films grown by a sputtering technique has recently been investigated [42, 52, 53]. Figure 4.20 shows the typical variation of transmittance and absorptance as a function of O_2/Ar gas flow ratio. It can be observed that the transmission increases monotonically whereas absorption decreases continuously with an increase in gas flow ratio. For gas flow ratio less than 4.2%, the films are reported to be yellowish and the transmission is poor. On the other hand, for higher gas flow rates, the films are highly transparent. Similar results have also been observed by other workers [42, 53]. The yellowish colour is characteristic of highly oxygen-deficient tin oxide films prepared either in pure argon [54] or in low oxygen gas flow rates [42]. Poor transmission at low oxygen gas rates is due to the presence of SnO phase, which has been confirmed by X-ray diffraction studies [42]. With an increase in oxygen partial pressure, the films possess predominantly SnO_2 phase instead of SnO phase.

$SnCl_4$ has also been used in place of $SnCl_2$ in the CVD process for the formation of SnO_2 films [55]. Figure 4.21 shows the effect of the flow rate of $SnCl_4$ vapour in the reaction process for films deposited at 600 °C. The oxygen flow rate was fixed at 4.1×10^{-2} mol min^{-1} and the $SnCl_4$ flow rates for a, b, c, and d samples were 2.32×10^{-3}, 4.62×10^{-3}, 6.96×10^{-3} and 9.28×10^{-3} mol min^{-1}, respectively. Figure 4.21 also shows the variation of absorption coefficient with wavelength for films with different deposition

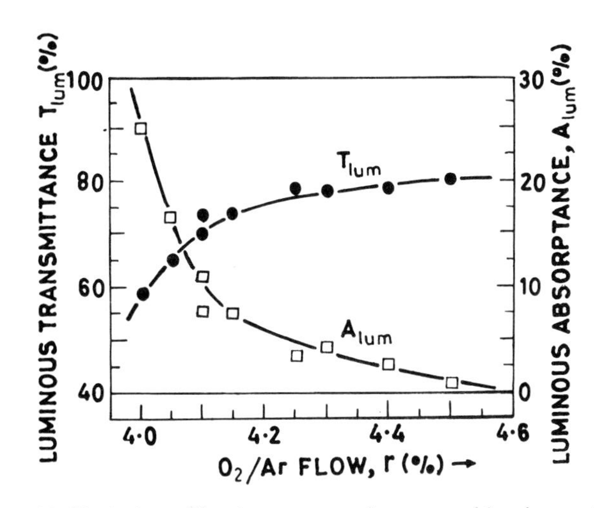

Figure 4.20 Variation of luminous transmittance and luminous absorptance as a function of O_2/Ar flow rate for RF sputtered tin oxide films (from [52]).

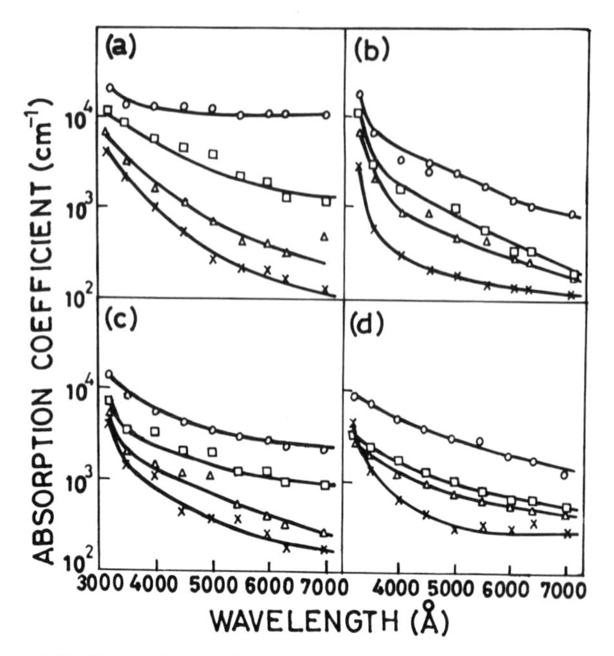

Figure 4.21 Dependence of absorption coefficient of CVD-grown SnO_2 films on $SnCl_4$ flow rates and deposition time. $SnCl_4$ vapour flow rate and deposition time for (a) 2.32×10^{-3} mol min^{-1}, (×, △, □ and ○) 1, 2, 3, 10 min; (b) 4.62×10^{-3} mol min^{-1}, 0.5, 1, 2, 4 min; (c) 6.96×10^{-3} mol min^{-1}, 1, 2, 3, 4 min; (d) 9.28×10^{-3} mol min^{-1}, 0.5, 1, 2, 4 min (from [55]).

times. It is evident that SnO_2 films deposited over a long duration and/or under a high $SnCl_4$ vapour flow rate give higher absorption coefficients and hence lower transparency in the short wave length range of the visible spectrum. The absorption coefficients are in the range 10^2–10^4 cm^{-1}. The effect of deposition time and thus thickness is more pronounced in series 'a' (figure 4.21(a)). X-ray diffraction and electrical studies revealed that the series 'a' and 'b' films, i.e. with low $SnCl_4$ vapour flow rates, were highly resistive and had very broad SnO_2 peaks. An additional SnO phase peak is also present. In addition to this the films show a lot of amorphous background. However, the thicker films deposited for longer durations have improved crystallinity and lower resistivity. On the other hand, films grown with higher flow rates (series c and d) are less resistive and show sharply defined diffraction patterns. This indicates that films formed at higher $SnCl_4$ flow rates have better crystallinity. The increase in grain size, crystallinity and conductivity with increased $SnCl_4$ flow rate and deposition time explains the higher values of absorption coefficient in these films. Agashe *et al* [56] observed a reduction in average transmission with increase in thickness of

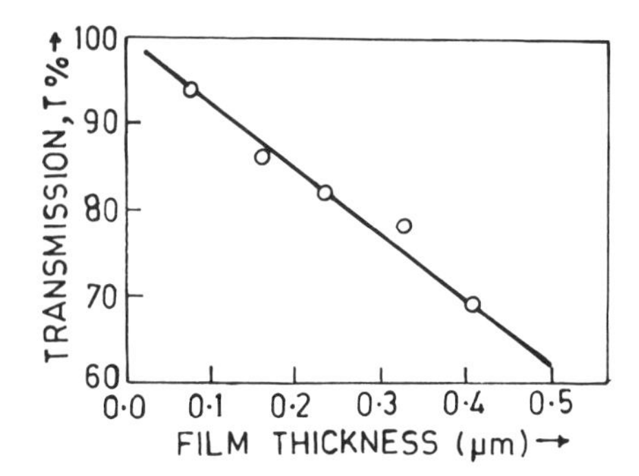

Figure 4.22 Dependence of average visible transmission on thickness for sprayed SnO_2 films (from [56]).

spray-deposited SnO_2 films. Figure 4.22 shows this variation for films formed from solutions of $SnCl_4$:methanol and distilled water sprayed at 425 °C. Melsheimer and Zifgler [45], while studying the effect of thickness on transmission, observed that increased thickness not only decreases the transmission but also shifts the absorption edge towards the longer wavelengths. Their electrical results show that thicker films are more resistive and hence have a lower number of free carriers. The decrease in the number of carriers is expected to shift the band edge towards longer wavelengths on the basis of Burstein Moss [57] theory.

Growth temperature is another important parameter that affects the transmission properties of SnO_2 films. Figure 4.23 shows this effect for CVD-grown tin oxide films. It can be observed that growth temperature influences transmission mainly in the low wavelength region. The absorption edge shifts towards the longer wavelength as the deposition temperature is increased. Electrical properties of these films revealed that the films

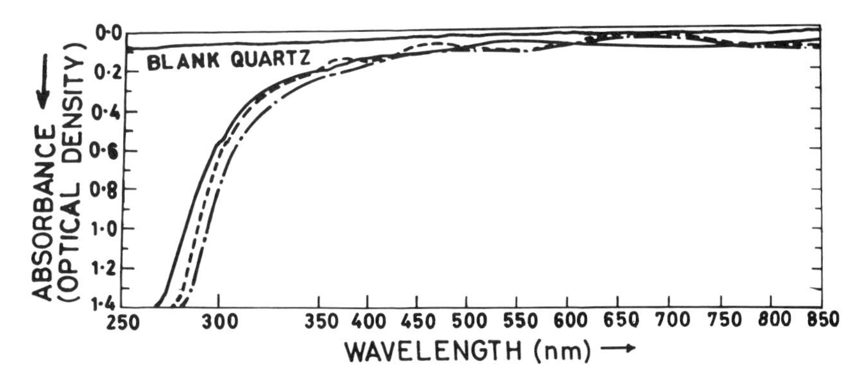

Figure 4.23 Optical absorption characteristics for CVD-grown SnO_2 films deposited at different temperatures: ———, 400 °C; – – – –, 450 °C; –·–·–, 500 °C (from [43]).

grown at 400 °C had higher resistivity than those grown at 500 °C. This indicates that the band gap decreases as conductivity increases. This result is not in agreement with the data of other workers [44, 45] for undoped as well as for doped films, to be discussed in section 4.4.2. This discrepancy is because of the presence of SnO phase in addition to SnO_2 phase at low growth temperatures, and as such Burstein Moss theory [57] cannot be applied to explain the results. Muranaka *et al* [58] studied the effect of crystallinity on the transmission properties of SnO_2 films. Films were prepared by reactive deposition in an oxygen atmosphere. Films deposited at 1–5 × 10^{-3} Torr and substrate temperatures up to 250 °C were amorphous in nature, whereas films deposited at the same pressure but at higher substrate temperature (>400 °C) were crystalline. It was further observed that the crystalline films produced at higher substrate temperatures showed very good transparency (~90%). However, films deposited at temperatures lower than 400 °C had relatively poor transmission. Figure 4.24 shows the typical transmission of as-grown amorphous SnO_2 films. It is quite evident from the figure that the as-grown films have poor transmission in the visible range. The same films, however, after annealing at 250 °C in air, had higher transparency. Although the films were amorphous even after annealing, the increase in transparency was presumably due to the fact that SnO phase oxidized to SnO_2. This was also quite evident from the electrical results. As-grown films were almost an insulator, whereas annealing at 250 °C resulted in a significant increase in conductivity.

The effect of annealing has also been studied by Beensh-Marchwicka *et al* [54] while investigating the transmission properties of SnO_2 films grown using a reactive ion sputtering technique. Figure 4.25 shows the typical variation of transmission with wavelength for as-grown films and films annealed at 625 K for 4 h in air. The curves are for different oxygen contents, K, which is defined as the ratio of partial pressure of O_2 to that of the total

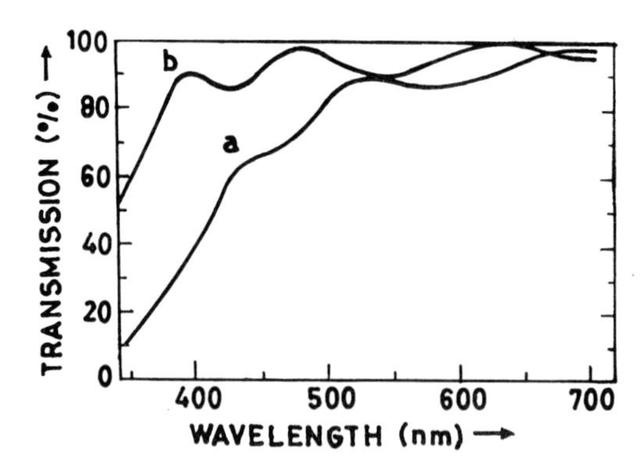

Figure 4.24 Dependence of transmission on wavelength for (a) as-deposited (b) annealed at 250 °C SnO_2 films grown by reactive deposition (from [58]).

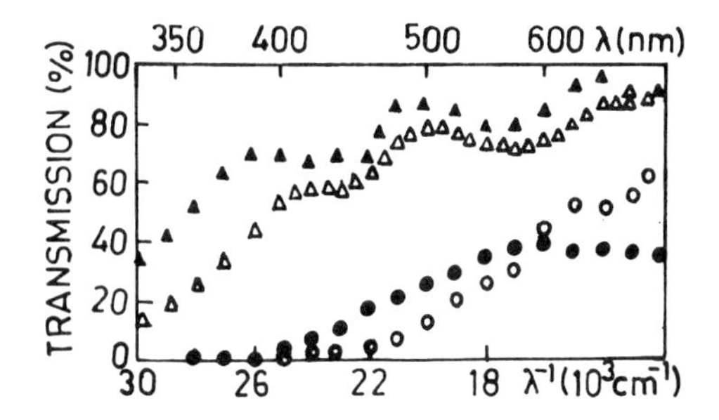

Figure 4.25 Spectral dependence for as-grown and annealed SnO_2 films with $K = 10\%$, ○ and ● before and after annealing, $K = 15\%$ △ and ▲ before and after annealing (from [54]).

pressure of O_2–Ar gases. It can be observed that for low value of K, transmission is very poor, which has been attributed to the presence of the amorphous SnO phase (also observed by Muranaka *et al* [58] in reactively deposited films). The effect of annealing in air on optical properties was observed to be dependent on sputtering parameters. Films deposited in a pure argon atmosphere underwent partial oxidation during annealing in air at 623 K. X-ray studies indicated the presence of β-Sn, SnO and SnO_2 phases. Films deposited at $K = 5\%$ oxidized to SnO which was stable and did not oxidize further during prolonged annealing at 623 K. Amorphous films with $K = 10\%$ crystallized, with SnO_2 as the prominent phase, whereas films with $K > 15\%$ always resulted only in SnO_2 phase after annealing. The maximum change was observed over the first thermal ageing cycle. Subsequent annealing cycles or increased duration did not have any significant effect on transmission characteristics. The effect of annealing SnO_2 films in a hydrogen atmosphere is more pronounced [59]. Figure 4.26 shows the average transmission (500–800 nm) as a function of annealing temperature

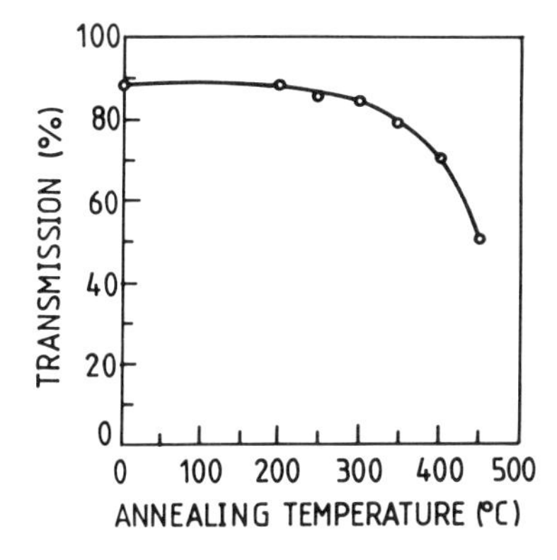

Figure 4.26 Effect of annealing in H_2 atmosphere on transmission at different annealing temperatures (from [59]).

in hydrogen for sprayed SnO_2 films. The flow rate of hydrogen was maintained at 500 ml min^{-1}. It can be observed from this figure that annealing beyond 300 °C decreased transmission significantly. The decrease in transmission can be correlated with the electrical measurements on the films, which suggest that hydrogen results in a greater number of oxygen vacancies. This results in an increase in free carrier concentration thus increasing free carrier absorption.

4.4.2 Doped tin oxide films

Attempts have been made to improve the properties of SnO_2 films by doping them with different dopants. Dopants such as phosphorus, arsenic, indium, thallium, tellurium, chlorine, etc. have not been found to be electrically effective and as such no studies have been carried out to observe the effect of these dopants on the optical properties. Antimony and fluorine are found to be the most useful dopants from a device application point of view. Many workers [1–5, 23, 36–39, 44, 47, 51, 54, 59, 60, 62, 67, 73, 77, 79, 82–87] have studied the effect of doping and other growth parameters on the optical properties of these films.

In general the optical properties, namely n, k, E_g, m_e^*, etc., are governed more by the carrier concentration of the films than the nature of the dopant itself. The main effects are to shift the fundamental absorption edge towards shorter wavelengths, to increase the effective mass, to decrease the refractive index, to shift the plasma edge towards shorter wavelengths and to increase IR reflectance.

(a) *Band-gap studies*
Figure 4.27 shows the effect of carrier concentration on the absorption coefficient in the lower wavelength range for Sb-doped SnO_2 films. It can be observed that an increase in carrier concentration shifts the band edge towards shorter wavelengths. The effect of carrier concentration is on both the direct and indirect band-gaps. Figure 4.28(a) and (b) show α^2 versus photon energy and $\alpha^{1/2}$ versus photon energy for SnO_2 films with different resistivities. These plots show that direct transitions lie between 4.17 and 4.45 eV for the different samples. On the other hand, the values of indirect transition lie between 2.25 and 3.05 eV. Figure 4.29 shows the dependence of band-gap on carrier concentration. The linear nature of the plot shows that band-gap widening depends on $N^{2/3}$, where N is the carrier concentration. The widening of the band-gap can be understood on the basis of Burstein-Moss shift [57]. This effect is observable in materials where the electron density far exceeds the Mott critical density [68], which is given as

$$N \gg (0.25 m_c^* e^2/\varepsilon_0 \hbar^2)^3 \tag{4.40}$$

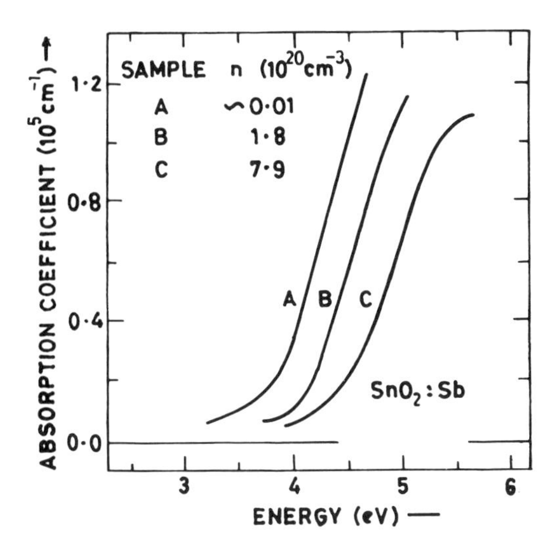

Figure 4.27 Spectral absorption coefficient for Sb-doped SnO_2 films with different electron densities (from [67]).

where e is the electron charge, ε_0 is the permittivity of free space, $\hbar$ is Planck's constant divided by 2π, and m_c^* is the conduction band effective mass. Under the condition that the carrier concentration greatly exceeds the Mott critical density, the conduction band is partly filled, i.e. its lowest states are blocked, which leads to widening of the optical band-gap. It should be mentioned here that this widening is partly compensated by a downward shift of the conduction band and an upward shift of the valence band, which occur as a consequence of electron–electron and electron–impurity scattering [69–71]. Band-gap narrowing because of heavy doping is quite common in silicon [72]. In addition, doping also affects the effective masses m_c^* and m_v^* where m_v^* is the valence band effective mass.

Quantitatively the band-gap E_g is given by [67]:

$$E_g = E_{go} + \Delta E_g^{BM} + \hbar \sum_c (k_F, \omega) - \hbar \sum_v (k_F, \omega) \qquad (4.41)$$

where E_{go} is the band-gap of the undoped semiconductor, k_F is the Fermi wave number, $\hbar\Sigma_v$ and $\hbar\Sigma_c$ are self-energies due to electron–electron and electron–impurity scattering. The Burstein-Moss shift ΔE_g^{BM} is given by

$$\Delta E_g^{BM} = \frac{\hbar^2}{2}\left(\frac{1}{m_v^*} + \frac{1}{m_c^*}\right)(3\pi^2 N_e)^{2/3}. \qquad (4.42)$$

The Fermi wave number k_F is given by

$$k_F = (3\pi^2 N_e)^{1/3}. \qquad (4.43)$$

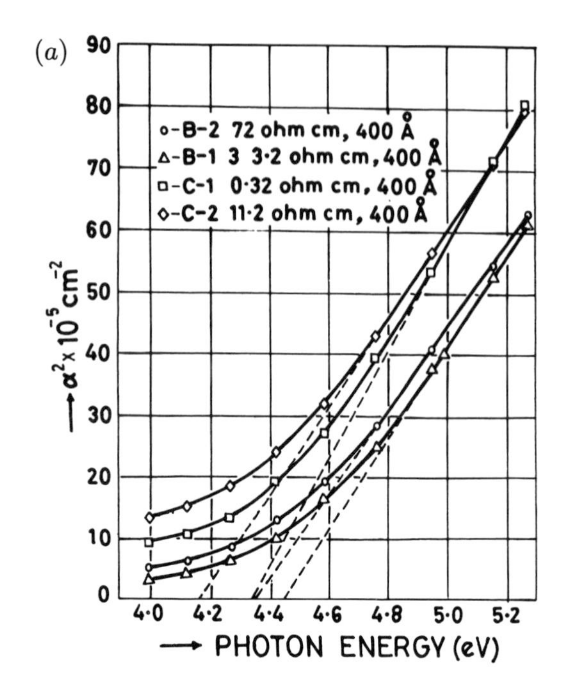

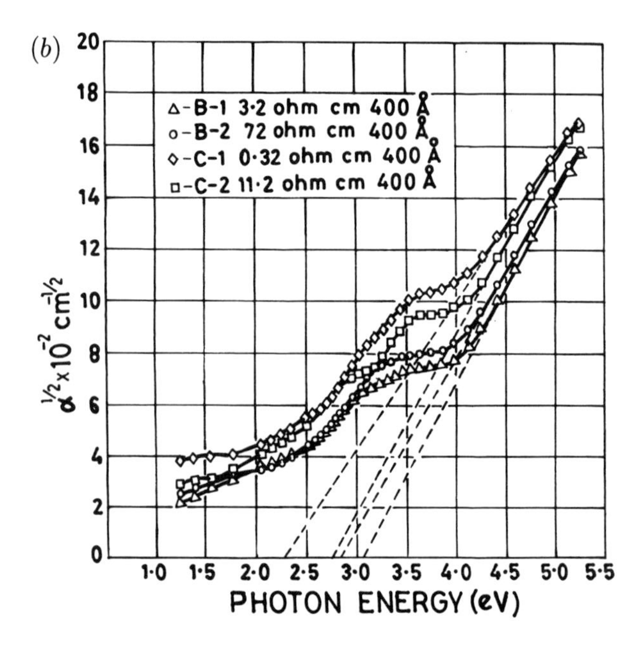

Figure 4.28 Photon energy dependence of (a) α^2 and (b) $\alpha^{1/2}$ for SnO_2 films with different resistivities (from [44]).

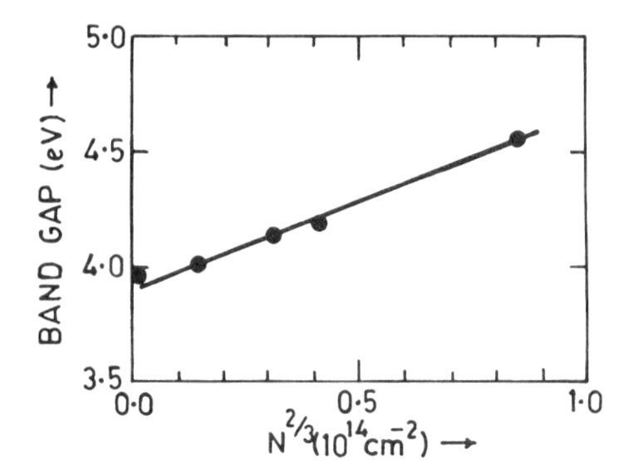

Figure 4.29 Band-gap dependence of carrier concentration in doped SnO_2 films (from [3]).

It is evident from figure 4.29 that in doped SnO_2 films, the band-gap is directly proportional to $N^{2/3}$, thereby suggesting that band-gap widening is the most prominent process, and the effect of electron–electron scattering and electron–impurity scattering is not very significant. Afify *et al* [39] studied the effect of film thickness on band-gap in fluorine-doped SnO_2 films. Figure 4.30 shows the variation of $(\alpha\Delta t)^2$ with $h\nu$ for films of different thicknesses. It can be observed that the band-gap is $\approx$4.21 eV and is independent of film thickness. This is expected because the resistivity in all these films is almost the same and is not a function of film thickness. Further, the value of band-gap 4.21 eV in these doped films is higher than the 4.11 eV for undoped films (figure 4.10). This may be attributed to the fact that the resistivity in undoped films is one order higher than that in doped films.

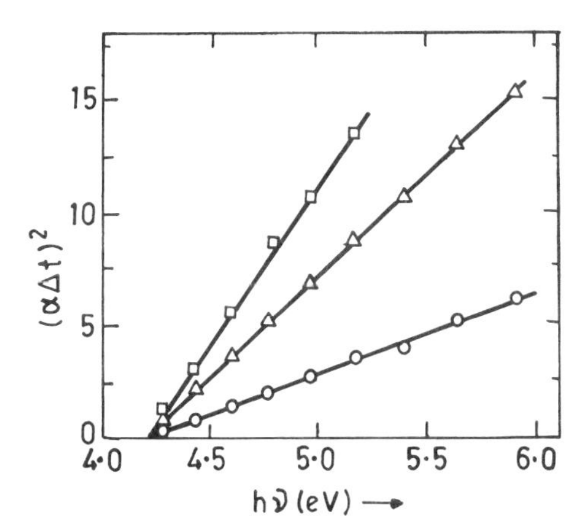

Figure 4.30 Variation of $(\alpha\Delta t)^2$ versus $h\nu$ for different thicknesses of F-doped SnO_2 films; t = (○) 85 nm, (△) 172 nm and (□) 350 nm (from [39]).

(b) *Optical constants n and k*

The variation of refractive index n and extinction coefficient k for fluorine-doped SnO_2 films in the wavelength range 0.3 to 1.5 μm for a typical thickness of 0.15 μm is shown in figure 4.31. It can be observed that the value of n is practically constant beyond 0.4 μm, whereas near the band edge a peak is observed. On the other hand, the value of k remains constant and is very near to zero for wavelengths beyond 0.4 μm, but in the wavelength range less than 0.4 μm, it increases significantly. The large increase in n and k at wavelengths less than 0.4 μm is due to fundamental band-gap absorption in the films. Afify *et al* [39] also observed a similar type of n versus λ variation for 10% fluorine-doped SnO_2 films of 85 nm thickness. However, as the thickness increases, this dispersion in the value of n is very weak.

The variation of n with wavelength also depends on the thickness of the films [39, 73]. Figure 4.32 shows the effect of thickness on the refractive index of films [73] for various wavelengths. The refractive index shows a marked dependence on film thickness. A maximum in the plot occurs for films of thickness about 0.8 μm. The peak at this particular thickness can be understood on the basis of the electrical results of these films. The reflectivity, which is related to refractive index, is also related to conductivity by the relation [74]

$$R = (1 + 2\varepsilon_0 c_0 R_S)^{-2} \tag{4.44}$$

where R_S is the sheet resistivity of the films and $\varepsilon_0 c_0 = 1/376\ \Omega^{-1}$. The electrical properties indicate [73] that films of 0.8 μm thickness have maximum conductivity. According to equation (4.44), these films should have maximum reflectivity and hence maximum refractive index at 0.8 μm.

The effect of doping SnO_2 films with antimony or fluorine on n and k near the plasma edge has been studied in detail by Shanthi *et al* [38, 75]. Figures 4.33 and 4.34 show the variation of n and k for different doping concen-

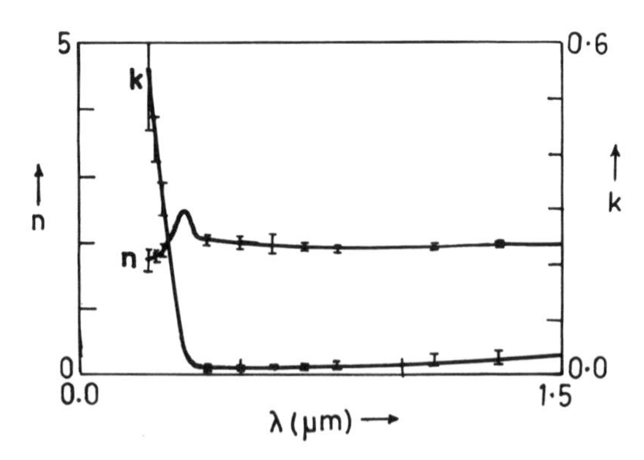

Figure 4.31 Refractive index n and extinction coefficient k against λ for SnO_2:F films (from [37]).

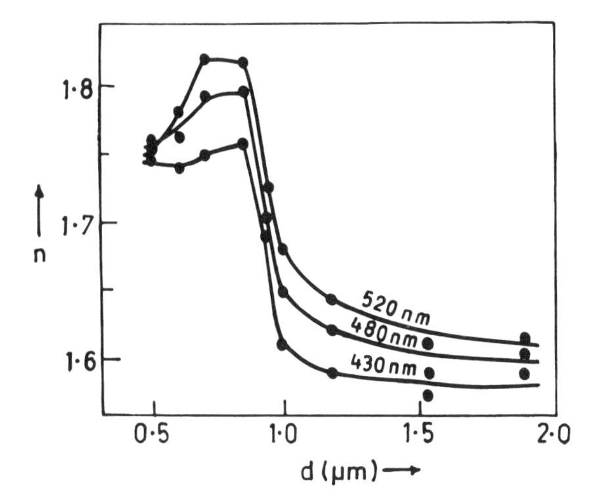

Figure 4.32 Variation in film refractive index with film thickness for different wavelengths (from [73]).

trations of Sb and F, respectively. It can be observed that, in general, the effect of doping is to decrease the refractive index and to shift the onset of increased extinction coefficient from longer wavelengths to shorter wavelengths. These results can be well understood on the basis of Drude's theory as discussed earlier. According to equation (4.25), the effect of doping, i.e.

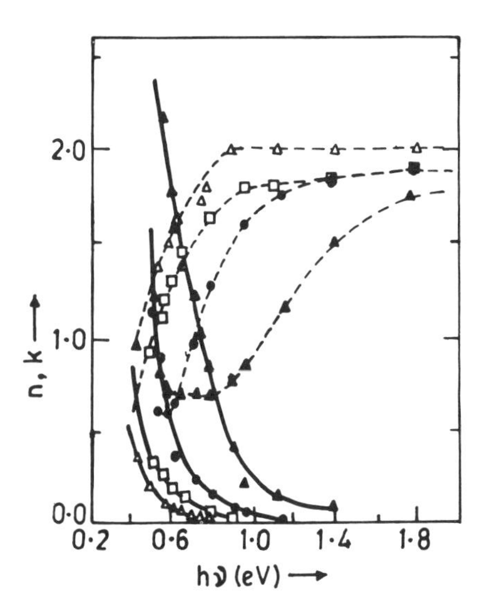

Figure 4.33 Refractive index n and extinction coefficient k variation for different antimony doping concentrations: △, 0; □, 0.7; ●, 1.4; ▲, 3 mol % (from [38]).

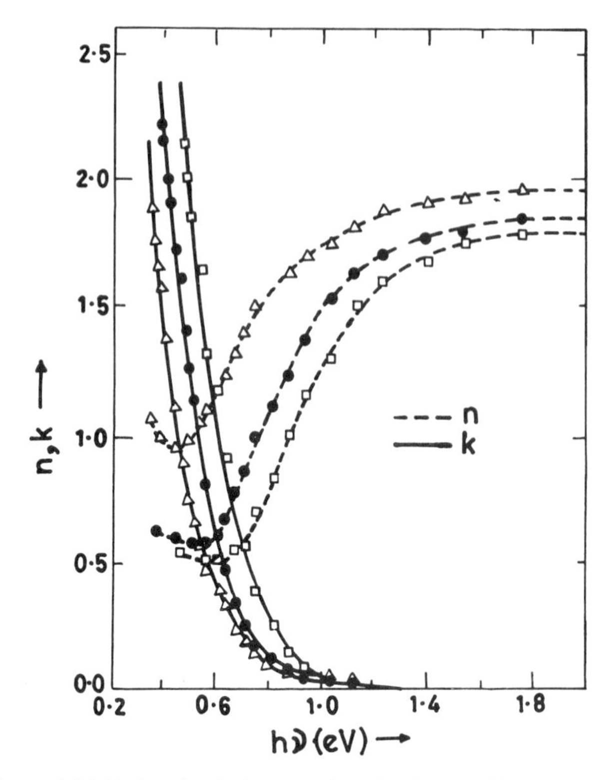

Figure 4.34 Refractive index n and extinction coefficient k of fluorine-doped SnO_2 films: △, 10; ●, 30; □, 65 at% fluorine (from [75]).

increased carrier concentration, is to increase the plasma frequency ω_p and hence reduce plasma wavelength.

(c) *Effective mass*

The effective mass of the free carriers is significantly influenced as a result of change in carrier concentration due to doping. The effective mass of the free carriers can be computed using equation (4.25), knowing the values of carrier concentration N and plasma frequency ω_p. The value of carrier concentration can be determined using Hall effect data as discussed in chapter 3. The value of ω_p can be obtained using equation (4.23). Near the plasma edge, the ε' versus $1/(\omega^2 + \gamma^2)$ plot is a straight line and the plasma frequency is estimated from the slope of the plot at ε'. Figure 4.35 shows the typical variation of effective mass with carrier concentration in antimony-doped SnO_2 films. It can be observed that the value of m_e^* varies from $0.1m_o$ for undoped tin oxide films to $0.29m_o$ for 3 mol% Sb-doped films, which corresponds to a carrier concentration of $\approx 6 \times 10^{20}$ cm^{-3}. Such a large change in effective mass indicates the nonparabolicity of the conduction

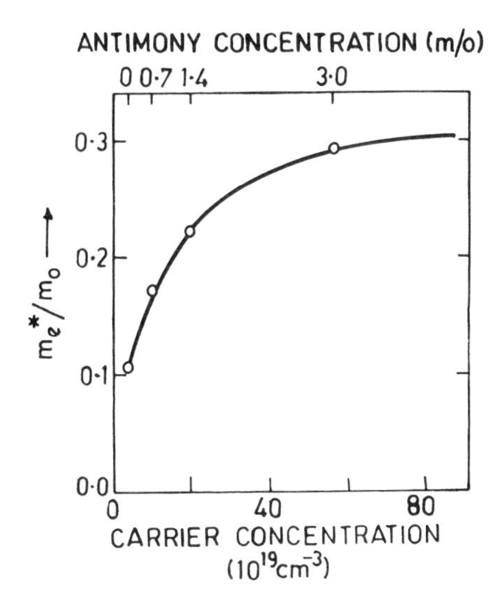

Figure 4.35 Variation of effective mass of electrons (m_e^*/m_o) as a function of antimony carrier concentration (from [38]).

band. In general, the value of effective mass depends only on free carrier concentration and not on the nature of the dopant.

(d) *Transmission, reflection and absorption*

The typical transmission, reflection and absorption spectra for SnO_2:F films are shown in figure 4.36. Absorption has been computed using the relation $A + T + R = 100\%$. Compared to figure 4.7, the plasma wavelength now shifts to around 1.5–1.6 μm. Figure 4.37 compares the variation of transmission with wavelength for undoped and fluorine-doped SnO_2 films. The value of plasma wavelength can easily be adjusted by varying the free carrier

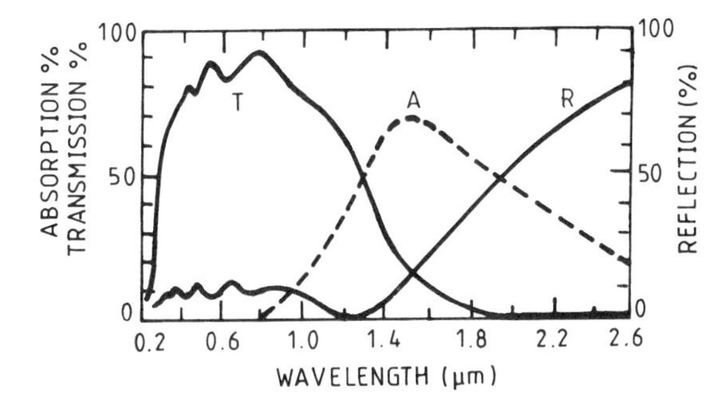

Figure 4.36 Typical transmission T, reflection R and absorption A for the fluorine-doped SnO_2 films (from [1]).

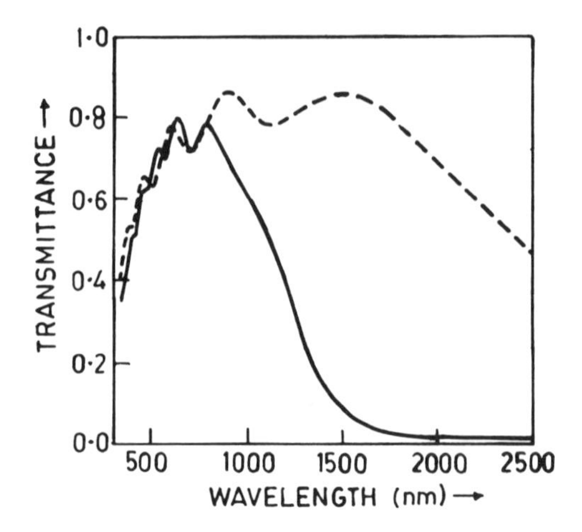

Figure 4.37 Transmittance versus wavelength for – – – –, undoped and ———, F-doped SnO_2 films (from [47]).

concentration; the higher the value of carrier concentration, the lower is the value of plasma wavelength. The free carrier concentration can be varied up to 10^{21} cm^{-3} in SnO_2 films [1]. Figure 4.38 shows the dependence of absorption on wavelength in the plasma wavelength region, for different dopants. It can be observed from this figure that the absorption, in general, increases, as a result of doping SnO_2 films. The most effective dopant is Sb_2O_3, which produces maximum shift of the plasma edge.

Figure 4.39 shows typical transmission curves for Sb:SnO_2 films obtained by spraying. It can be observed from this figure that an increase in the thickness of the film (volume of the sprayed solution) results in a significant

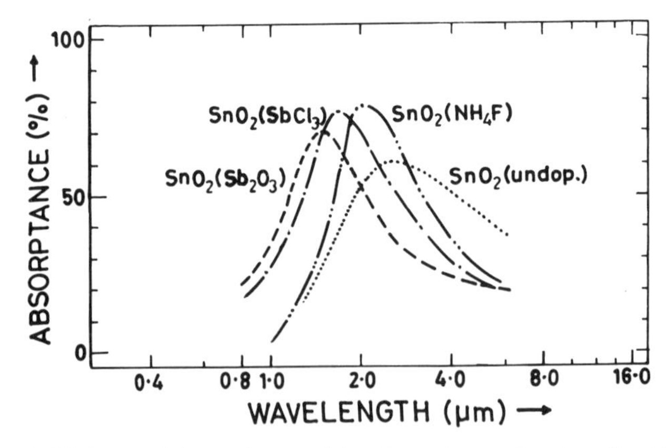

Figure 4.38 Spectral dependence of the absorptance $A = 1 - R - T$ for different dopants (from [36]).

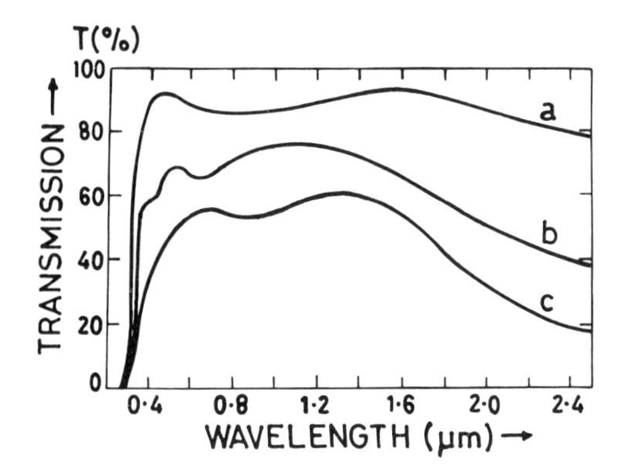

Figure 4.39 Dependence of the transmission T on the wavelength for SnO_2:Sb films obtained by spraying various volumes of solution (a) 0.25 ml, (b) 0.5 ml, (c) 0.55 ml (from [76]).

reduction in transmission. This can be explained on the basis of the electrical properties of the film. It has been observed [76] that the electrical sheet resistivity for films sprayed using 0.25, 0.5 and 0.55 ml of solution are 501.5, 111.8 and 69.8 $\Omega\,\square^{-1}$, respectively. This significant decrease in resistivity with volume of solution is responsible for the reduction in transmission of these films. Figure 4.40 shows the dependence of transmission on thickness for F–SnO_2 films. In these F-doped films an increase in thickness up to 1.5 μm reduces the transmission to 70%, whereas in Sb-doped films [76], the transmission corresponding to 0.55 ml solution ($\sim$0.4 μm) is about 50%. This suggests that F doping is more useful than Sb doping provided similar electrical properties can be achieved.

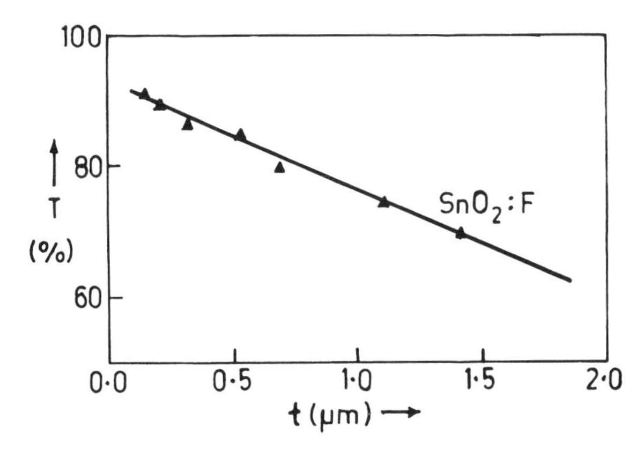

Figure 4.40 Variation of transmission (T) with thickness (t) for SnO_2:F films (from [77]).

A comparison of the experimental data of the dependence of reflectance on wavelength as a function of carrier concentration can be made with theoretical curves based on Drude's theory. Such theoretical curves are depicted in figure 4.41. The typical layer thickness assumed is 0.5 μm and the mobility and effective mass of the charge carriers are taken as $15 \text{ cm}^2 \text{ V}^{-1} \text{ s}^{-1}$ and $0.25m_o$, respectively, for calculation. The effect of increasing carrier concentration is to shift the plasma edge towards shorter wavelengths and to increase the overall reflectance in the infrared region. Figures 4.42 and 4.43 show the transmission and reflection spectra for different doping concentrations of antimony and fluorine in spray-deposited SnO_2 films. It can be observed from these figures that the plasma edge shifts towards the shorter wavelengths with an increase in carrier concentration. The IR reflectance, in general, increases with the increase in carrier concentration. These results are in accordance with those predicted by Drude's theory as shown in figure 4.41. Further, the experimentally observed reflectance at 10 μm for a carrier concentration of $6 \times 10^{20} \text{ cm}^{-3}$ is ≈80% in Sb-doped SnO_2 films, which is in good agreement with the theoretical value. It can also be observed from figure 4.42 that the transmission of these films remains high (>80%) over a large wavelength range from 0.4–1.56 μm. The transmission initially increases for doping levels up to 1.4 mol% Sb, but thereafter decreases with further addition of antimony. The reflection in the infrared region increases with increasing Sb concentration up to 3 mol%. The reflectivity, however, decreases with further addition of antimony (10 mol%). The decrease in reflection in 10 mol% Sb-doped films may be due to the low mobility and coarse nature of the films obtained with these concentrations. Murty and Jawalekar [51], while studying CVD-grown Sb–SnO_2 films, observed that the average transmission in the range 0.4–1.1 μm remains practically constant at about 88% for up to 3 mol% of antimony doping. From 3 mol% to 4 mol% doping level, trans-

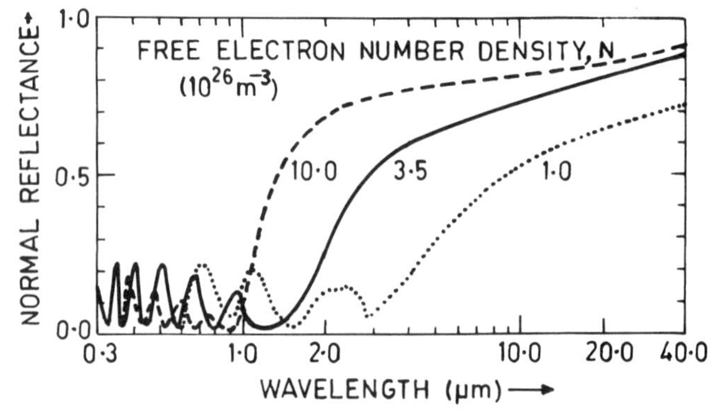

Figure 4.41 Calculated normal reflectance for different carrier concentrations in doped SnO_2 films (from [23]).

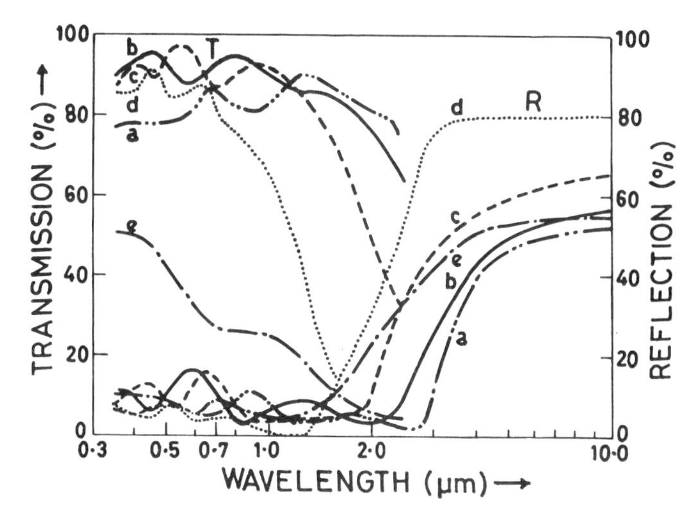

Figure 4.42 Transmission and reflection spectra of Sb-doped SnO_2 films with different antimony concentrations: (a) 0 mol%; (b) 0.7 mol%; (c) 1.4 mol%; (d) 3 mol%; (e) 10 mol% (from [38]).

mission falls to about 85%, but with further doping transmission falls abruptly and the films become dark.

Figure 4.44 shows transmission and reflection spectra at substrate temperatures of 15 °C, 300 °C and 460 °C for Sb–SnO_2 films deposited by an RF magnetron-sputtering technique. It is observed that the transmittance de-

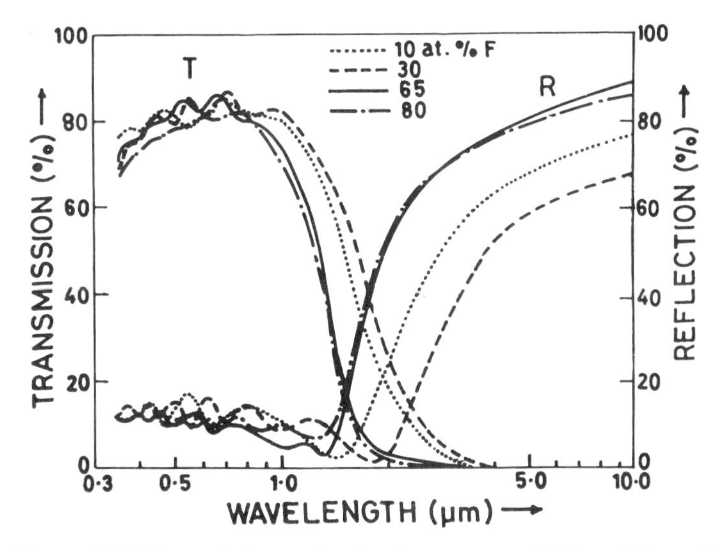

Figure 4.43 Transmission and reflection spectra of F-doped tin-oxide films at different fluorine concentrations (from [75]).

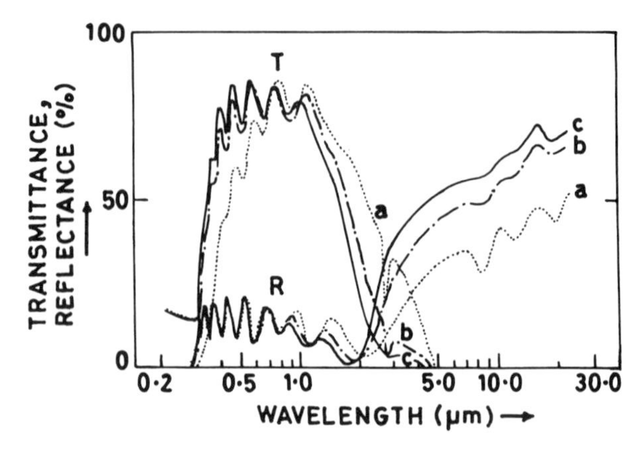

Figure 4.44 Transmittance and reflectance curves for Sb:SnO_2 films deposited at various substrate temperatures: (a) 15 °C; (b) 300 °C; (c) 460 °C (from [62]).

creases, whereas reflectance in the IR region increases, with increase of substrate temperature. The oscillations in the visible region are due to interference effects. The average transmission in the visible range is always >80% and the absorption is very low. The effect of free carrier absorption is clearly observed as reduced transmission and increased reflection in the near-IR region. A sharp decrease in transmittance and shift of reflectance minimum, which are obvious for higher substrate temperatures, are mainly due to the increase in carrier density. The steep rise in the reflectance curve for higher substrate temperature is mainly due to higher electron mobilities. The properties in the UV region can be explained by the Burstein shift [57]. The values of fundamental absorption edge estimated from α versus $h\nu$ plots are 3.07, 3.58 and 3.78 eV for films grown at 15, 300 and 460 °C, respectively. The low value of absorption edge for films grown at the lowest substrate temperature can be explained by assuming the presence of SnO phase, which has a fundamental absorption edge at a lower energy than that for SnO_2. High values of 3.7–3.8 eV for highly conducting films have been reported by Spence [44] and Arai [78] using α versus $h\nu$ plots, whereas similar plots yield a value of 3.5 eV for bulk SnO_2. The shift in the absorption edge from 3.5 eV to 3.7–3.8 eV is a result of the Burstein effect, as discussed earlier.

Figure 4.45 shows the spectral transmittance in the visible and near-infrared regions for Sb-doped SnO_2 films deposited on different substrates. Curves (a) and (a′) are for films deposited on fused quartz substrates whereas (b) and (b′) are for films deposited on soda lime glass. As shown in the figure, films on fused quartz generally have a higher spectral transmittance over all wavelengths when compared to the films grown on soda lime glass. It has been further observed by Kaneko and Miyake [79] that the

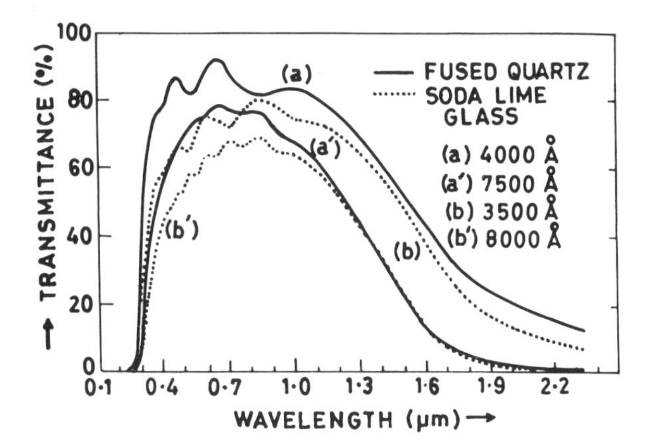

Figure 4.45 Spectral transmittance of Sb:SnO_2 films deposited on fused quartz and soda lime substrates (from [79]).

spectral transmittance of films on borosilicate glass substrates is also nearly equal to that of films on fused quartz. The direct optical band-gap of the films prepared on different substrates has been derived from a plot of $(\alpha h\nu)^2$ versus $h\nu$. The optical band-gap of 2250–13000 Å thick films deposited on fused quartz and borosilicate glass substrates is found to be independent of the film thickness and is almost the same ($\approx$3.75 eV). However, the band-gap of films deposited on soda lime glass substrates varies with film thickness, and ranges from 2.85 to 3.08 eV. The difference in these physical properties of films deposited on soda lime glass with respect to films deposited on fused quartz or borosilicate glass substrates, is presumed to result from the out-diffusion of active ions in the soda lime glass into the Sb-doped SnO_2 films. This is quite probable, because preparation of films is usually carried out at high substrate temperatures (>500 °C), which is near the softening point of soda lime glass. Similar substrate effects on the properties of films has also been reported by Kane *et al* [80] and Mizuhashi [81]. The relatively low value of transmission in films deposited on soda lime glass has also been attributed to the comparatively poor surface finish of these films. Scanning electron micrograph studies have revealed that the surface of thin films on soda lime glass substrates differs from that of films deposited on fused quartz or borosilicate glass substrates; many sharp-cornered pits on the smooth surface have been observed. Furthermore, the shape and size of the grains in thick films on soda lime glass differ from those of thick films on fused quartz or borosilicate glass. In films thicker than 8000 Å deposited on soda lime glass, fine cracks are often visible.

Post-deposition annealing of doped tin oxide in different ambients affects their transmission properties significantly. Figure 4.46 shows the optical transmission spectra of DC reactively sputtered Sb-doped SnO_2 films

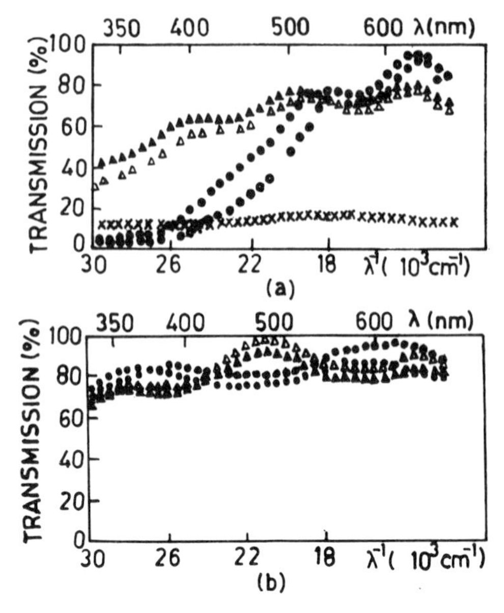

Figure 4.46 Optical transmission as a function of wavelength for Sb:SnO_2 films for various oxygen concentrations and annealed for 4 h at 623 K: (a) (i) $K = 0\%$, ×, after annealing; (ii) $K = 10\%$, ○, before and ●, after annealing; (iii) $K = 15\%$, △, before and ▲, after annealing; (b) (i) $K = 40\%$, ○, before and ●, after annealing; (ii) $K = 50\%$, △, before and ▲, after annealing (from [54].

deposited at various oxygen concentrations and subsequently annealed in air at 623 K for 4 h. It can be observed that annealing improves the transmittance of the as-deposited films. X-ray studies revealed that the films prepared only in argon, i.e. $K = 0\%$ (ratio of oxygen to argon and oxygen) or very low values of oxygen, i.e. $K \leqslant 15\%$, and post-oxidized in air contain β-Sn, SnO, SnO_2 and Sb_2O_4 phases, which are responsible for the overall low transmission of these films. Moreover, films which are initially amorphous remain amorphous even after post-annealing in air. The transmittance observed in films deposited with $K = 40$–50% was always about 80% over the complete wavelength range studied, i.e. 350–600 nm. Post-annealing of these films resulted in a slight reduction in transmittance as shown in figure 4.46(b).

Annealing of fluorine-doped SnO_2 films in a hydrogen atmosphere is also found to affect the optical transmission properties. Figure 4.47 shows typical average transmission curves as a function of annealing temperature in a hydrogen atmosphere for doped and undoped SnO_2 films. It can be observed that the decrease in transmission as a result of annealing in hydrogen is more pronounced in F-doped films than in undoped SnO_2 films. This is

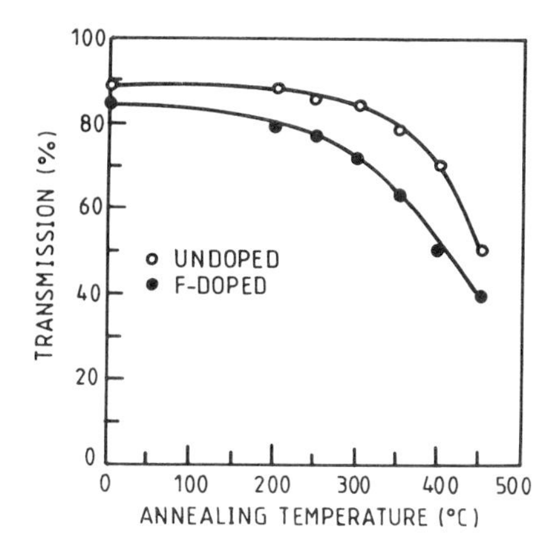

Figure 4.47 Variation of average optical transmission (%) with annealing temperature in undoped and F-doped SnO_2 films (from [59]).

possibly due to the fact that hydrogen not only removes oxygen but also fluorine from these F–SnO_2 films. This result has been further supported by electrical studies (figure 3.32) where it is observed that hydrogen annealing results in a significant decrease in carrier concentration suggesting the removal of fluorine from the F–SnO_2 films.

4.5 Optical Properties of Indium Oxide Films

4.5.1 Undoped indium oxide films

During the last decade, indium oxide has become the favoured material for the preparation of transparent, conducting and highly IR reflecting thin films on glass. These films are superior to SnO_2 films because they can be prepared with higher conductivity and IR reflectivity. In general, the films are transparent in the range 0.4–2.0 μm. This transparency, and the other optical parameters such as refractive index n, extinction coefficient k, band-gap E_g, etc., depend upon many factors such as the growth process itself, thickness and conductivity of the films. Some workers [27, 89, 91–94, 97–102] have studied the effect of these parameters on the optical properties of In_2O_3 films.

(a) *Band-gap studies*

Figures 4.48 and 4.49 show the α^2 versus $h\nu$ variation for In_2O_3 films grown by e-beam evaporation and CVD techniques, respectively. The direct band-gap, as found by extrapolation, is $\approx$3.60 eV and 3.7 eV, respectively, which

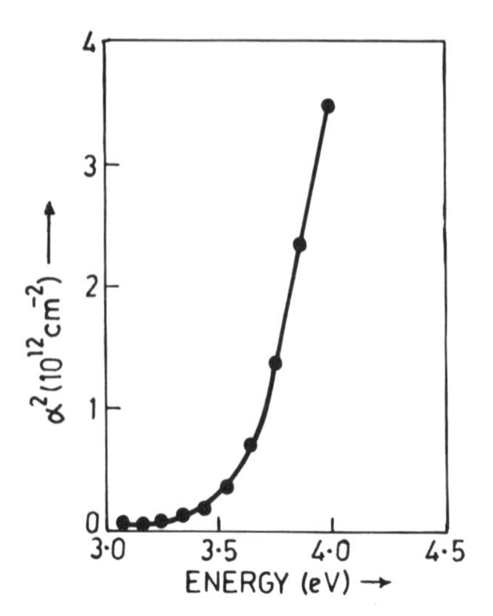

Figure 4.48 α^2 versus $h\nu$ graph for In_2O_3 films grown by electron beam evaporation (from [88]).

is very close to the value of 3.75 eV observed in bulk samples of In_2O_3 [90]. Bellingham *et al* [91] reported a value of 3.34–3.58 eV for the direct band-gap of amorphous In_2O_3 films, depending on the carrier concentration in the samples. These optical studies indicate that interband absorption is similar in crystalline and amorphous In_2O_3. Ryabova *et al* [92] studied the variation of α with wavelength for In_2O_3 films prepared by a pyrolysis technique. Figure 4.50 shows such a spectral dependence of the absorption coefficient.

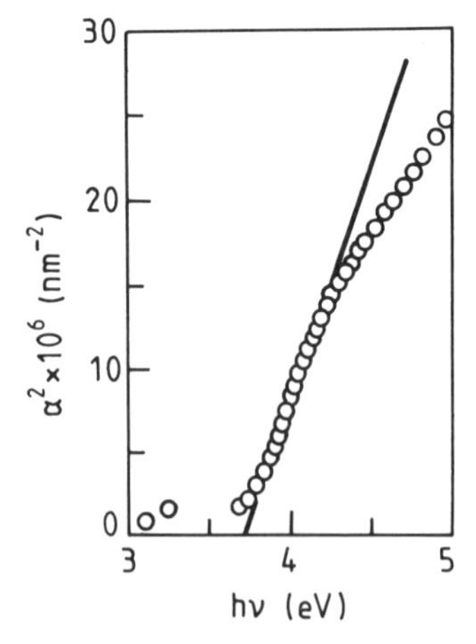

Figure 4.49 α^2 versus $h\nu$ graph for indium oxide films grown by a CVD technique (from [89]).

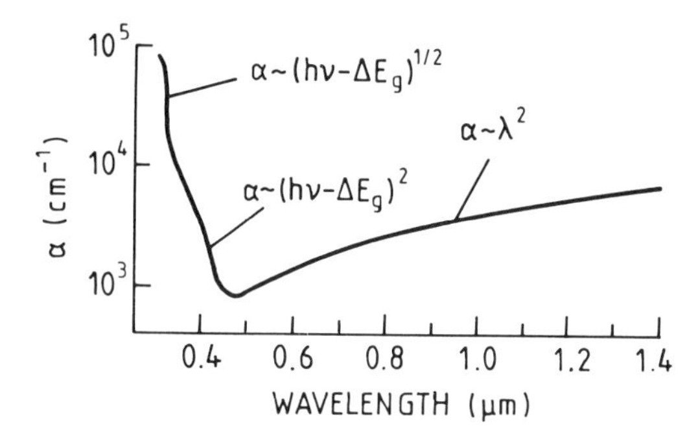

Figure 4.50 Spectral dependence of the absorption coefficient of In_2O_3 films (from [92]).

The value of α was calculated using the transmission data for two films of different thicknesses t_1 and t_2 in accordance with the relation

$$\alpha = \frac{1}{t_2 - t_1} \ln \left(\frac{T_1}{T_2} \right). \tag{4.45}$$

The thickness was particularly chosen such that

$$\exp(2\alpha t) \gg R^2 \tag{4.46}$$

where R is reflectance. It can be observed in figure 4.50 that light is absorbed by both direct and indirect interband transitions. The values of direct and indirect band-gaps are 3.78 and 2.6 eV, respectively. For wavelengths in the range 0.55–1.1 μm, α is proportional to λ^2. A dependence of this type is characteristic of light absorption by free carriers. Pan and Ma [93] used electrolytic electro-reflectance measurements for the determination of the band-gap of thermally evaporated In_2O_3 films. In this technique, the relative differential reflectance $\Delta R/R$ modulated by an AC field, is plotted as a function of optical wavelength. Figure 4.51 shows the spectral dependence of the electro-reflectance for some In_2O_3 films. Four critical energies are clearly visible, marked in the figure. The second and sharpest peak, centred around 3.56 eV, corresponds to the direct band-gap. The fourth critical peak is at about 2.69 eV, which corresponds to the indirect band-gap for In_2O_3. The first and third peaks at 3.88 eV and 3.18 eV, respectively, depend on the free carrier concentration and are attributable to Burstein shift [57].

(b) *Optical constants* n *and* k

Very little work has been reported on the study of n and k for indium oxide films. Figure 4.52 shows the variation of refractive index n as a function of wavelength in the range 0.45–1.8 μm for RF sputtered indium oxide films. In this figure n_1 and n_2 represent the refraction coefficients at the upper

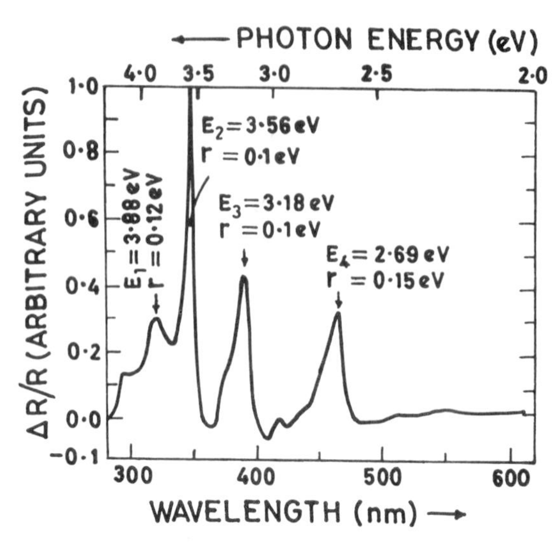

Figure 4.51 Electro-reflectance as a function of wavelength for In_2O_3 films (from [93]).

surface, and the film–substrate interface, respectively. It is worth noting that the refractive indices are slightly different and are nearly independent of energy. The slight difference in the values of n_1 and n_2 suggest that the films are optically inhomogeneous. Such inhomogeneities have also been reported on the basis of the electrical properties of indium tin oxide films [95]. Figure 4.53 presents the variation of both refractive indices, n_1 and n_2, as a function of oxygen partial pressure during deposition. The value of n_2 at the film–substrate interface is always constant and has a value of $\approx$2.08,

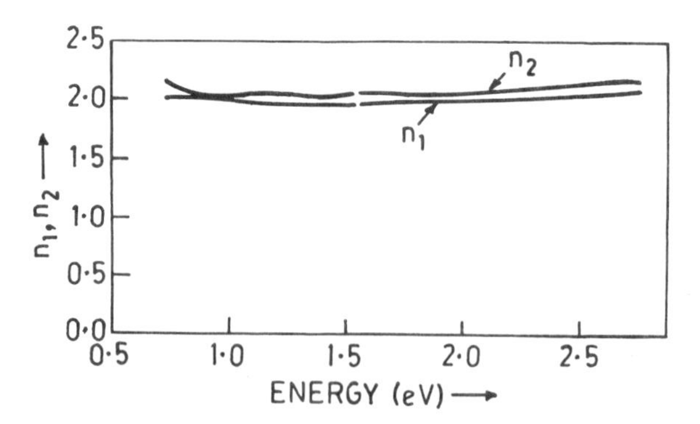

Figure 4.52 Refractive indices n_1 and n_2 for In_2O_3 films grown by RF sputtering (from [94]).

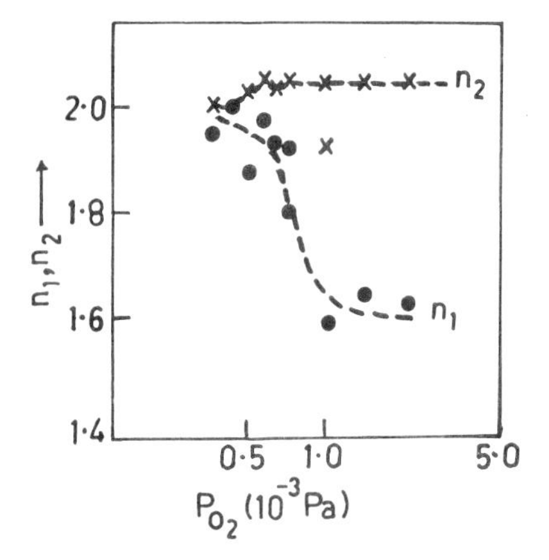

Figure 4.53 Dependence of refractive indices n_1 and n_2 on oxygen partial pressure for In_2O_3 films; $h\nu = 2$ eV (from [94]).

which is very near to the value observed in bulk material [96]. On the other hand, the refractive index n_1 at the upper surface of the film shows a clear dependence on the oxygen partial pressure of the sputter gas. Electrical measurements suggest that the oxygen partial pressure of the sputter gas mainly affects the carrier concentration in the interior of the films. This change in carrier concentration should be observed in the optical properties in the regions of plasma reflection, but should not significantly affect the optical constants in the region of intrinsic absorption. The results shown in figure 4.53 for n_2 support this argument. The strong dependence of n_1 on the oxygen partial pressure could be due to the presence of a depletion layer, formed after deposition by the chemisorption of oxygen from the atmosphere [94].

(c) *Transmission, reflection and absorption studies*

Figure 4.54 illustrates typical transmission, reflection and absorption curves for magnetron-sputtered thin films of In_2O_3. The transmission is greater than 80% throughout the visible range and extends up to wavelengths of 1 μm. The reflectivity curve shows the plasma edge at around 2 μm, where reflectivity is about 10%, but increases to more than 80% beyond 10 μm. Sundaram and Bhagavat [98] and Chen *et al* [99] have shown that the electrical parameters also influence the optical characteristics of indium oxide films. Figure 4.55 depicts the influence of sheet resistivity on the transmission of In_2O_3 films. Films were prepared by post-oxidation of indium films at about 300–400 °C. The sheet resistivity for the films was controlled by the duration of oxidation. It can be observed from figure 4.55 that transmission varies from 76 to 91%, depending on the thickness and

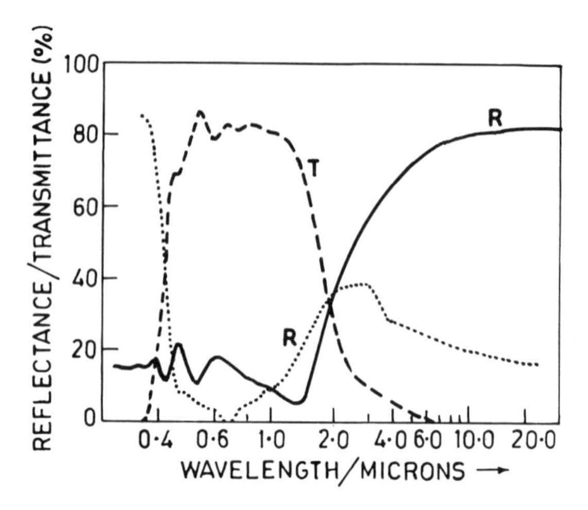

Figure 4.54 Transmission T, reflection R and absorption A for magnetron sputtered In_2O_3 films (from [97]).

resistivity of the films. Films with the highest sheet resistivity (10 kΩ $\square^{-1}$) exhibit the highest transmission. This is expected as discussed earlier in section 4.3, since transmission depends on the conductivity of the films, which in turn is a function of the free carrier absorption. Figure 4.56 shows transmission and reflection response in the infrared region (2.5–12 μm) for samples $T_{27}R_{27}$, $T_{49}R_{49}$ with sheet resistivities of 27.8 and 49 Ω $\square^{-1}$ respectively. Films were deposited on ZnS substrates because of their transparency in the IR region. Reflection and transmission curves for the ZnS substrate are also shown in this figure. It can be observed that the transmission decreases while reflection increases appreciably for films with lower sheet resistivity. This is expected on the basis of Drude's theory, as discussed earlier.

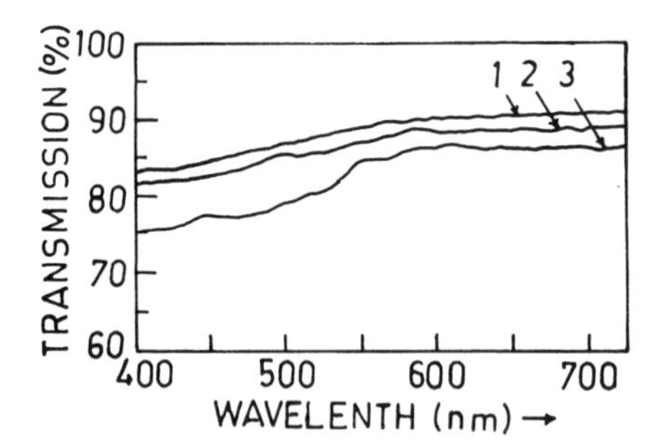

Figure 4.55 Transmission curves in visible region for In_2O_3 films of different sheet resistivities: (1) 10 kΩ $\square^{-1}$; (2) 4.6 kΩ $\square^{-1}$; (3) 1 kΩ $\square^{-1}$ (from [98]).

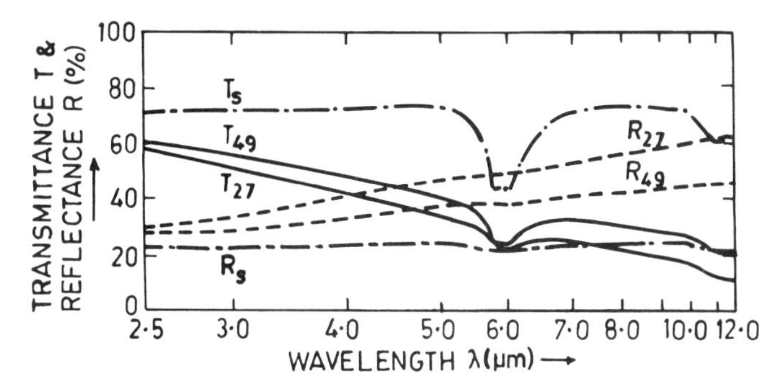

Figure 4.56 Transmission and reflection curves in IR region for films of different sheet resistivity. T_S and R_S represent substrate transmission and reflection, respectively. Subscripts represent the sheet resistivity (from [99]).

4.5.2 Doped indium oxide films

(a) *Indium tin oxide (ITO)*

As in the case of tin oxide, the optical properties of In_2O_3 films are also greatly influenced by doping. Though various dopants have been tried, the most commonly used dopant for In_2O_3 films has been tin or tin oxide and the films are generally known as ITO (indium tin oxide). Indium tin oxide has been the subject of great interest due to its excellent properties, both electrical and optical, suitable for device applications. Many workers [67, 74, 77, 83, 84, 86, 88, 89, 91–93, 97, 99–104, 106, 108, 109, 112–118, 120–138] have studied the optical properties of ITO films as a function of tin doping and other growth parameters.

Band-gap studies. The effect of tin doping in indium oxide is to increase the band-gap, both direct and indirect. Figure 4.57 shows the variation of absorption with photon energy as a result of change in carrier concentration due to tin doping. It can be observed that the increase in carrier concentration shifts the absorption edge towards higher energies. The effect of doping is on both direct and indirect band-gaps. Figures 4.58 and 4.59 show the α^2 versus $h\nu$ and $\alpha^{1/2}$ versus $h\nu$ curves for different carrier concentrations of RF reactively sputtered ITO films. These plots show that the direct band-gap shifts from $\approx$3.7 eV to about 4.2 eV when the carrier concentration increases from 3.2×10^{19} to 8.23×10^{20} cm^{-3}. The indirect band-gap, on the other hand, shifts from about 2.8 eV to 3.4 eV when the carrier concentration increases from 5.38×10^{19} to 6.54×10^{20} cm^{-3}. The shift of band-gap with change in carrier concentration is further elaborated in figure 4.60. The linear nature of the curves shows that band-gap widening is

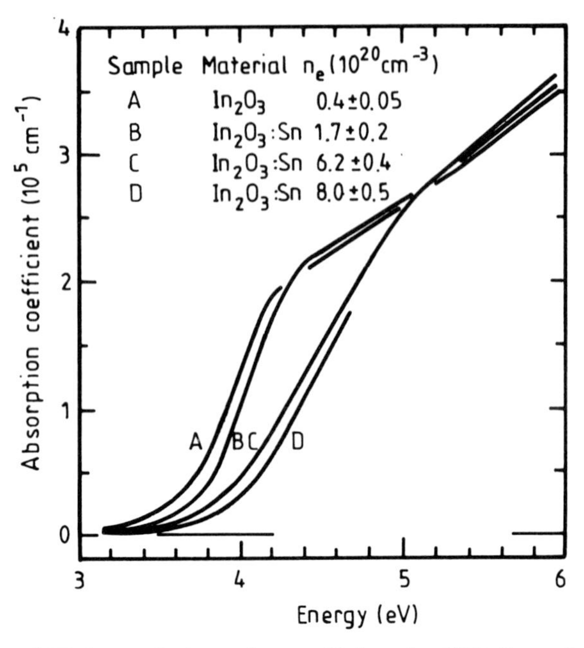

Figure 4.57 Spectral absorption coefficient for ITO films of different carrier concentrations (from [67]).

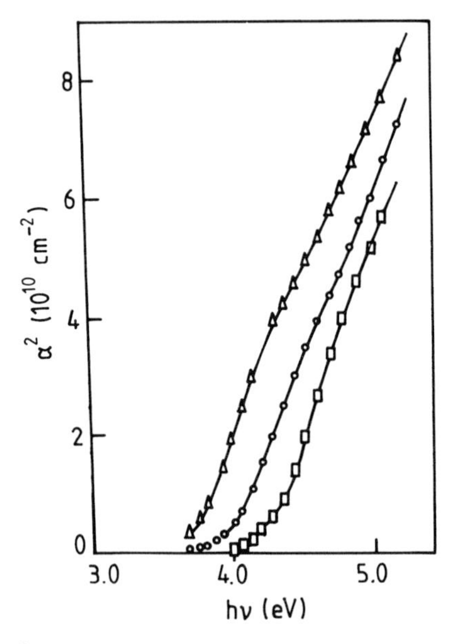

Figure 4.58 α^2 versus $h\nu$ curves for ITO films of carrier concentration: △, 3.2×10^{19} cm^{-3}; ○, 3.24×10^{20} cm^{-3}; □, 8.23×10^{20} cm^{-3} (from [103]).

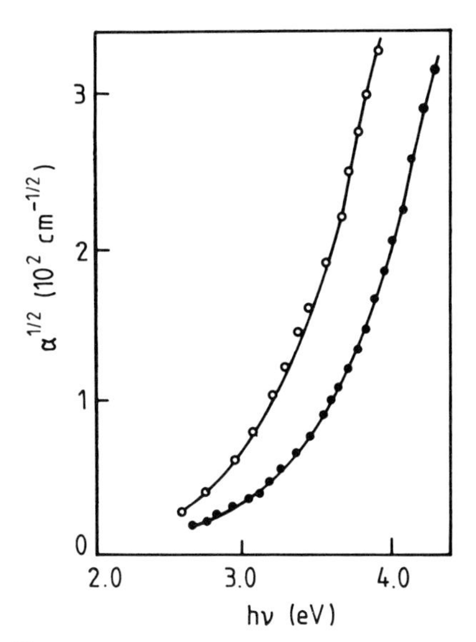

Figure 4.59 $\alpha^{1/2}$ versus $h\nu$ curves for ITO films of carrier concentration: ○, $5.38 \times 10^{19}\ cm^{-3}$ and ●, $6.54 \times 10^{20}\ cm^{-3}$ (from [103]).

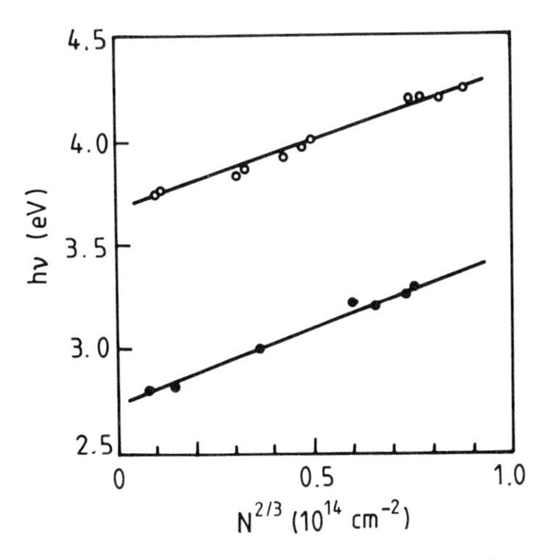

Figure 4.60 Variation of direct band-gap (○) and indirect band-gap (●) as a function of $N^{2/3}$ (from [103]).

proportional to $N^{2/3}$. A similar variation of band-gap with carrier concentration (figure 4.61) has also been reported by Bellingham *et al* [91] for amorphous ITO films. The results have been discussed on the basis of Burstein–Moss shift (as discussed earlier in section 4.4.2). Assuming that Burstein–Moss shift is the predominant effect, the band-gap can be written using a simplified expression (4.41), as

$$E_g = E_{go} + \Delta E_g^{BM} \tag{4.47}$$

where

$$\Delta E_g^{BM} = \frac{\hbar^2}{2m_{vc}^*}(3\pi^2 N)^{2/3}$$

where m_{vc}^* is the reduced effective mass and is given by

$$\frac{1}{m_{vc}^*} = \frac{1}{m_v^*} + \frac{1}{m_c^*}. \tag{4.48}$$

Figures like 4.60 give the value of m_{vc}^* from the slope. Values of m_{vc}^* reported by different workers vary from 0.5 to 1.2 m_o [104–106] where m_o is the rest mass of the electron. The value of m_{opt}, which is assumed to be m_c^*, is usually determined using plasma frequency data. The value of effective mass is related to the plasma frequency by the relation (4.25).

$$\omega_p^2 = \frac{4\pi Ne^2}{\varepsilon_o \varepsilon_\infty m_c^*} \tag{4.25}$$

where ε_o is the permitivity of vacuum and ε_∞ is assumed to be the optical

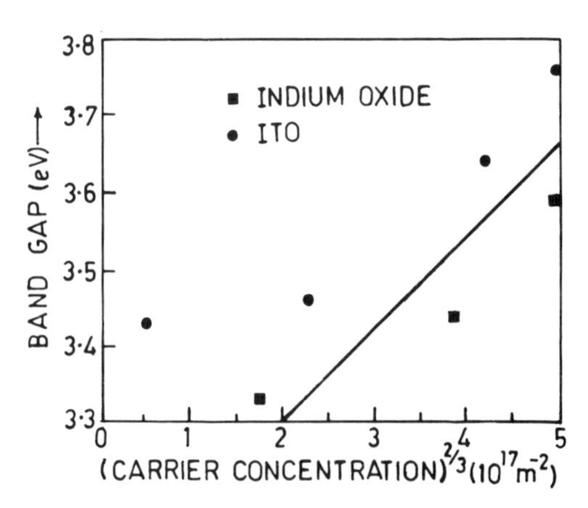

Figure 4.61 Band-gap as a function of carrier concentration ($N^{2/3}$) for amorphous indium oxide and ITO films. The solid line shows the expected curve for an effective mass of $0.3m_o$ (from [91]).

dielectric constant, ε_{opt}. The plasma frequency is usually determined experimentally from the frequency corresponding to the maximum of the plasma resonance absorption. The slopes of the plots of ω_p^2 versus N give the value of $\varepsilon_{opt} m^*_{opt}$. In order to determine m^*_{opt}, ε_{opt} must be obtained as a function of carrier concentration. ε_{opt} is determined either by the extrapolation of the real part of the dielectric constant ε_r to $\lambda = 0$ [105, 106] or by calculating the slope of ε_r versus λ^2 curves [107]. Substituting the values of ε_{opt} in $\varepsilon_{opt} m^*_{opt}$, one can find the variation of m^*_{opt} as a function of carrier concentration. Figure 4.62 shows such a variation for ITO films [103]. Various workers [103, 108, 109] have reported values of optical effective mass between 0.3 and 0.4 m_o.

Assuming a nominal value of $m^*_{vc} = 0.55\, m_o$ and $m^*_c = m_{opt} = 0.3\, m_o$, equation (4.48) gives the value of m^*_v, which is negative. This negative value indicates that the curvature of the valence band concerned with the direct transition is negative, which is not possible. This anomaly has been removed by taking into consideration the band-gap narrowing effects in ITO. As discussed earlier in equation (4.41) and also suggested by Hamberg and Granqvist [6], and Roth *et al* [110], in addition to band-gap widening, there are effects of band-gap narrowing because of electron–electron and electron–impurity scattering. These effects are quite significant in ITO and should be taken into consideration. These effects have been studied and their contributions have been calculated by Roth *et al* [110], Stern and Tolley [111] and Hamberg and Granqvist [6]. Band-gap narrowing has also been reported to be proportional to $N^{2/3}$ and if these effects are also taken into consideration, the slope of the line shown in figure 4.60 reduces significantly, typically from $0.49m_o$ to $0.22m_o$ [104]. After putting the value of $m^*_{vc} = 0.22m_o$, one obtains the value of m^*_v as positive, thus suggesting that

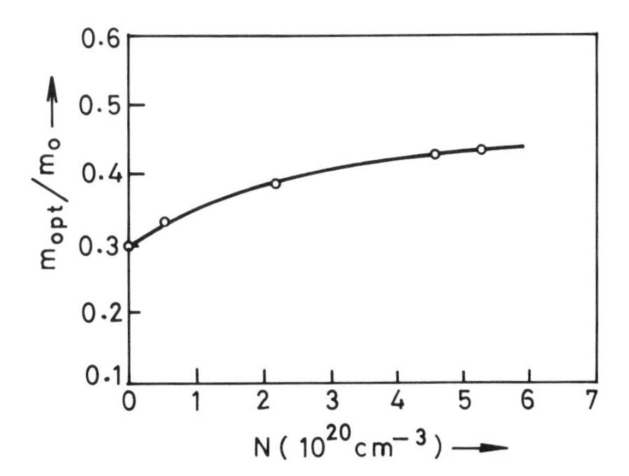

Figure 4.62 Optical effective mass as a function of carrier concentration for ITO films (from [103]).

one can assume a parabolic structure for ITO with conventional band structure, i.e. the conduction band curves upwards and the valence band curves downwards.

Optical constants n *and* k. A reduction in refractive index with increase in carrier concentration usually occurs in indium tin oxide films [103, 112]. The reduction in refractive index is expected in accordance with the relation

$$n^2 = \varepsilon_{\text{opt}} - \frac{4\pi Ne^2}{m^* \omega_o^2}. \tag{4.49}$$

The increase of tin content results in an increase in carrier concentration and thus reduces the value of n. Such an effect of tin concentration on refractive index in ITO films is shown in figure 4.63. Ohhata *et al* [103] observed that the value of n varies from 2 to about 1.70 when the carrier concentration increases from 0.2×10^{20} to 5×10^{20} cm^{-3}.

The typical variation of refractive index n and extinction coefficient k, as a function of wavelength for ITO films prepared by a spray deposition technique is shown in figure 4.64. It can be observed that the value of k is $\leqslant 0.03$ for the wavelength range 0.5–1.0 μm. However, there is a steep rise in the value of k for wavelengths less than 0.5 μm. This behaviour is similar to that observed in the case of undoped SnO_2 (figure 4.15) and fluorine-doped SnO_2 (figure 4.31) films. This increase in the value of k at the lower wavelength range is due to interband absorption. It can be observed from figures 4.15, 4.31 and 4.64 that the value of k starts increasing very slowly for wavelengths exceeding 1 μm, because of the onset of the contribution from

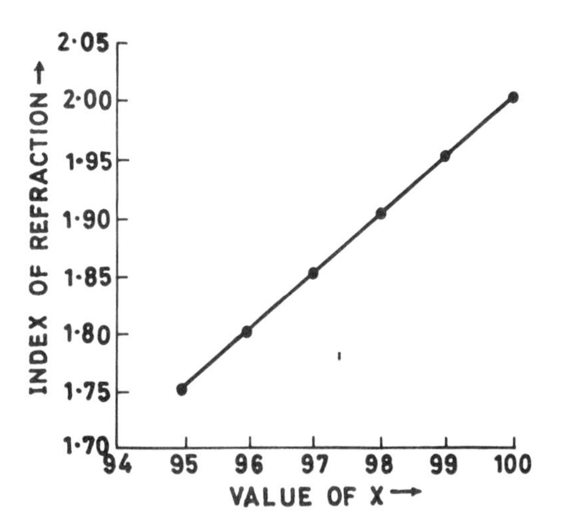

Figure 4.63 Variation of refractive index of $(In_2O_3)_x$:Sn films with x (from [112]).

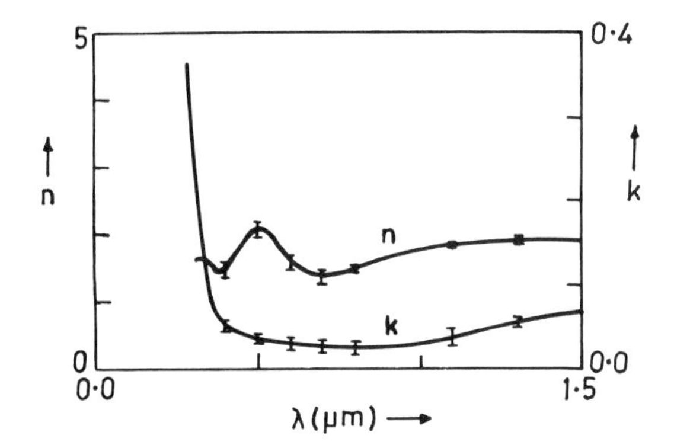

Figure 4.64 Variation of n and k with wavelength for ITO films prepared by a spray deposition technique (from [37]).

the plasma region. The value of n is almost constant and has a value of about 2 in this region, where dispersion effects due to interband absorption are not dominant. It can further be seen from these figures that the value of n increases with decrease of wavelength near the absorption edge. Theuwissen and Declerck [113] studied, in more detail, the variation of refractive index with wavelength in the dispersion region for ITO films. They explained their results on the basis of the general relation

$$n^2 = A + \frac{C}{\lambda^2 + B} \tag{4.50}$$

where A, B and C are constants. They also studied the effect of annealing ITO films in forming gas. Figure 4.65 shows the variation of refractive index

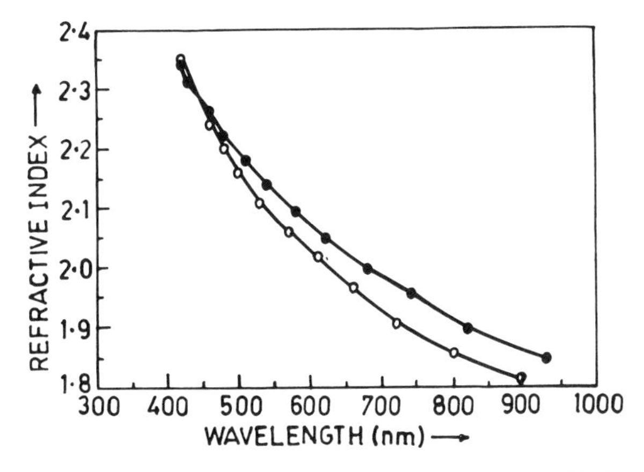

Figure 4.65 Variation of n with wavelength for sputtered ITO films (●) before and (○) after annealing in forming gas (from [113]).

with wavelength for sputtered ITO films and for films annealed in forming gas. The effect of annealing is to decrease the refractive index, because annealing in forming gas increases the free carrier concentration and conductivity which in turn decreases the refractive index [103]. Pommier *et al* [77] studied the effect of substrate temperature on the refractive index of spray-deposited ITO films. Their results are given in figure 4.66. It can be observed that refractive index is not an intrinsic property of ITO, but depends upon fabrication parameters. The decrease in refractive index can be understood on the basis of the electrical results. The resistivity of the films decreases as the substrate temperature increases, possibly because of the removal of the other suboxides of indium, which are highly resistive. The decrease in resistivity suggests that a large number of free carriers are generated, which are responsible for the reduction in the value of refractive index.

Reflection and transmission studies. In general, ITO films are superior even to fluorine-doped tin oxide films. Figure 4.67 compares the transmission and reflection characteristics of tin-doped indium oxide films and fluorine-doped tin oxide films. It can be observed that, although the transmission characteristics of these films in the visible region are comparable, the reflection properties in the infrared region are far better for ITO films than for doped SnO_2 films. Figure 4.68 shows typical transmission and reflection curves for spray-deposited indium oxide films and films doped with 4 mol % SnO_2. There are significant changes that occur as a result of doping. Although transmission is reduced typically only from 90% to 80%, the increase in reflectance is many-fold. In the infrared range, the reflectance increases from about 30% to about 90%. It is this property which makes ITO films superior even to doped SnO_2 films for infrared reflector applications. The plasma wavelength edge also shifts towards shorter wavelengths as a result of doping. Figure 4.69 shows the transmission and reflection characteristics

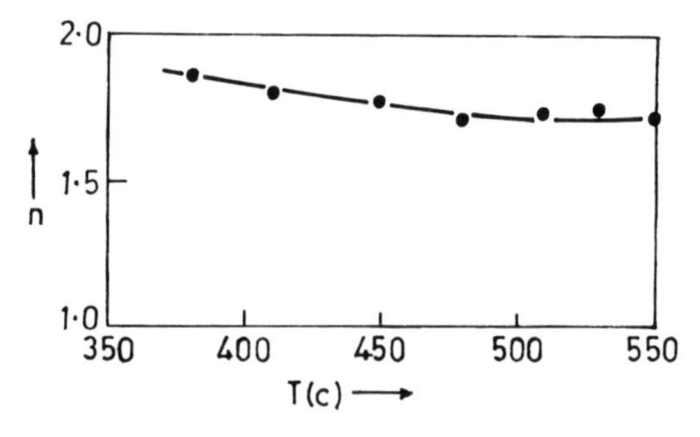

Figure 4.66 Variation of *n* with substrate temperature for sprayed ITO films (from [77]).

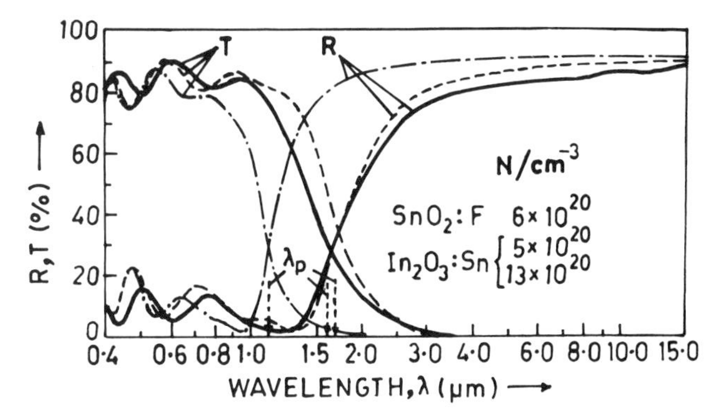

Figure 4.67 *R* and *T* spectral dependence for SnO_2:F (———) and ITO films (– – – –) 5×10^{20} and (–·–·–) 13×10^{20} cm^{-3} (from [83]).

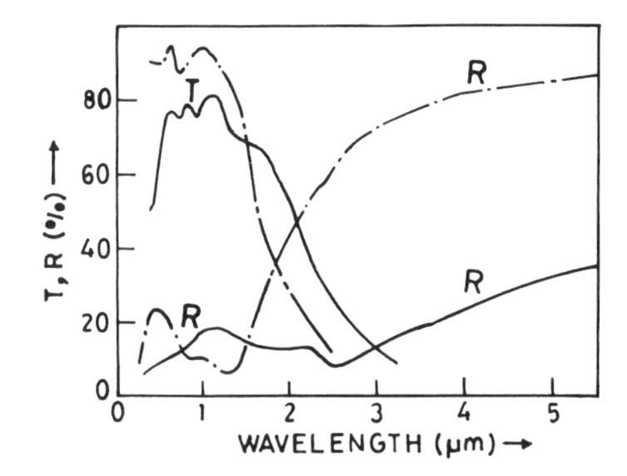

Figure 4.68 Typical variation of transmittance *T*, and reflectance *R* for sprayed ITO films (–·–·–) and for In_2O_3 (———) films (from [88]).

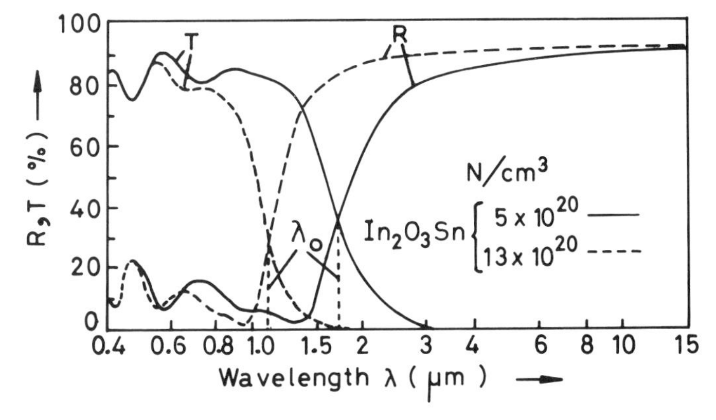

Figure 4.69 Spectral dependence of *R* and *T* on λ for ITO films of two carrier concentrations (———) 5×10^{20} and (– – – –) 13×10^{20} cm^{-3} (from [83]).

of spray-deposited ITO films for two different carrier concentrations: 5×10^{20} and $13 \times 10^{20}\,\text{cm}^{-3}$. It can be observed that transmission decreases with increase in carrier concentration, but the decrease is not very significant. Similarly the effect of increasing the carrier concentration is to improve the reflectance marginally in the infrared region (>5 μm). However, in the near-infrared region, the effect of increase in carrier concentration is significant. Typically at 1.5 μm, the reflectance increases from less than 10% to more than 80%. These results suggest that the effect of increasing the carrier concentration beyond $10^{20}\,\text{cm}^{-3}$ can shift the plasma edge to about 1 μm, which can be exploited in their use as efficient IR reflectors.

The reduction in sheet resistivity, which is due to increase in either thickness and/or carrier concentration, results in a decrease in transmission. Figure 4.70 shows the effect of increasing sheet resistivity on the transmission of ITO films. The results are for ion-plated films, and transmission is the average in the visible range. It can be observed that for low values of sheet resistivity, transmission is very low. Typically for films with sheet resistivity of $5\,\Omega\,\square^{-1}$, it is about 40%. Figure 4.71 shows the effect of increasing thickness on transmission in ITO films. It can be observed from this figure that increasing the thickness from 0.1 μm to about 2 μm reduces transmission from about 90% to only 70%. Comparison of figure 4.71 with figures 4.22 and 4.40 reveals that the thickness effect on transmission is less prominent in ITO films than in undoped and fluorine-doped SnO_2 films. It must be noted that the dependence of transmission on thickness is not exponential in any of these cases. This is contrary to the expected behaviour of transmission as per the relation

$$T = (1 - R)^2 \exp^{-\alpha t}. \tag{4.51}$$

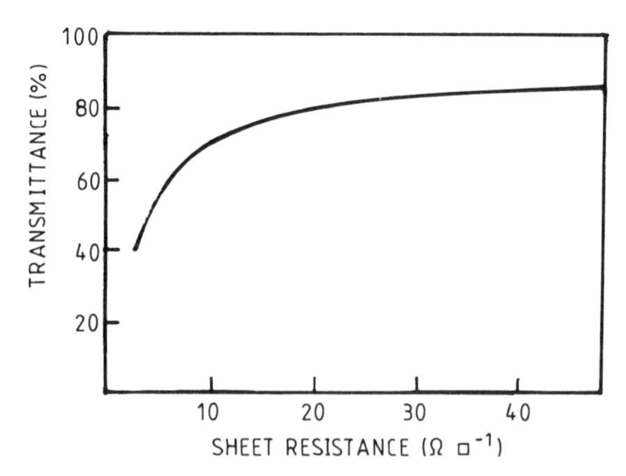

Figure 4.70 Variation of transmission with sheet resistance for ITO films deposited by an ion-plating technique (from [114]).

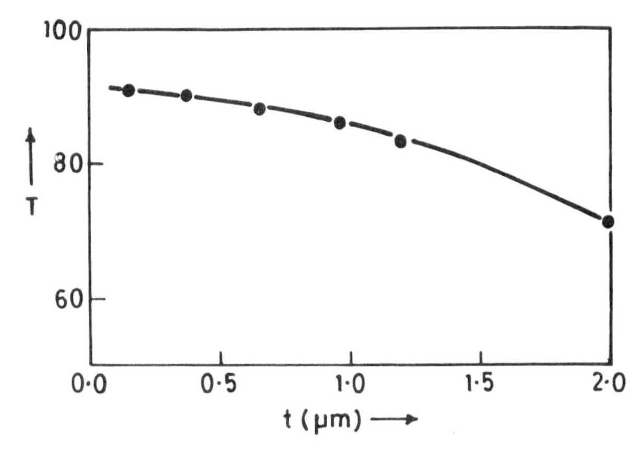

Figure 4.71 Variation of transmission with thickness for sprayed ITO films (from [77]).

The non-exponential variation of transmission with thickness suggests that the physical properties of these films change with thickness. Figure 4.72 shows the effect of an increase in the Sn/In ratio on the transmission of ITO films. It can be observed from this figure that the addition of Sn to indium oxide does not change transmission significantly up to 15% Sn addition. However, if tin is added beyond 20%, transmission decreases, mainly because of higher free carrier absorption. Geoffroy *et al* [26] also studied the effect of sheet resistivity on transmittance and reflectance of fluorine-doped ITO films. Typical spectral curves for films with sheet resistance varying from 31 to 145 $\Omega\,\square^{-1}$ are shown in figure 4.73. These results also indicate that the decrease in sheet resistance results in a decrease in transmittance and an increase in reflectance. These effects are more pronounced near and

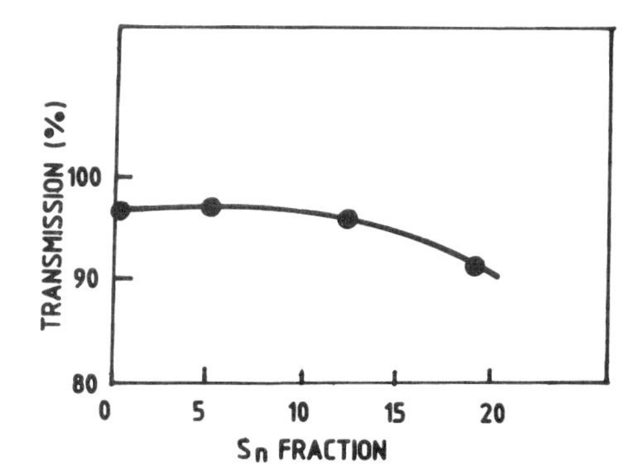

Figure 4.72 Variation of transmission with Sn concentration in ITO films prepared by a reactive evaporation technique (from [115]).

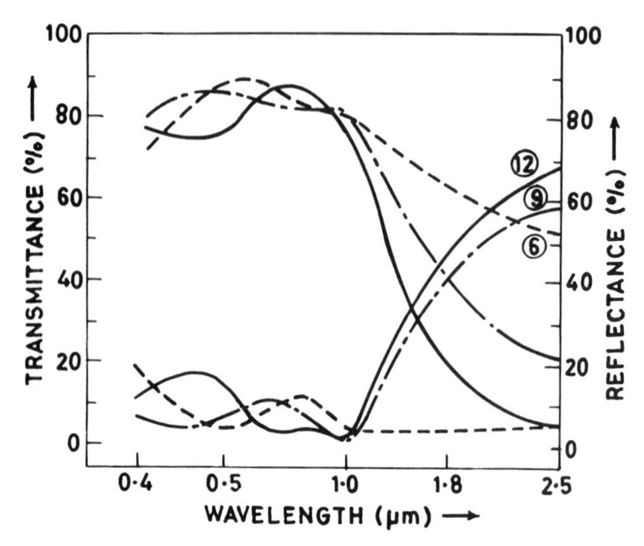

Figure 4.73 Variation of transmittance and reflectance with wavelength for fluorine-doped ITO films with different sheet resistivities. Sample (6) 145 $\Omega\,\square^{-1}$, (9) 36 $\Omega\,\square^{-1}$ and (12) 31 $\Omega\,\square^{-1}$ (from [26]).

beyond the plasma edge, and are in agreement with the studies of Frank *et al* [83] on tin-doped indium oxide films, as discussed earlier (figure 4.69). Studies by Yuanri *et al* [117] on transmission in the visible range near the band edge for ITO films revealed that transmission increased with increase in carrier concentration. This is understandable from the fact that an increase in carrier concentration results in shifting the absorption band-edge towards shorter wavelengths, as discussed earlier. This Burstein–Moss shift [57] reduces the contribution of interband absorption, thus increasing the overall transmission at shorter wavelengths.

Substrate temperature affects the optical properties of ITO films significantly, particularly when the films are grown in the presence of oxygen. Figure 4.74 shows the effect of substrate temperature on transmission and reflection for reactively evaporated films of In_2O_3 doped with Sn/In contents of 5–9 at%. Films evaporated at a substrate temperature of about 100 °C have a transmission nearly equal to 16%. Transmission becomes gradually higher as the substrate temperature rises, and finally a transmission of over 80% is observed for substrate temperatures greater than 400 °C. This suggests that oxidation begins at about 100 °C and is almost complete at a substrate temperature of 400 °C. Electrical measurements also suggest that the number of free carriers decreases with increase of substrate temperature. Typically, the carrier concentration decreases from 6×10^{21} to about 6×10^{20} cm^{-3} when the substrate temperature increases from 100 to 400 °C. This change in transmission through film oxidation is less abrupt than that

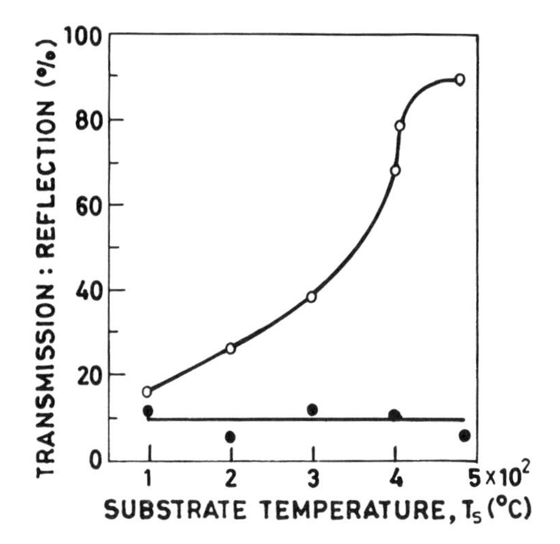

Figure 4.74 Effect of substrate temperature on transmission (○) and reflection (●) of ITO films (from [118]).

observed for undoped In_2O_3 films [119], since undoped films oxidize over a narrower range of temperatures (100–200 °C). This suggests that the higher oxidation temperature for the ITO films is due to the doped tin atoms, which disturb the oxidation of In_2O_3. Figure 4.74 also shows that the reflectivity is almost constant at ≈10% for films grown in the substrate temperature range 100–500 °C. Even at 100 °C, no metallic reflections are observed. This indicates that films formed at this temperature range consist not of metals but of imperfectly oxidized, non-stoichiometric indium tin oxide. A similar effect of substrate temperature on the optical properties of sprayed and reactively evaporated ITO films has also been reported by other workers [77, 115, 120]. ITO films have recently been fabricated at room temperature using a magnetron sputtering technique [121, 122]. Latz *et al* [121] used a hot pressed ITO target instead of indium tin target for the growth of these films. They observed that there is no appreciable increase in transmittance (85%) of these films even when the substrate temperature is increased to 300 °C. Harding and Window [122] produced ITO films at room temperature by reactively sputtering an indium tin target using a gas mixture of argon–oxygen–hydrogen. These deposition conditions are particularly suitable for producing ITO layers on substrates like polyester, which cannot be heated to higher temperatures during growth or post-deposition annealing. Figure 4.75 shows the variation of transmission with wavelength for ITO films grown on polyester at room temperature. It can be observed that the average transmittance of these films is greater than 80% over most of the visible range.

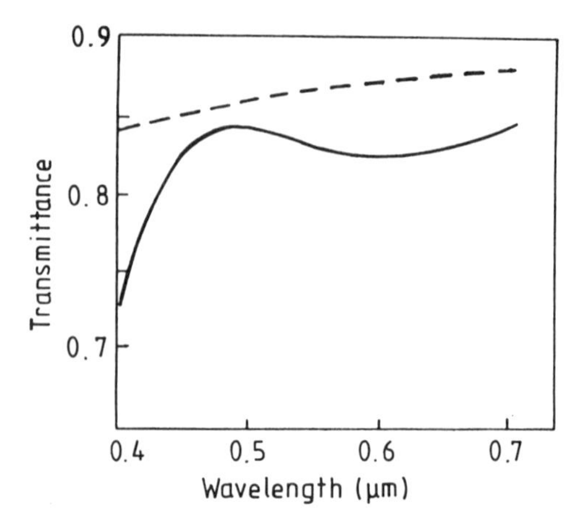

Figure 4.75 Transmittance versus wavelength for uncoated polyester (– – – –) and for ITO-coated polyester (———) (from [122]).

Oxygen partial pressure plays a crucial role in controlling the optical properties of ITO films. Figure 4.76 shows transmission and reflection as a function of oxygen pressure $P(O_2)$ for ion beam sputtered ITO films grown using a hot pressed ITO target. It can be observed that for very low values of oxygen pressure, transmission is very poor (10% at 1×10^{-5} Torr). With increasing oxygen pressure, transmission first increases rapidly (>80% at 3×10^{-5} Torr) and then becomes almost constant at 90%. The inset in the

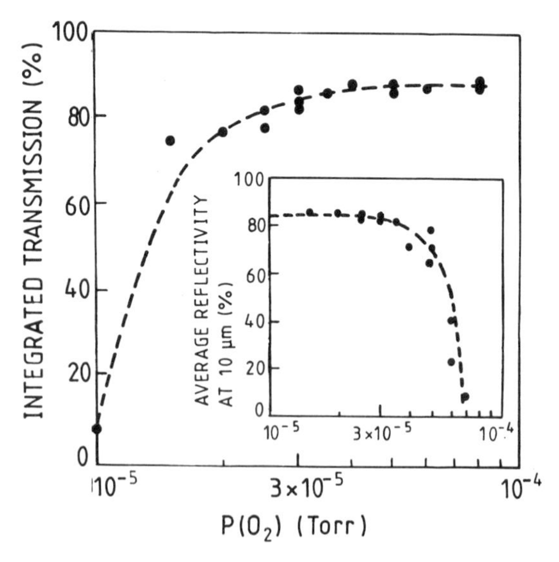

Figure 4.76 Transmission versus oxygen pressure for ITO films. Inset is reflection at ≈10 μm as a function of oxygen pressure (from [123]).

figure shows the dependence of IR reflectivity of the films, centred at a wavelength around 10 μm. For $P(O_2) < 3 \times 10^{-5}$ Torr, R remains constant at about 0.84, but at higher oxygen pressures R begins to decrease and drops sharply at 5×10^{-5} Torr. These results suggest that oxygen pressure control is effective for producing high transparency in the visible range and high reflectivity in the IR range only over a narrow range of pressures. The reduction in the value of R for higher oxygen pressures is due to the fact that the number of carriers and the mobility decreases for these films. Similar results have also been reported by Latz *et al* [121]. Theuwissen and Declerck [113] studied the effect of oxygen partial pressure on the transmission of ITO films (figure 4.77) grown by a DC reactive sputtering technique using an In–Sn target. It was observed that transmission remained at about 10% for oxygen partial pressures up to 1.2×10^{-3} Torr, after which transmission sharply increased to about 90% and remained in the range 80–90% for oxygen partial pressures up to 3×10^{-3} Torr. These results suggest that a minimum of 0.16 Pa (1.2×10^{-3} Torr) oxygen partial pressure is required for complete oxidation of the metallic components. For pressures less than 1.2×10^{-3} Torr, oxidation of the target is not possible because all the oxygen atoms are consumed in oxidation of the sputtered metal ions which are oxidized partially during their transport to the substrate. As such, the deposited films are also non-stoichiometric. It has also been observed that for very high values of oxygen partial pressures ($>1.7 \times 10^{-3}$ Torr), the target is completely oxidized and there is no further scope for the sputtered particles to be oxidized. In this range ($>1.7 \times 10^{-3}$ Torr), although the transmission is very good, the conductivity is relatively poor (figure 3.46). This low conductivity for films grown with oxygen partial pressure greater than 1.7×10^{-3} Torr makes these films unsuitable for infrared reflective

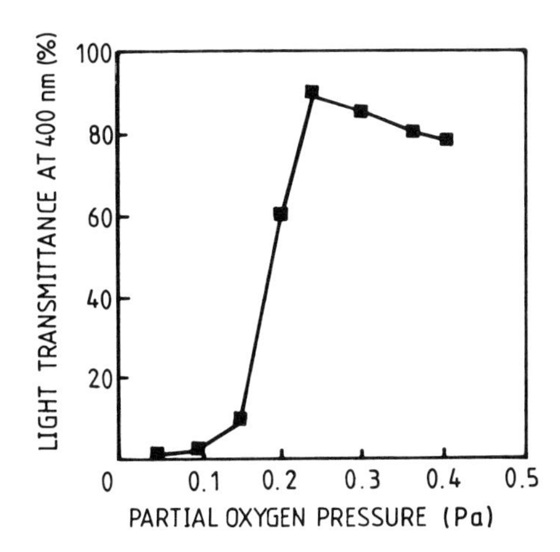

Figure 4.77 Variation of transmission with oxygen partial pressure (from [113]).

applications. Comparison of figures 4.76 and 4.77 reveals that, to obtain high quality ITO films, the oxygen partial pressure required in the case of an In–Sn target is more than that required for an ITO target. This is expected because in the case of In–Sn targets some oxygen is used to convert In–Sn to ITO.

Sputtering power is also found to affect the optical properties of sputtered ITO films. Figure 4.78 shows the dependence of transmission on sputtering power for a given oxygen partial pressure of 1.13×10^{-3} Torr. It can be observed that for power levels up to ≈900 W, transmission is about 80%, and decreases rapidly with further increase in sputter power; its value drops to about 20% for a sputtering power of 2 kW. This curve can be explained as follows: for $P < 900$ W, the target surface is non-stoichiometrically oxidized. The sputtered particles can be oxidized further during transport or during growth of the films. If the power is increased, the number of sputtered species also increases which means that a higher consumption of oxygen atoms is needed to oxidize the target surface and the sputtered atoms. With increasing power, the number of oxygen atoms available to oxidize the target decreases, the surface of the target becomes more metallic and the films become highly conducting, as can be seen from the electrical data (figure 3.45). The composition of the deposited films deviates more and more from stoichiometry. This deviation from stoichiometry results in a reduction in transmission. Ishibashi *et al* [124] observed that the sputtering voltage (discharge voltage) affects the optical properties of ITO films deposited by a DC magnetron sputtering technique. Transmission decreases with increase of sputtering voltage due to the formation of a black InO phase in these films.

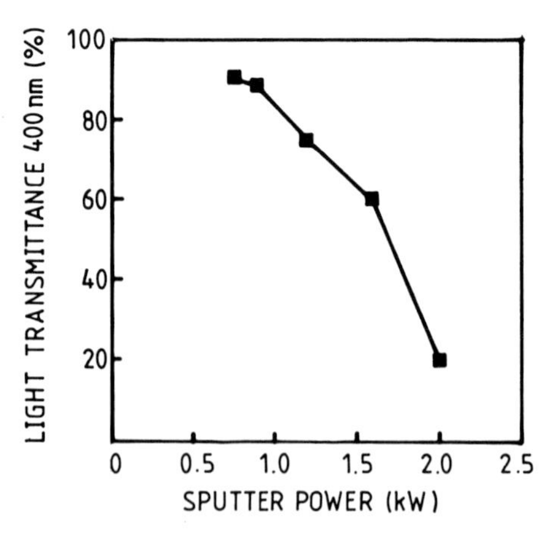

Figure 4.78 Transmission versus sputtering power for ITO films prepared by DC magnetron sputtering (from [113]).

Post-deposition annealing in various ambients influences the optical properties of ITO films by removing sub-oxides and by changing the carrier concentration. Weijtens and Vanloon [109] observed that the annealing of reactively DC magnetron sputtered ITO films in nitrogen resulted in an increase in carrier concentration from $2.7 \times 10^{20}\,\mathrm{cm}^{-3}$ for as-deposited films to $8.1 \times 10^{20}\,\mathrm{cm}^{-3}$ for nitrogen annealed films. The band-gap increased from 3.75 eV for as-deposited ITO films to 4.21 eV for nitrogen annealed films. Theuwissen and Declerck [113] studied the effect of annealing in forming gas on the transmission properties of ITO films and their results are shown in figure 4.79. There are two main effects of annealing on optical properties: an increase in transmission at the shorter wavelength range; and a reduction in optical transmission at the higher wavelength region ($\approx 1\,\mu$m). The results for lower wavelengths can be explained on the basis of Burstein–Moss shift [57]. The effect of annealing in forming gas is also to increase carrier concentration thereby shifting the absorption edge towards shorter wavelengths. In order to explain the reduction of transmission at longer wavelengths, one can make use of the following relation [125]:

$$\alpha = \frac{C\lambda^2 N}{\mu} \tag{4.52}$$

where C is a constant and μ is the mobility of the charge carriers. The increase in carrier concentration on annealing results in an increase in the value of α. However, as α depends on the square of wavelength λ, the wavelength dependence of α becomes stronger. That is why transmission

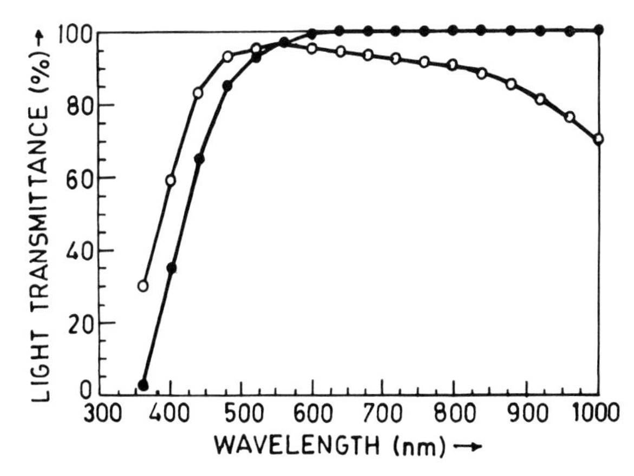

Figure 4.79 Transmission versus wavelength for unannealed (●) and annealed in forming gas (○) films of ITO (from [113]).

decreases more rapidly with increasing wavelength in the near-IR spectrum. Krokoszinski and Oesterlein [126] observed a monotonic increase in transmittance up to 90% with annealing temperature and time in air-annealed, electron beam evaporated ITO films. The rate of increase in transmission is greater for films annealed at higher temperatures. Further, for a given annealing temperature, the increase in transmission is faster for thinner films than for thicker films.

(b) *Other dopants*

In addition to tin, other dopants such as fluorine [89, 139, 140], titanium [141], antimony [141, 142] and lead [142] have been tried in In_2O_3 films. However, no systematic studies of these dopants have been reported. Maruyama and Fukui [89] and Avaritsiotis and Howson [139, 140] studied the optical properties of indium oxide films doped with fluorine. Maruyama and Fukui [89] reported that fluorine doping improved the transmission of CVD-grown In_2O_3 films marginally, without affecting the band-gap. Avaritsiotis and Howson [139, 140] observed that annealing of fluorine-doped In_2O_3 films grown by ion-plating sharpened the plasma edge and shifted the absorption edge towards shorter wavelengths. Their results are depicted in figure 4.80. The films were annealed in air at 450 °C for 1 h. The shift of the absorption edge on annealing can be described in terms of the Burstein effect [57] in relation to increased carrier concentration as the films become more degenerate as a result of the heat treatment. The sharpening of the plasma edge may be explained using Drude's model.

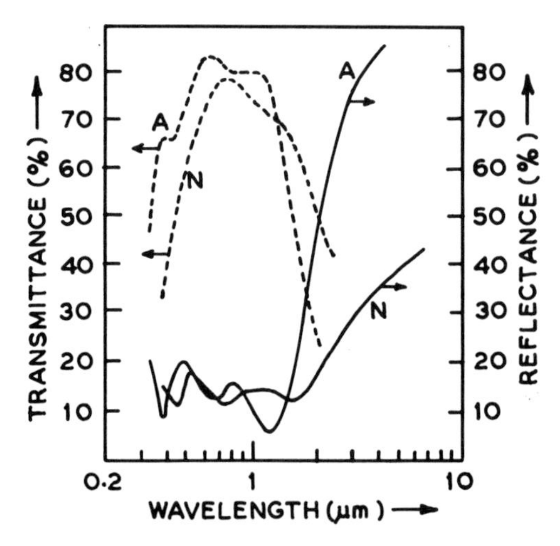

Figure 4.80 Reflectance (———) and transmittance (– – – –) for as-deposited (curve N) and air-annealed (curve A) fluorine-doped In_2O_3 films (from [140]).

4.6 Optical Properties of Cadmium Stannate Films

Although the most commonly used transparent conducting materials are doped tin oxide and indium tin oxide, cadmium stannate is an alternative material, since it has comparable properties. The potential utility of cadmium stannate for transparent electrodes was first indicated by Nozik [143], who measured electrical and optical properties of amorphous cadmium stannate (CTO) films. Following this, extensive work has been reported [114, 135, 143–159] on the study of the optical properties of CTO films.

(a) *Band-gap studies*
Figure 4.81 shows the $(\alpha h\nu)^{1/2}$ versus $h\nu$ plot for as-grown CTO films deposited using a DC reactive sputtering technique from Cd–Sn alloy in the presence of $Ar–O_2$ mixture. The absorption edge of these films is estimated to be 2.4 eV. Figure 4.81 also shows an $(\alpha h\nu)^{1/2}$ versus $h\nu$ plot for annealed CTO films. The annealing was done in an argon atmosphere at 300 °C for 6 h. It is observed that annealing shifts the absorption edge from 2.4 eV to 2.7 eV. The shift in the absorption edge towards shorter wavelengths can be understood on the basis of Burstein–Moss shift [57] as described earlier. An idea of the increase in carrier concentration can be gleaned from the fact that the resistivity of the annealed films was 8×10^{-4} compared to 4×10^{-2} Ω cm for unannealed films. Similar results have also been reported by Leja *et al* [145] for CTO films grown by reactive DC sputtering of Cd–Sn alloy. The

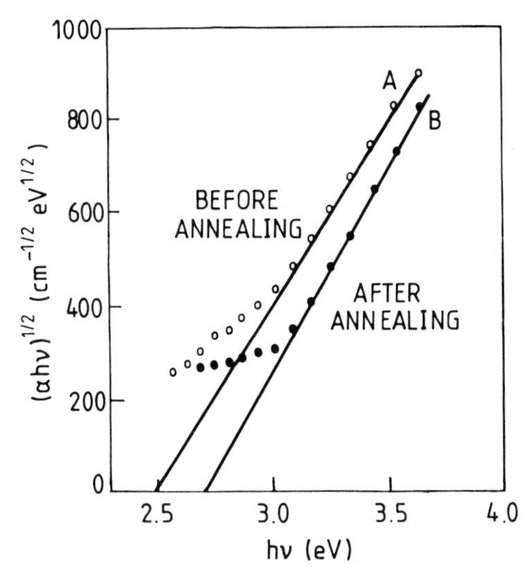

Figure 4.81 $(\alpha h\nu)^{1/2}$ versus $h\nu$ plots for CTO films; (○) before and (●) after annealing (from [144]).

maximum value of direct band-gap observed by these workers was 3.04 eV for films having carrier concentrations 3×10^{20} cm^{-3}. The value of fundamental edge (3.04 eV) observed by Leja *et al* [145] was slightly higher than the 2.7 eV, observed by Miyata *et al* [144]. However, the carrier concentration in the CTO films in the latter case [144] was 2.0×10^{21} cm^{-3}, which is significantly higher than the 3×10^{20} cm^{-3} reported for the films grown by Leja *et al* [145]. These results suggest that there is a significant contribution of electron–electron scattering and impurity scattering in the band-gap narrowing of CTO films. The presence of defect scattering is well established [146] on the basis of electrical measurements.

(b) *Effective mass*

Like all other transparent semiconducting films, CTO films have high reflectivity in the infrared frequency range. The reflectivity minimum in the case of CTO films is observed [145] in the energy range 0.55–0.72 eV, and depends on the carrier concentration of the samples tested. The highest value of effective mass estimated using reflectivity data obtained by Leja *et al* [145] is $0.25m_o$. The effective mass as reported by Miyata *et al* [147] shows a variation from $0.11m_o$ to $0.22m_o$ when carrier concentration increases from 4.7×10^{19} to 2.8×10^{20} cm^{-3}.

(c) *Transmission and reflection properties*

Typical transmission and reflection spectra of DC reactively sputtered CTO films are depicted in figure 4.82. The films are transparent in the range 0.5–2.5 eV. Transparency is reduced beyond 2.5 eV because of interband absorption. On the other hand, in the range less than 0.5 eV, the contribution of plasma reflectivity is quite evident. Films have an average transmission greater than 80% throughout the visible range. CTO films deposited by planar magnetron sputtering onto substrates kept at room temperature have also been reported [114] to have comparable properties. Transmission in the visible range is observed to be greater than 80% beyond

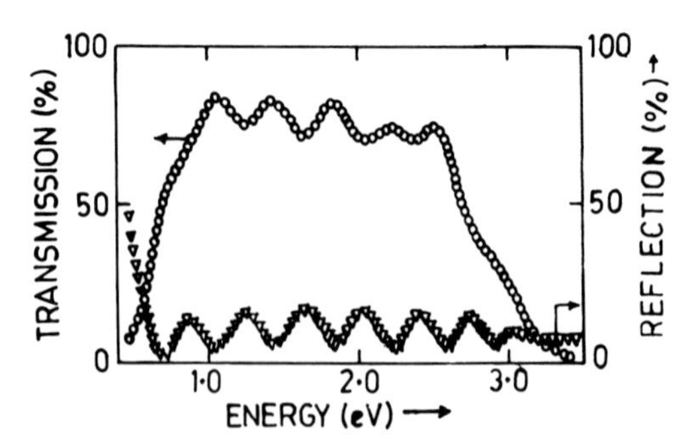

Figure 4.82 Typical T and R curves for CTO films (from [145]).

0.5 μm and the reflectivity minimum is observed at 2.5 μm. The plasma edge observed in these CTO films is at longer wavelengths than in ITO and doped SnO_2 films mainly because the carrier concentration in these films is less than that in doped SnO_2 and ITO films.

The effect of sheet resistivity on the transmission of CTO films is shown in figure 4.83 for cadmium stannate films RF sputtered in a pure oxygen atmosphere. In order to increase free carrier concentration and thus sheet conductivity, films were annealed in an Ar/CdS atmosphere at temperatures ranging from 600–700 °C. Typically, annealing for 10 minutes increased the carrier concentration from 10^{19} to 10^{21} cm^{-3}. It can be observed from figure 4.83 that the absorption edge shifts towards longer wavelengths with the decrease in sheet resistivity. It can also be observed that the average transmission in the visible range decreases from more than 90% for 3.4 $\Omega\,\square^{-1}$ films to less than 80% for 1.2 $\Omega\,\square^{-1}$ films. This decrease in transmission is due to the significant contribution of free carrier absorption. Similar results have been reported by Miyata and Miyake [149, 150] for DC reactively sputtered CTO films. The average transmission of films in the visible region ranged from 70 to 95% depending on the sheet resistivity of the films. Typically it was 90% for films with sheet resistivity of 100 $\Omega\,\square^{-1}$. Figure 4.84 shows the effect of increasing sheet resistivity on the optical transmission of CTO films prepared by an ion-plating technique. It can be observed that transmission is almost independent of sheet resistivity. On comparing the results shown in figure 4.84 with those shown in figure 4.70, it can be observed that the transmission properties of ITO films are more sensitive to sheet resistivity than are those of CTO films. Less dependence of transmission on sheet resistivity down to 5 $\Omega\,\square^{-1}$ renders CTO films superior to ITO films in the visible range.

The transmission properties of spray-deposited CTO films depend on the

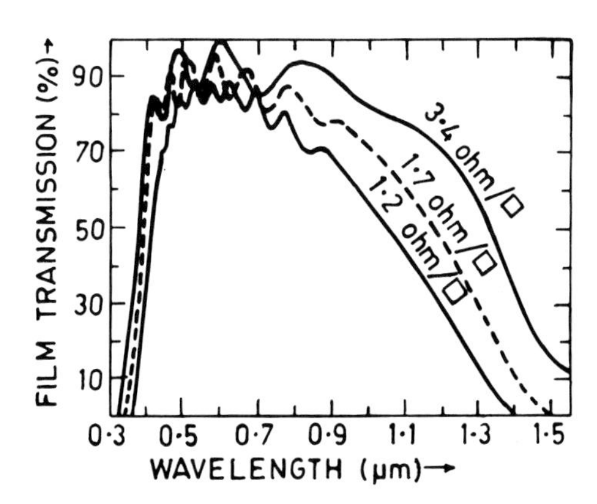

Figure 4.83 Transmission of CTO films with different sheet resistivity (from [148]).

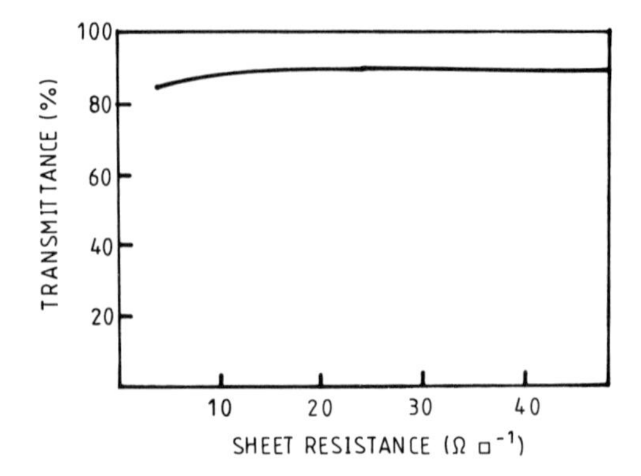

Figure 4.84 Effect of sheet resistivity on transmission of CTO films (from [114]).

flow rates of $SnCl_4$ and $CdCl_2$ solutions. Figure 4.85 shows transmission versus wavelength curves for various flow rates. The effect of increasing the flow rate is essentially to increase the thickness. Increase in thickness, in turn, reduces sheet resistivity. Figure 4.85 also shows that the effect of increasing thickness or decreasing sheet resistivity on the transmission properties of CTO films is not very significant. This result is in agreement with those reported by Howson *et al* [114] as discussed earlier.

Unlike flow rate, substrate temperature has a significant effect on the transmission properties of CTO films. Figure 4.86 shows transmission curves

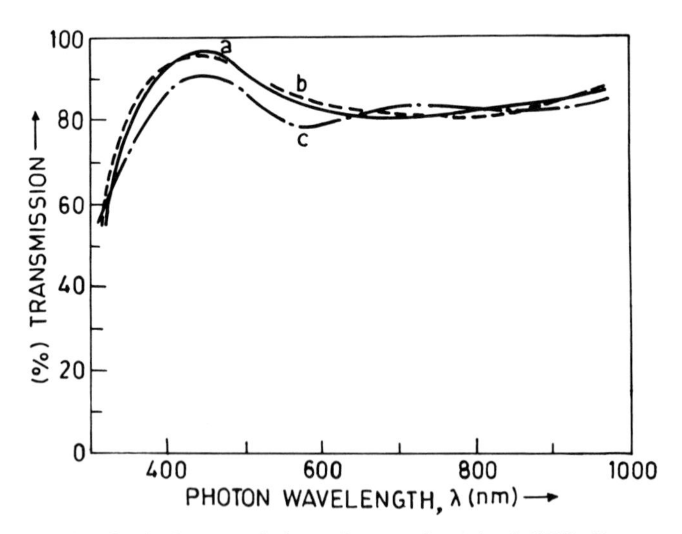

Figure 4.85 Optical transmission of spray-deposited CTO films as a function of wavelength for different solution flow rates: (a) 1.5 cc min^{-1}; (b) 1.8 cc min^{-1}; (c) 2.1 cc min^{-1} (from [151]).

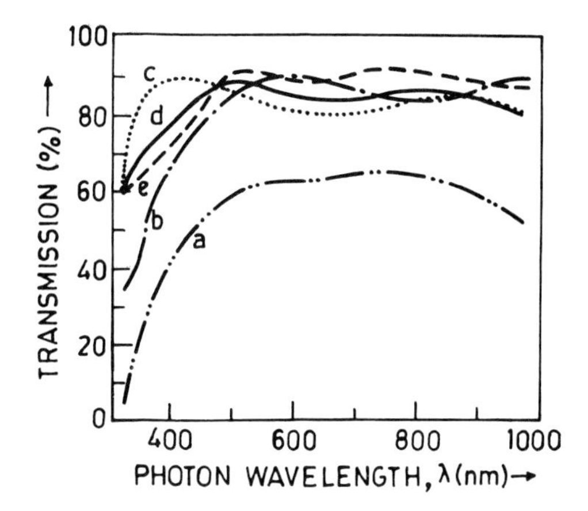

Figure 4.86 Optical transmission curves of spray-deposited CTO films deposited at different substrate temperatures: (a) 370 °C; (b) 390 °C; (c) 410 °C; (d) 430 °C; (e) 450 °C (from [151]).

for films grown at substrate temperatures of 370, 390, 410, 430 and 450 °C. Transmission in CTO films increases as the substrate temperature increases to 430 °C and then remains constant. The increase in substrate temperature usually results in higher conductivity and higher carrier concentration, however, the films become thinner. As discussed earlier, in CTO films the effect of thickness and sheet resistivity is not very significant. The observed reduced transmission (figure 4.86) in films grown at lower substrate temperature suggests the presence of sub-oxides and other surface inhomogeneities. Scanning electron microscopy (SEM) studies of these samples shows the existence of clusters in films grown at lower substrate temperatures. With increasing substrate temperature the clusters disappear and the surface becomes almost smooth. A similar improvement in transmission when substrate temperature is increased has also been observed by Schiller *et al* [152] for magnetron-sputtered films. It can be further observed from figure 4.86 that there is a shift in the direct absorption edge as a result of the increase in substrate temperature. This may be attributed to an increase in carrier concentration which results in band-gap widening.

An important parameter in sputtered films, which changes their properties appreciably, is the oxygen partial pressure. Figure 4.87 shows the transmission properties of CTO films sputtered from a $CdO–SnO_2$ target, in (A) pure Ar, (B) Ar–2% O_2 and (C) Ar–13% O_2. The average transmission of all these films is more than 85% in the wavelength range 0.5–0.7 μm. However, the transmission of films deposited in a pure Ar atmosphere drops rapidly in the near-infrared region. The reason for this behaviour is the presence of higher carrier density, $\sim 5 \times 10^{20}$ cm^{-3}, in films produced in

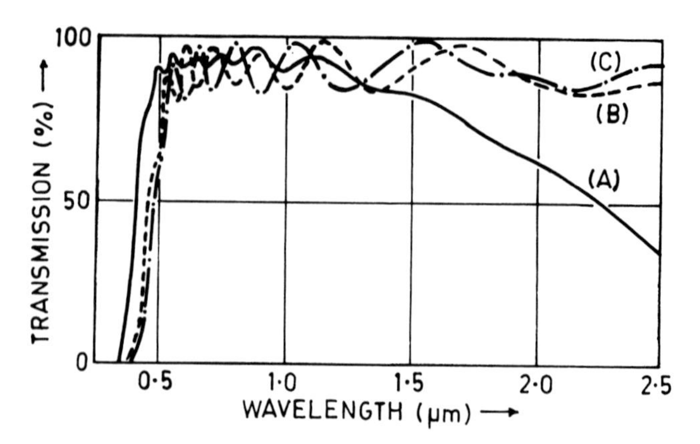

Figure 4.87 Transmission of sputtered CTO films in (A) pure argon; (B) Ar–25% O_2; (C) Ar–13% O_2 (from [153]).

pure argon, compared to 8.8×10^{18} cm^{-3} in films grown in Ar–13% O_2 atmosphere. Optical band-gap studies of these films suggest that the absorption edge shifts towards shorter wavelengths with decreasing oxygen concentration. Typically, the band-gap of films formed in a pure Ar was estimated to be 2.71 eV and those of films formed in 2% O_2 and Ar–13% O_2 were determined to be 2.45 eV and 2.39 eV, respectively. The required partial pressure is greater when films are sputtered using metallic targets, namely Cd–Sn instead of oxide targets. It has been reported [154, 155] that tin and cadmium form two stable compounds with oxygen, namely $CdSnO_3$ and Cd_2SnO_4. During the sputtering process Cd-Sn alloys with various oxygen concentrations in the mixture, and secondary phases such as CdO, SnO_2, SnO and the metallic phase of cadmium and tin, especially for low oxygen concentrations, can be observed. For low oxygen concentrations in the reactive mixture, sputtering assumes a metallic nature, i.e. the sputtering rate of the target is greater than the oxidation rate. With increasing oxygen concentration in the mixture the oxidation rate increases and the sputtering process assumes a reactive nature. The critical value of oxygen concentration in the Ar–O_2 mixture for metallic targets is usually greater than 20%.

In general, the effect of heat treatment in a reducing atmosphere at high temperatures is to increase the conductivity, which may influence the optical properties because of a change of carrier concentration. Figure 4.88 shows the effect of annealing on transmission for DC reactively sputtered CTO films. The films were sputtered in pure oxygen, but annealed in a reducing atmosphere at 500 °C. Annealing the films results in significant improvement in the transmission and shifts the absorption edge towards shorter wavelengths. The shift in the absorption edge can be understood on the basis of Burstein–Moss shift [57] caused by the increase in carrier concentration.

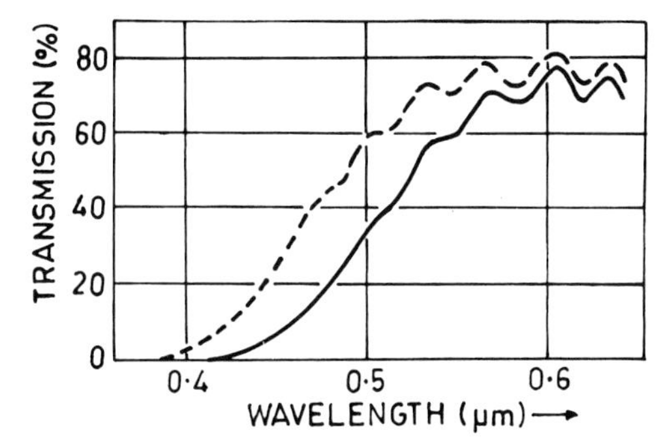

Figure 4.88 Optical transmission for sputtered CTO films: (———) as sputtered, (– – – –) heat treated in a reducing atmosphere (from [156]).

However, the increase in overall transmission suggests that the properties of these films also improve as a result of annealing at a high temperature of the order of 500 °C. Optical transmission values of the order of 95% have been reported [157] for CTO films prepared by RF sputtering froma Cd_2SnO_4 powder target combined with post-deposition heat treatment in a reducing atmosphere.

4.7 Optical Properties of Zinc Oxide Films

Recently, thin films of zinc oxide (ZnO) have generated immense interest as a transparent semiconducting material because coatings of this material are relatively inexpensive and have a sharp UV cut-off (figure 4.89]. Although

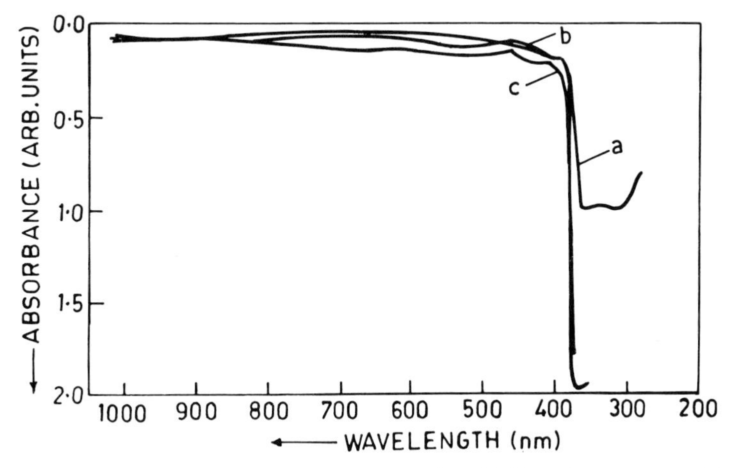

Figure 4.89 Optical absorption spectra of ZnO films deposited at (a) 500 °C, (b) 550 °C, and (c) 600 °C (from [160]).

at present, ITO is the most commonly used material for optoelectronic devices due to its excellent properties, its main disadvantage is that the indium source material is expensive. Thin films of ZnO are being actively investigated as an alternative candidate for applications as transparent, conductive and IR-reflective coatings. Films are, in general, transparent in the wavelength range 0.3–2.5 μm, and the plasma edge lies in the range 2–4 μm, depending upon the carrier concentration in the films. The optical properties of undoped and doped ZnO films have been studied by many workers [110, 160–163, 167, 169, 171–189]. Sarkar *et al* [161] studied the effect of carrier concentration on the optical band-gap of ZnO films. Carrier concentration was changed by indium doping and subsequent annealing in various ambients. Figure 4.90 shows $(\alpha h\nu)^2$ versus $h\nu$ plots for these films. Samples with different carrier density on which the data has been taken and shown in figure 4.75 are

(i) ZIO–A: as-deposited ZnO:In film ($N = 2.7 \times 10^{19}$ cm^{-3});
(ii) ZIO–A–O: sample (i) annealed in air ($N = 4.7 \times 10^{18}$ cm^{-3});
(iii) ZIO–A–O–H: sample (ii) annealed in hydrogen for 30 min ($N = 4.0 \times 10^{19}$ cm^{-3});
(iv) ZIO–A–O–H–V: sample (iii) annealed in vacuum for 30 min ($N = 4.3 \times 10^{19}$ cm^{-3});

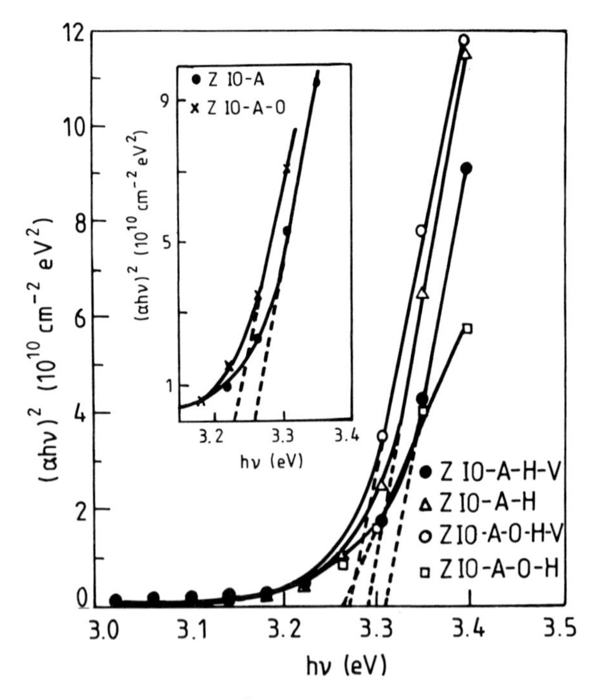

Figure 4.90 Plots of $(\alpha h\nu)^2$ versus $h\nu$ for ZnO films (from [161]).

(v) ZIO–A–H: sample (i) annealed in hydrogen for 30 min ($N = 7.2 \times 10^{19}\,\text{cm}^{-3}$).

The values of band-gap estimated from the plots of figure 4.90 vary from 3.232 to 3.290 eV. It may be noted that as carrier density increases, the band edge shifts towards lower wavelengths. This variation of band-gap with carrier concentration is shown in figure 4.91. It can be observed from this figure that the band-gap varies linearly with $N^{2/3}$, indicating that the Burstein–Moss model [57] can explain the results on ZnO films. Similar results have also been reported by Caporaletti [162] and Roth *et al* [110, 163]. However, these workers observed a drop in the value of band-gap shift at $N = 3 \times 10^{19}\,\text{cm}^{-3}$. According to Roth *et al* [110, 163], for carrier concentrations greater than this, there are two competing phenomena which affect the absorption edge in heavily doped semiconductors. The first phenomenon is Burstein–Moss band-filling, which shifts the band edge positively. The second phenomenon, which affects the optical absorption edge with increasing donor density, is due to a change in the nature and strength of the interaction potentials between donors and host materials. This latter effect increases the tailing of the absorption edge and hence reduces the band-gap. The onset of gap shrinkage can be related to a semiconductor–metal transition accompanied by a merging of donor and conduction bands. The dependence of gap shrinkage on carrier concentration has been discussed by many workers [69, 71, 164]. For very high carrier concentrations where $N \gg N_c$ (the carrier concentration at which a semiconductor–metal transition is expected to occur), the gap shrinkage ΔE_g is proportional to $N^{1/3}$. Values of ΔE_g have been estimated by subtracting the band-gap for the insulating ZnO and the expected Burstein–Moss

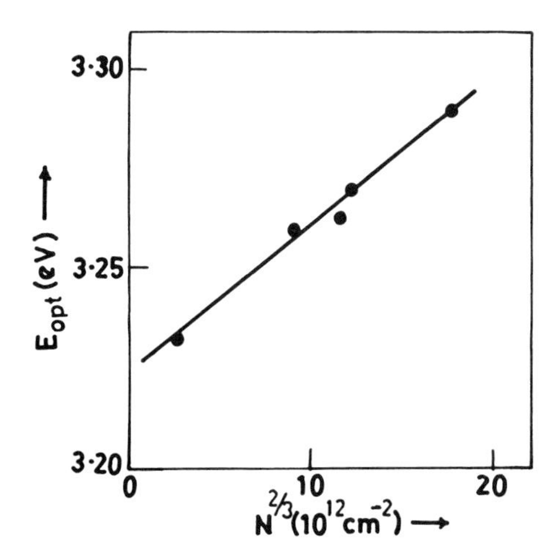

Figure 4.91 Plot of band-gap (E_{opt}) versus $N^{2/3}$ (from [161]).

shift from the experimentally observed band-gap. The estimated values of ΔE_g at 300 K for ZnO films grown using different techniques (organometallic CVD (OCVD), reactive RF magnetron sputtering (RRFMS), bias sputtering (BS) and reactive evaporation (RE)) are shown in figure 4.92. The dashed line in the figure shows the curve in accordance with the relation

$$\Delta E_g = AN^{1/3} \tag{4.53}$$

where A is a constant. Although this relation has been successfully used for n-Ge [165] and p-GaAs [166], for ZnO films the agreement does not seem to be very good.

(a) *Optical constants* n *and* k

Figure 4.93 shows the variation of n and k for In-doped ZnO films prepared using a spray pyrolysis technique. The value of refractive index in the visible range is ≈2, which is very close to the value reported for bulk material [168]. It can be observed that the value of k increases for wavelengths less than 0.5 μm as well as for wavelengths greater than 1.0 μm. The sharp rise in the value of k near the band edge (<0.5 μm) is due to interband absorption. On the other hand, the increase in the value of k behond 1 μm is due to free carrier absorption. A similar type of behaviour has been observed by Sarkar *et al* [161] for magnetron-sputtered In-doped ZnO films. It can be further observed from figure 4.93 that the value of n reaches a minimum (≈1) near 2.3 μm for 3% In-doped ZnO films. This behaviour of n with λ is a

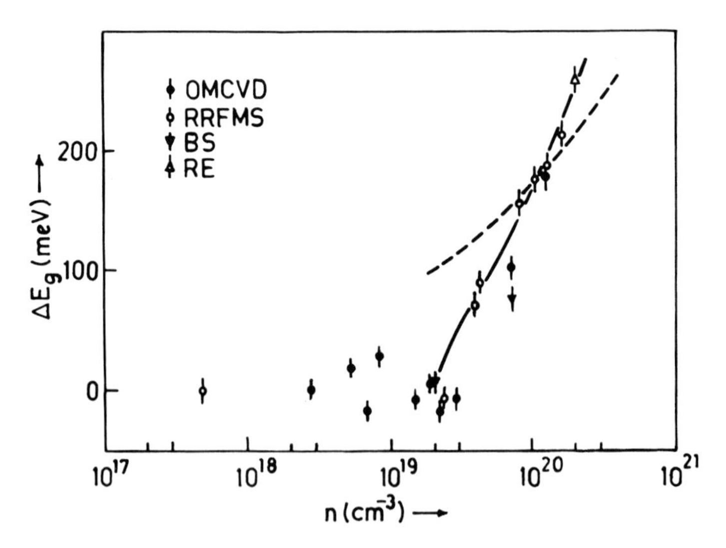

Figure 4.92 Band-gap shrinkage in ZnO films prepared by different techniques as a function of carrier concentration (from [110]).

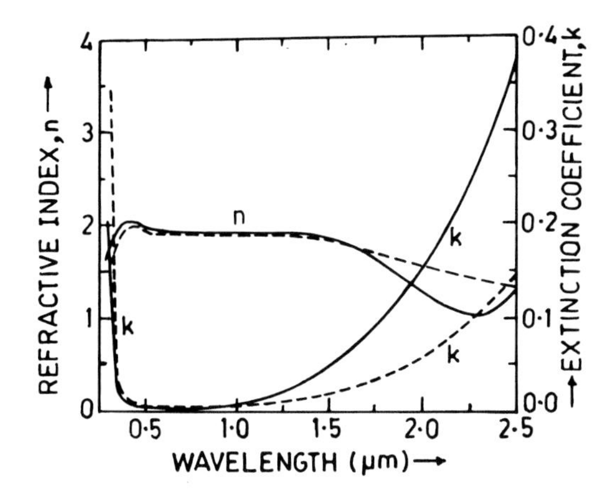

Figure 4.93 Variation of n and k with wavelength for (– – – –) 1% and (———) 3% In-doped ZnO films (from [167]).

characteristic feature near the plasma resonance edge. It has also been observed [167] that the effect of increasing carrier concentration is to shift the plasma edge towards shorter wavelengths and to increase the value of extinction coefficient in the infrared region.

(b) *Transmission and reflection studies*

Both oxygen partial pressure and substrate temperature play a crucial role [169] in controlling the optical as well as electrical properties of these films (see figure 3.61). For oxygen concentration >2%, the films are highly transparent, but non-conductive. On the other hand, for oxygen concentration <2%, the films are conducting but opaque up to a substrate temperature ~473 K; thereafter the films are conducting as well as transparent up to a substrate temperature of 523 K. For substrate temperatures higher than 523 K, the films are transparent and non-conducting irrespective of oxygen concentration. The transmission of films grown under optimum conditions is ~90%.

Figure 4.94 shows the spectral dependence of transmission and reflection for undoped and In-doped ZnO films. All the films exhibit high transmittance (>80%) in the visible range. Transmission, however, falls very sharply in the UV region due to the onset of fundamental absorption. The absorption edge is also observed to shift slightly towards shorter wavelengths. This observed small shift instead of the expected large Burstein–Moss shift [57] is further indication of band-gap narrowing in ZnO films. It can be further observed from figure 4.94 that the reflectance depends strongly on the indium concentration; the higher the indium concentration, the greater is

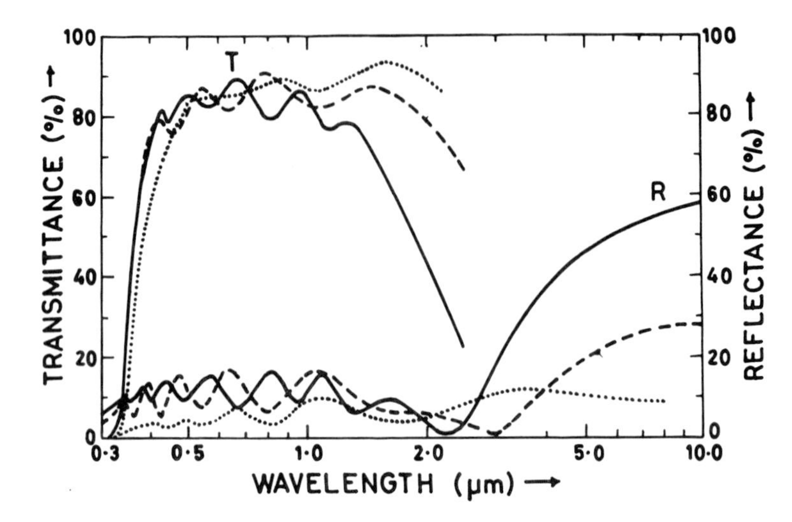

Figure 4.94 Spectral dependence of transmittance and reflectance for undoped (······), 1% (– – – –) and 3% (———) In-doped ZnO films (from [167]).

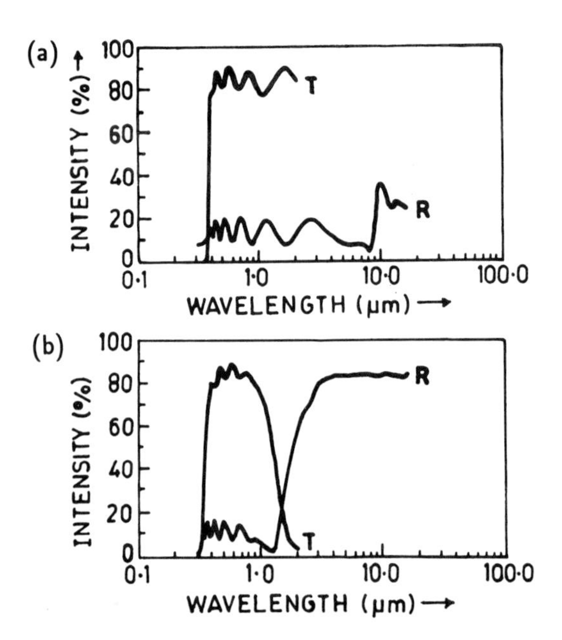

Figure 4.95 Transmittance and reflectance of (a) undoped and (b) Al-doped ZnO films (from [28]).

the value of reflectance. Typically, the value of reflectance is about 60% for 3% In-doped ZnO films. The optical transmission is greater than 80% in ZnO films, which is comparable with SnO_2, ITO and CTO films, but the reflectance is significantly lower in indium-doped ZnO than in other materials. This is mainly due to the fact that the maximum carrier concentration attainable in these films is $\sim 10^{20}\ cm^{-3}$, which is an order lower than that which can be obtained in doped SnO_2 and ITO films. Recently, aluminium-doped zinc oxide (ZnO:Al) films have been investigated as a transparent conducting material [28, 170–176]. Figure 4.95 shows the transmission and reflection response for undoped and aluminium-doped ZnO films prepared by a CVD technique at 367 °C. The undoped film is transparent throughout the visible and near-infrared range and has a plasma wavelength at $\approx 8.2\ \mu m$. The doped films are still transparent in the visible range but the plasma edge shifts to $1.28\ \mu m$. The optical properties of doped samples are dominated by high free carrier concentration in the films. The maximum infrared reflectivity observed in doped films is 82%. An increase in substrate temperature to 420 °C is reported to further improve the infrared reflectivity to 90% [28].

Figure 4.96 shows the transmission spectra for undoped and Al-doped sputtered ZnO films. Aluminium has been incorporated in the films in the form of Al_2O_3. It can be observed from the figure that there is a considerable shift in the absorption edge in films with higher Al content. Electrical measurements on these films show that doping the films with Al increases the concentration up to $10^{21}\ cm^{-3}$, which is an order higher than the In-doped ZnO films. Moreover, the mobility observed in Al–ZnO films is also significantly higher. Both the higher value of carrier concentration and the higher value of mobility make Al–ZnO films superior to In–ZnO films for IR reflector applications.

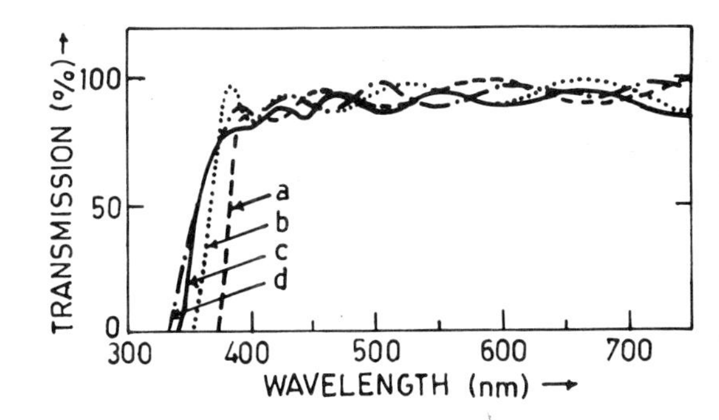

Figure 4.96 Optical transmission spectra for undoped, 0.5, 1.0 and 2 wt% Al-doped ZnO films (a, b, c, d) (from [170]).

4.8 Conclusions

The optical properties of these films (transmission, reflection, band-gap, refractive index) are strongly dependent on dopant concentration and growth conditions. Table 4.2 summarizes the optical properties of some prominent oxide films. Semiconducting transparent oxide films exhibit the following important features:

(i) There are both direct and indirect allowed transitions in these materials. The values of these band-gaps strongly depend on the carrier concentration, which in turn depends on the deposition conditions. The absorption edge shifts towards higher energy with an increase in carrier concentration. In general, the shift in band edge can be explained on the basis of the Burstein–Moss model. However, for films with relatively higher carrier concentration ($\approx 10^{20}\ cm^{-3}$), band-gap narrowing due to electron–electron scattering and electron–impurity scattering also plays an important role.

(ii) The substrate temperature maintained during growth of the films influences the optical properties significantly. This is due to the fact that substrate temperature decides the extent of oxidation and hence free carrier concentration.

(iii) ITO and fluorine-doped tin oxide films are widely recommended for practical device applications. ITO films are usually preferred for high grade applications, i.e. opto-electronics, while fluorine-doped tin oxide films are normally used to coat large surfaces, e.g. heat mirrors, because of the lower cost of the tin oxide. Transmission characteristics of doped SnO_2 films are comparable to those of ITO films in the visible region. However, the reflection properties of ITO films are superior to those of doped SnO_2 films in the infrared region. Moreover, transmission properties of ITO films are less sensitive to film thickness. This means that thicker films of ITO with higher sheet conductivities can be used without compromising the optical quality of these films. The better features of ITO films are mainly due to the fact that these films can be prepared with higher carrier concentrations and mobilities.

(iv) Coatings of CTO and ZnO have also shown promising optical properties. Although these films have transparencies comparable to SnO_2 and ITO films in the visible range, the reflection properties are usually inferior.

(v) Undoped ZnO films are not of practical interest as these films are not stable at high temperatures.

(vi) Aluminium-doped ZnO films, which have stable electrical and optical properties, are emerging as an alternative candidate for transparent conducting applications. These films have shown

Table 4.2 Optical properties of some important transparent conducting oxide films.

Material	Process	Transmission T (visible region)	Reflection R (infrared region)	Bandgap E_g (eV)	Refractive index n	Effective mass (m^*/m_o)	Reference
SnO_2	Spray	82	—	3.69	—	—	65
SnO_2	Spray	90	—	—	1.8–2.2	—	45
SnO_2	Spray	97	—	4.11	2.2	—	39
SnO_2	Spray	80	—	3.95	—	—	38
SnO_2	Sputtering	75	55	—	—	0.38	63
SnO_2	Sputtering	95	—	4.13	—	—	42
SnO_2	Sputtering	80	—	—	2.0	—	64
SnO_2:Sb	Spray	85	—	3.75	—	—	79
SnO_2:Sb	Spray	80	84	—	—	—	36
SnO_2:Sb	Spray	88	80	—	—	—	84
SnO_2:Sb	Spray	>90	85	—	—	—	23
SnO_2:Sb	Spray	80	—	4.62	—	(0.1–0.3)	38
SnO_2:Sb	Sputtering	80	—	3.87	—	—	62
SnO_2:F	Spray	80–90	—	—	1.9	—	191
SnO_2:F	Spray	88	85	—	—	—	36
SnO_2:F	Spray	91	80	—	—	—	1
SnO_2:F	Spray	>90	90	—	—	—	23
SnO_2:F	Spray	>80	90	4.41	—	0.33	75
SnO_2:F	Spray	90	—	4.21	1.85	—	39
SnO_2:Mo	Reactive evaporation	>85	—	4.10	—	—	40
SnO_2:As	CVD	90	—	3.87	—	—	190
In_2O_3	Spray	>90	—	—	1.8–2.1	—	199

continued

Table 4.2 *Continued.*

Material	Process	Transmission T (visible region)	Reflection R (infrared region)	Bandgap E_g (eV)	Refractive index n	Effective mass (m^*/m_o)	Reference
In_2O_3	Thermal evaporation	>90	—	3.56	—	—	93
In_2O_3	Sputtering	80	—	—	2	—	94
In_2O_3:F	CVD	>85	—	3.5	—	—	101
ITO	Spray	>85	>85	—	2	—	74
ITO	Spray	>85	—	—	2	—	196
ITO	Spray	82	—	3.64	—	—	197
ITO	Spray	92	—	—	1.8	—	77
ITO	Spray	>85	85	—	—	—	83
ITO	Spray	88	90	—	—	—	84
ITO	Spray	90	—	—	2	—	193
ITO	Sputtering	90	90	—	—	—	195
ITO	Sputtering	90	84	(3.0–3.5)	—	—	123
ITO	Sputtering	95	90	—	—	—	100
ITO	Sputtering	85	80	—	—	—	135
ITO	Sputtering	95	—	(3.8–4.2)	2	—	133
ITO	Sputtering	90	—	—	2	—	113
ITO	Sputtering	80	—	3.75	—	0.30	91
ITO	Sputtering	92	—	3.66	—	—	194
ITO	Sputtering	—	—	4.21	—	0.31	109
ITO	Sputtering	85	>80	—	—	—	200
ITO	CVD	90–95	90	—	—	—	92
ITO	CVD	90	—	3.9	—	—	89
ITO	e-beam evaporation	90	>80	(3.6–4.1)	1.9–2.1	0.33	88

ITO	e-beam evaporation	95	>90	(3.9–4.6)	—	—	128
ITO	e-beam evaporation	>90	—	4.1	—	—	192
ITO	e-beam evaporation	90	—	—	—	0.35	108
ITO:F	Sputtering	>80	70	—	—	—	26
Cd_2SnO_4	Spray	83	—	2.83	—	—	151
Cd_2SnO_4	Sputtering	>80	80	—	(2.05–2.10)	—	152
Cd_2SnO_4	Sputtering	90	60	—	—	—	135
Cd_2SnO_4	Sputtering	85	90	—	—	—	148
Cd_2SnO_4	Sputtering	93	—	2.7	—	—	153
Cd_2SnO_4	Sputtering	>80	—	3.04	—	—	145
Cd_2SnO_4	Sputtering	90	—	3.0	—	0.55	158
ZnO	Spray	70	—	3.1	—	—	198
ZnO	CVD	>90	—	3.23	—	—	160
ZnO	Reactive evaporation	88	—	3.3	—	—	177
ZnO:In	Sputtering	>80	—	3.29	1.85	0.35	161
ZnO:Al	Sputtering	90	—	3.52	—	—	173
ZnO:Al	CVD	85	90	—	—	—	28
ZnO:Ga	Sputtering	>85	—	3.59	—	—	179

extremely promising results: typically, transmission and reflection values of the order of 85 and 90%, respectively, have been achieved. These results are quite comparable with those achieved with the best ITO and SnO_2:F films. The use of ZnO has other advantages: zinc is a cheap, abundant and non-toxic material, and as such will have lower cost than tin oxide and ITO, which are being currently used in a variety of applications.

References

[1] Manifacier JC 1982 *Thin Solid Films* **90** 297
[2] Jarzebski M 1982 *Phys. Status Solidi* a **71** 13
[3] Chopra KL, Major S and Pandya DK 1983 *Thin Solid Films* **102** 1
[4] Vossen JL 1977 *Phys. Thin Films* **9** 1
[5] Dawar AL and Joshi JC 1984 *J. Mater. Sci.* **19** 1
[6] Hamberg I and Granqvist CG 1986 *J. Appl. Phys.* **60** R123
[7] Heavens OS 1955 *Optical Properties of Thin Solid Films* (London: Butterworth)
[8] Bennett JM and Booty MJ 1966 *J. Appl. Opt.* **5** 41
[9] Nilsson PO 1988 *J. Appl. Opt.* **7** 435
[10] Campbell RD 1968 *PhD Thesis* University of Adelaide, Australia
[11] Tomlin SG 1968 *J. Phys. D: Appl. Phys.* **1** 1667
[12] Hall JF and Ferguson WFC 1955 *J. Opt. Soc. Amer.* **45** 714
[13] Lyashenko SP and Milosolavskii VK 1964 *Opt. Spectrosc.* **16** 80
[14] Manifacier JC, Gasiot J and Fillard JP 1976 *J. Phys. E: Sci. Instrum.* **9** 1002
[15] Azzam, RMA and Bashara NM 1977 *Ellipsometry and Polarized light* (Amsterdam: North-Holland)
[16] Borgogno JP, Lazairdes B and Pelletier E 1982 *Appl. Opt.* **21** 4020
[17] Borgogno JP and Lazairdes B 1983 *Thin Solid Films* **102** 209
[18] Pelletier E, Roche P and Vidal B 1976 *Nouv. Rev. Opt.* **7** 353
[19] Valeer AS 1963 *Opt. Spectrosc.* **15** 269
[20] Nestell JE Jr and Christy RW 1972 *Appl. Opt.* **11** 643
[21] Case WE 1983 *Appl. Opt.* **22** 1832
[22] Arndt DP *et al* 1984 *Appl. Opt.* **23** 3571
[23] Simonis F, Leij MV and Hoogendoorn CJ 1979 *Solar Energy Mater.* **1** 221
[24] Abeles F 1972 *Optical Properties of Solids* (Amsterdam: North-Holland)
[25] Hagen E and Rubens R 1903 *Ann. Phys.* **11** 873
[26] Geoffroy C, Campet G, Portier J, Salardenne J, Couturier G, Bourrel M, Chabagno JM, Ferry D and Quet C 1991 *Thin Solid Films* **202** 77
[27] Bellingham JR, Phillips WA and Adkins CJ 1991 *Thin Solid Films* **195** 23
[28] Hu J and Gordon RG 1992 *J. Appl. Phys.* **71** 880
[29] Hamberg I, Hjortsberg A and Granqvist CG 1982 *Appl. Phys. Lett.* **40** 362
[30] Yoshida S 1978 *Appl. Opt.* **17** 145
[31] Fan JCC and Bachner FJ 1975 *J. Electrochem. Soc.* **122** 1719
[32] Fan JCC and Bachner FJ 1976 *Appl. Opt.* **15** 1012

[33] Gerlach E and Grosse P 1977 *Fest Korperprobleme* **17** 157
[34] Frank G, Kauer E, Kostlin H and Schmitte FJ 1982 *Proc. SPIE* **324** 58
[35] Grosse P, Schmitte FJ, Frank G and Kostlin H 1982 *Thin Solid Films* **90** 309
[36] Karlsson T, Roos A and Rilebing CG 1985 *Solar Energy Mater.* **11** 469
[37] Demichelis F, Mezzetti EM, Smurro V, Tagliaferro A and Tresso E 1985 *J. Phys. D: Appl. Phys.* **18** 1825
[38] Shanthi E, Dutta V, Banerjee A and Chopra KL 1980 *J. Appl. Phys.* **51** 6243
[39] Afify HH, Momtaz RS, Badawy WA and Nasser SA 1991 *J. Mater. Sci.: Mater. Electron.* **2** 40
[40] Casey V and Stephenson MI 1990 *J. Phys. D: Appl. Phys.* **23** 1212
[41] Demiryont H, Nietering KE, Surowiec R, Brown FT and Platts DR 1987 *Appl. Opt.* **26** 3803
[42] De A and Ray S 1991 *J. Phys. D: Appl. Phys.* **24** 719
[43] Sundram KB and Bhagavat GK 1981 *J. Phys. D: Appl. Phys.* **14** 921
[44] Spence W 1967 *J. Appl. Phys.* **38** 3767
[45] Melsheimer J and Zifgler D 1983 *Thin Solid Films* **109** 71
[46] Roos A and Hedenqvist P 1990 *Proc. SPIE* **1275** 148
[47] Gottlieb B, Koropecki R, Arce R, Crisalle R and Ferron J 1991 *Thin Solid Films* **199** 13
[48] Omar OA, Ragaie HF and Fikry WF 1990 *J. Mater. Sci.: Mater. Electron.* **1** 79
[49] Sanon G, Rup R and Mansingh A 1990 *Thin Solid Films* **190** 287
[50] Srinivasamurty N, Bhagavat GK and Jawalekar SR 1982 *Thin Solid Films* **92** 347
[51] Murty SN and Jawalekar SR 1983 *Thin Solid Films* **108** 277
[52] Stjerna B and Granqvist CG 1990 *Solar Energy Mater.* **20** 225
[53] Howson RP, Barankova H and Spencer AG 1991 *Thin Solid Films* **196** 315
[54] Beensh-Marchwicka G, Krol-Stepniewska L and Misiuk A 1984 *Thin Solid Films* **120** 215
[55] Lou JC, Lin MS, Chyl JI and Shieh JH 1983 *Thin Solid Films* **106** 163
[56] Agashe C, Takwale MG, Marathe BR and Bhide VG 1988 *Solar Energy Mater.* **17** 99
[57] Burstein E 1954 *Phys. Rev.* **93** 632; Moss TS 1964 *Proc. Phys. Soc.* **B67** 775
[58] Muranaka S, Bando UY and Takada T 1981 *Thin Solid Films* **86** 11
[59] Jagadish C, Dawar AL, Sharma S, Shishodia PK, Tripathi KN and Mathur PC 1988 *Mater. Lett.* **6** 149
[60] Blandenet G, Court M and Lagarde Y 1981 *Thin Solid Films* **77** 81
[61] Geurts J, Rau S, Richter W and Schmitte FJ 1984 *Thin Solid Films* **121** 217
[62] Suzuki K and Mizuhashi M 1982 *Thin Solid Films* **97** 119
[63] Stejerna B and Granqvist CG 1990 *Thin Solid Films* **193/194** 704
[64] Howson RP, Barankova H and Spencer AG 1990 *Proc. SPIE* **1275** 75
[65] Vasu V and Subrahmanyam A 1991 *Thin Solid Films* **202** 283
[66] Pijolat C, Bruno L and Lalauze R 1991 *J. Physique* IV **1** 303
[67] Hamberg I, Granqvist CG, Berggren KB, Sernelius BE and Engstrom L 1985 *Solar Energy Mater.* **12** 479
[68] Mott NF 1974 *Metal Insulator Transitions* (London: Taylor and Francis)
[69] Berggren KF and Sernelius BE 1981 *Phys. Rev.* B **24** 1971
[70] Abram RA, Ress GJ and Wilson BLH 1978 *Adv. Phys.* **27** 799

[71] Mahan GD 1980 *J. Appl. Phys.* **51** 2634
[72] Vina L, Umbach C, Candona M, Compann A and Axman A 1983 *Solid State Commun.* **48** 457
[73] Mavrodiev G, Gajdardziska M and Novkovski N 1984 *Thin Solid Films* **113** 93
[74] Frank G, Fauer E and Kostlin H 1981 *Thin Solid Films* **77** 107
[75] Shanthi E, Banerjee A, Dutta V and Chopra KL 1982 *J. Appl. Phys.* **53** 1615
[76] Kulaszewicz S 1980 *Thin Solid Films* **74** 211
[77] Pommier R, Gril C and Marucchi J 1981 *Thin Solid Films* **77** 91
[78] Arai T 1960 *J. Phys. Soc. Japan* **15** 916
[79] Kaneko H and Miyake K 1982 *J. Appl. Phys.* **53** 3629
[80] Kane J, Schweizer HP and Kern W 1976 *J. Electrochem. Soc.* **123** 270
[81] Mizuhashi M 1980 *J. Non. Cryst. Solids* **38** 329
[82] Abass AK and Mohammad MT 1986 *J. Appl. Phys.* **59** 1641
[83] Frank G, Kauer E, Kostlin H and Schmitte FJ 1983 *Solar Energy Mater.* **8** 387
[84] Siefert W 1984 *Thin Solid Films* **120** 275
[85] Randhawa HS, Mathews MD and Bunshah RF 1981 *Thin Solid Films* **83** 267
[86] Manifacier JC and Fillard JP 1981 *Thin Solid Films* **77** 67
[87] Unaogu AL and Okeke CE 1990 *Solar Energy Mater.* **20** 29
[88] Agnihotry SA, Saini KK, Saxena TK, Napgal KC and Chandra S 1985 *J. Phys. D: Appl. Phys.* **18** 2087
[89] Maruyama T and Fukui K 1991 *J. Appl. Phys.* **70** 3848
[90] Weiher RL and Ley RP 1966 *J. Appl. Phys.* **37** 299
[91] Bellingham JR, Phillips WA and Adkins CJ 1990 *J. Phys.: Condens. Matter* **2** 6207
[92] Ryabova LA, Salun VS and Serbinov IA 1982 *Thin Solid Films* **92** 327
[93] Pan CA and Ma TP 1981 *J. Electron. Mater.* **10** 43
[94] Szczyrbowski J, Dietrich A and Hoffmann H 1982 *Phys. Status Solidi* a **69** 217
[95] Hoffmann H, Dietrich A, Pickl J and Krause D 1978 *Appl. Phys.* **16** 381
[96] Staritzky E 1956 *Analyt. Chem.* **28** 553
[97] Howson RP, Ridge MI and Suzuki K 1983 *Proc. SPIE* **428** 14
[98] Sundaram KB and Bhagavat GK 1981 *Phys. Status Solidi* a **63** K15
[99] Chen TC, Ma TP and Barker RC 1983 *Appl. Phys. Lett.* **43** 901
[100] Bawa SS, Sharma SS, Agnihotry SA, Biradar AM and Chandra S 1983 *Proc. SPIE* **428** 22
[101] Maruyama T and Fukui K 1990 *Japan. J. Appl. Phys.* **29** L1705
[102] Hoheisel M, Heller S, Mrotzek C and Mitwalsky A 1990 *Solid State Commun.* **76** 1
[103] Ohhata Y, Shinoki F and Yoshida S 1979 *Thin Solid Films* **59** 255
[104] Gupta L, Mansingh A and Srivastava PK 1989 *Thin Solid Films* **176** 33
[105] Kostlin H, Jost R and Lems W 1975 *Phys. Status Solidi* a **9** 87
[106] Clanget R 1973 *Appl. Phys.* **2** 247
[107] Muller HK 1968 *Phys. Status Solidi* **27** 733
[108] Robusto PF and Braunstein R 1990 *Phys. Status Solidi* a **119** 155
[109] Weijtens CHL and Vanloon PAC 1991 *Thin Solid Films* **196** 1
[110] Roth AP, Webb JB and Williams DF 1982 *Phys. Rev.* B **25** 7836
[111] Stern F and Tolley RM 1955 *Phys. Rev.* **100** 1638
[112] Chen RT and Robinson D 1992 *Appl. Phys. Lett.* **60** 1541
[113] Theuwissen AP and Declerck GJ 1984 *Thin Solid Films* **121** 109

[114] Howson RP, Ridge MI and Bishop CA 1981 *Thin Solid Films* **80** 137
[115] Nath P, Bunshah RF, Basol BM and Staffsud M 1980 *Thin Solid Films* **72** 463
[116] Christian KDJ and Shatynski SR 1983 *Thin Solid Films* **108** 319
[117] Yuanri C, Xinghao X, Zhaoting J, Chuancai P and Shuyun X 1984 *Thin Solid Films* **115** 195
[118] Noguchi S and Sakata H 1981 *J. Phys. D: Appl. Phys.* **14** 1523
[119] Noguchi S and Sakata H 1980 *J. Phys. D: Appl. Phys.* **13** 1129
[120] Kulaszewicz S 1981 *Thin Solid Films* **76** 89
[121] Latz R, Michael K and Scherer M 1991 *Japan. J. Appl. Phys.* **30** L149
[122] Harding GL and Window B 1990 *Solar Energy Mater.* **20** 367
[123] Fan JCC 1981 *Thin Solid Films* **80** 125
[124] Ishibashi S, Higuchi Y, Ota Y and Nakamura K 1990 *J. Vac. Sci. Technol.* A **8** 1403
[125] Smith RA 1978 *Semiconductors* (Cambridge: Cambridge University Press) p 294
[126] Krokoszinski HJ and Oesterlein R 1990 *Thin Solid Films* **187** 179
[127] Bellingham JR, Phillips WA and Adkins CJ 1992 *J. Mater. Sci. Lett.* **11** 263
[128] Hamberg I and Granqvist CG 1984 *Solar Energy Mater.* **11** 239
[129] Harding GL, Hamberg I and Granqvist CG 1985 *Solar Energy Mater.* **12** 187
[130] Hjortsberg A, Hamberg I and Granqvist CG 1982 *Thin Solid Films* **90** 323
[131] Oesterlein R and Krokoszinski HJ 1989 *Thin Solid Films* **175** 241
[132] Kulaszewicz S, Jarmoc W and Turowska K 1984 *Thin Solid Films* **112** 313
[133] Sreennivas K, Sudersena Rao T and Mansingh A 1985 *J. Appl. Phys.* **57** 384
[134] Enjouji K, Murata K and Nishikawa S 1983 *Thin Solid Films* **108** 1
[135] Howson RP and Ridge MI 1981 *Thin Solid Films* **77**, 119
[136] Jachimowski M, Brudnik A and Czternastek H 1985 *J. Phys. D: Appl. Phys.* **18** L145
[137] Maruyama T and Kojima A 1988 *Japan. J. Appl. Phys.* **27** L 1829
[138] Nagatomo T, Maruta Y and Omoto O 1990 *Thin Solid Films* **192** 17
[139] Avaritsiotis JN and Howson RP 1981 *Thin Solid Films* **77** 351
[140] Avaritsiotis JN and Howson RP 1981 *Thin Solid Films* **80** 63
[141] Groth R 1966 *Phys. Status Solidi* **14** 69
[142] Vossen JL 1971 *RCA Rev.* **32** 269
[143] Nozik AJ 1972 *Phys. Rev.* B **6** 453
[144] Miyata N, Miyake K and Nao S 1979 *Thin Solid Films* **58** 385
[145] Leja E, Stapinski T and Marszalek K 1985 *Thin Solid Films* **125** 119
[146] Stapinski T, Leja E and Pisarkiewicz T 1984 *J. Phys. D: Appl. Phys.* **17** 407
[147] Miyata N, Miyake K, Fukushima T and Koga K 1979 *Appl. Phys. Lett.* **35** 542
[148] Haacke G 1976 *Appl. Phys. Lett.* **28** 622
[149] Miyata N and Miyake K 1978 *Japan. J. Appl. Phys.* **17** 1673
[150] Miyata N and Miyake K 1979 *Surf. Sci.* **86** 384
[151] Ortiz RA 1982 *J. Vac. Sci. Technol.* **20** 7
[152] Schiller S, Beister G, Buedke E, Becker HJ and Schicht H 1982 *Thin Solid Films* **96** 113
[153] Miyata N, Miyake K, Koga K and Fukushima T 1980 *J. Electrochem. Soc.* **127** 918
[154] Smith AJ 1960 *Acta Crystallogr.* **13** 749
[155] Tromel M 1969 *Z. Anorg. Allg. Chem.* **341** 237

[156] Leja E, Budzynska K, Pisarkiewicz T and Stapinski T 1983 *Thin Solid Films* **100** 203
[157] Haacke G 1982 *Proc. Soc., Photo-Opt. Instrum. Eng.* **324** 10
[158] Pisarkiewicz T, Zakrzewska K and Leja E 1987 *Thin Solid Films* **153** 479
[159] Enoki H, Satoh T and Echigoya J 1991 *Phys. Status Solidi* a **126** 163
[160] Ogawa MF, Natsume Y, Hirayama T and Sakata H 1990 *J. Mater. Sci. Lett.* **9** 1354
[161] Sarkar A, Ghosh S, Chaudhuri S and Pal AK 1991 *Thin Solid Films* **204** 255
[162] Caporaletti O 1982 *Solar Energy Mater.* **7** 65
[163] Roth AB, Webb JB and Williams DF 1981 *Solid State Commun.* **39** 1269
[164] Schmid PE 1981 *Phys. Rev.* B **23** 5531
[165] Rogachev AA and Sablina NI 1966 *Sov. Phys.–Solid State* **8** 691
[166] Casey HC and Stern F 1976 *J. Appl. Phys.* **47** 631
[167] Major S, Banerjee A and Chopra KL 1985 *Thin Solid Films* **125** 179
[168] Bond WL 1965 *J. Appl. Phys.* **36** 1674
[169] Tsuji N, Komiyama H and Tanaka K 1990 *Japan. J. Appl. Phys.* **29** 835
[170] Minami T, Nanto H and Takata S 1985 *Thin Solid Films* **124** 43
[171] Ferrari A, Qian LS, Quaranta F and Valentine A 1990 *J. Mater. Res.* **5** 1929
[172] Igasaki Y. and Saito H 1991 *J. Appl. Phys.* **70** 3613
[173] Ghosh S, Sarkar A, Bhattacharya S, Chaudhuri S and Pal AK 1991 *J. Cryst. Growth* **108** 534
[174] Aktaruzzuman AF, Sharma GL and Malhotra LK 1991 *Thin Solid Films* **198** 67
[175] Harding GL, Window B and Horrigan EC 1991 *Solar Energy Mater.* **22** 69
[176] Minami T, Oohashi K, Takata S, Mouri T and Ogawa N 1990 *Thin Solid Films* **193/194** 721
[177] Swamy GHG and Reddy PJ 1990 *Semicond. Sci. Technol.* **5** 980
[178] Choi BH, Im HB and Song JS 1990 *J. Am. Ceram. Soc.* **73** 1347
[179] Choi BH, Im HB, Song JS and Yoon KH 1990 *Thin Solid Films* **193/194** 712
[180] Roth AP and Williams DF 1981 *J. Appl. Phys.* **52** 6685
[181] Webb JB, Williams DF and Buchanan M 1981 *Appl. Phys. Lett.* **39** 640
[182] Minami T, Nanto H, Shooji S and Takata S 1984 *Thin Solid Films* **111** 167
[183] Minami T, Sato H, Sonoda T, Nanto H and Takata S 1989 *Thin Solid Films* **171** 307
[184] Wu P, Gao YM, Baglio J, Kershaw R, Dwight K and Wold A 1989 *Mater. Res. Bull.* **24** 905
[185] Schropp R and Madan A 1989 *J. Appl. Phys.* **66** 2027
[186] Ruth M, Tuttle J, Goral J and Noufi R 1989 *J. Cryst. Growth* **96** 363
[187] Minami T, Tamura Y, Takata S, Mouri T and Ogawa N 1994 *Thin Solid Films* **246** 86
[188] Jin ZC, Hamberg I and Granqvist CG 1988 *J. Appl. Phys.* **64** 5117
[189] Tang W and Cameron DC 1994 *Thin Solid Films* **238** 83
[190] Vishwakarma SR, Upadhyay JP and Prasad HC 1989 *Thin Solid Films* **176** 99
[191] Calderer J, Esta J, Luquet H and Savelli M 1981 *Solar Energy Mater.* **5** 337
[192] Banerjee R, Das D, Ray S, Batabyal AK and Barua AK 1986 *Solar Energy Mater.* **13** 11
[193] Vasu V and Subrahmanyam A 1990 *Thin Solid Films* **193/194** 696

[194] Sberveglieri G, Benussi P, Coccoli G, Groppelli S and Nelli P 1990 *Thin Solid Films* **186** 349
[195] Croitoru N and Bannett E 1981 *Thin Solid Films* **82** 235
[196] Gouskov L, Saurel JM, Gril C, Boustani M and Oemry A 1983 *Thin Solid Films* **99** 365
[197] Kulaszewicz S, Jarmoc W, Lasocka I, Lasocki Z and Turowska K 1984 *Thin Solid Films* **117** 157
[198] Tomar MS and Garcia FJ 1982 *Thin Solid Films* **90** 419
[199] Lee CH and Huang CS 1994 *Mater. Sci. Eng.* B **22** 233
[200] Wen-Fa W and Bi-Shiou C 1994 *Thin Solid Films* **247** 201

5. Applications of Transparent Conducting Oxide Films

5.1 Introduction

Owing to unique combined electrical and optical properties, transparent conducting oxide films have found a variety of applications. Coatings of In_2O_3, SnO_2 and ITO are widely used in many applications related to their function as heat mirrors. These materials have found specific uses in various fields such as electrodes for liquid crystals, transparent heat reflectors in sodium or incandescent lamps, heated windscreens for cars, protective coatings on glass containers, gas sensors exploiting the variation in the conductivity of SnO_2 layers in various ambient gases, in hetero structure solar cells, etc.The selection of a suitable material for a specific application depends on its electrical, optical and mechanical properties as well as chemical stability. However, optimization of electrical and optical parameters is required in all transparent conductor applications.

5.2 Figure of Merit for Transparent Conducting Applications

Both optical transmission and electrical conductivity of these coatings should exceed a certain minimum value. Ideally, both parameters should be as large as possible. However, the simultaneous accomplishment of maximum transmission and conduction is not possible in most cases.

Fraser and Cook [1] defined a figure of merit F_{TC} to compare the performance of different transparent conductor materials. This figure of merit F_{TC} is given by

$$F_{TC} = T/R_S \tag{5.1}$$

where T is the optical transmission and R_S the electrical sheet resistance. The optical transmission T is given by the expression

$$T = I/I_0 = \exp(-\alpha t) \tag{5.2}$$

where I_0 is the radiation entering the coating on one side and I is the radiation leaving the coating on the other side, α is the optical absorption coefficient and t is the coating thickness.

The electrical sheet resistance R_S, as already defined in chapter 3, is given by

$$R_S = \frac{1}{\sigma t} \tag{5.3}$$

where σ is the electrical conductivity.

Substituting the values of T and R_S from equations (5.2) and (5.3), respectively, in equation (5.1), we obtain

$$F_{TC} = \sigma t \exp(-\alpha t). \tag{5.4}$$

It is evident from this equation that the figure of merit of a coating is a function of its thickness. The maximum value of F_{TC} corresponds to t_{max} which is calculated from

$$\frac{\partial F_{TC}}{\partial t} = \frac{\sigma \exp(\alpha t) - \sigma t \alpha \exp(\alpha t)}{\exp(2\alpha t)} = 0$$

$$t_{max} = \frac{1}{\alpha}. \tag{5.5}$$

Putting this value in equation (5.2), the optical transmission at maximum F_{TC} is given by

$$T = 1/\mathrm{e} = 0.37. \tag{5.6}$$

This shows that the optical transmission reduces to 37% for transparent conducting films with thicknesses corresponding to the maximum figure of merit. This value of transmission is too low for most transparent conductor applications.

In evaluating transparent conducting films based on F_{TC}, one particular film could be considered superior to others although its optical transmission is too low: this figure of merit F_{TC} is thus too much in favour of sheet resistance. A better balance between optical transmission and sheet resistance has been accomplished by Haacke [2] who redefined the figure of merit as

$$\phi_{TC} = \frac{T^{10}}{R_s}$$
$$= \sigma t \exp(-10\alpha t). \tag{5.7}$$

The maximum value of thickness that corresponds to the maximum value of the revised figure of merit ϕ_{TC} is given by

$$t_{max} = \frac{1}{10\alpha}. \tag{5.8}$$

The difference between ϕ_{TC} and F_{TC} is that maximum ϕ_{TC} occurs at 90% optical transmission, whereas maximum F_{TC} occurs at only 37% optical transmission. This revised figure of merit ϕ_{TC}, proposed by Haacke [2], is widely used to compare the performance of various transparent conductors over a wide thickness range.

Table 5.1 compares the values of ϕ_{TC} for different films prepared by different techniques.

Figure 5.1 shows the effect of thickness on ϕ_{TC} for transparent semiconductor oxide films of Cd_2SnO_4 and ITO. It is observed that the figure of merit is a strong function of thickness.

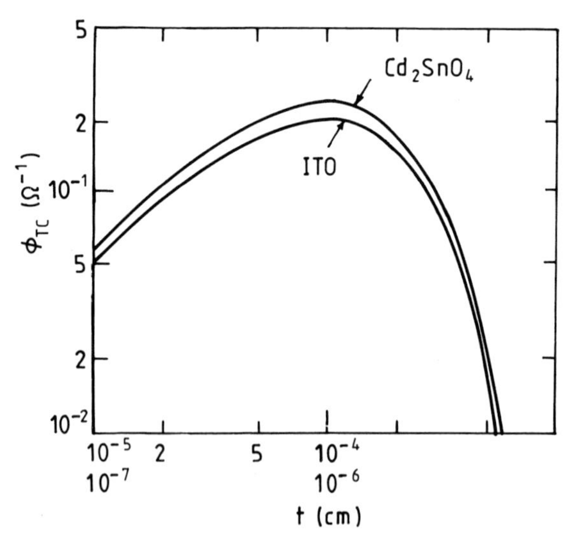

Figure 5.1 Variation of ϕ_{TC} versus thickness for Cd_2SnO_4 and ITO films (from [2]).

Table 5.1 Comparison of values of figure-of-merit ϕ_{TC} for different films prepared by different techniques.

Material	Process	Sheet resistance ($\Omega\,\square^{-1}$)	Transmission T (%]	Figure-of-merit, ϕ_{TC} (Ω^{-1}) $\times 10^{-3}$	Reference
SnO_2	CVD	4.4	87	56.4	3
SnO_2	CVD	420	95	1.43	4
SnO_2	Spray	8.5	80	12.6	7
SnO_2	Spray	43	80	2.5	8
SnO_2	Spray	85	80	1.26	9
SnO_2	Spray	509	97	1.45	5
SnO_2	Spray	13	89	24.0	5
SnO_2	Spray	63	82	2.18	6
SnO_2	Sputtering	600	75	0.1	10
SnO_2	Sputtering	98	75	0.57	11
SnO_2:Sb	CVD	72	87	3.45	12
SnO_2:Sb	CVD	55	90.6	6.78	4
SnO_2:Sb	Spray	160	90	2.18	13
SnO_2:Sb	Spray	86	84	2.03	14
SnO_2:Sb	Spray	14	80	7.6	8
SnO_2:Sb	Spray	39.6	88	7.0	15
SnO_2:Sb	Reactive evaporation	1.5	85	131.2	16
SnO_2:F	Spray	18	88	15.4	8
SnO_2:F	Spray	4	86	55.3	17
SnO_2:F	Spray	10.6	90	32.9	7
SnO_2:F	Spray	14.5	95	41.2	18

continued

Table 5.1 *Continued.*

Material	Process	Sheet resistance ($\Omega\,\square^{-1}$)	Transmission T (%)	Figure-of-merit, ϕ_{TC} (Ω^{-1}) $\times 10^{-3}$	Reference
SnO_2:F	Spray	9.2	93	52.6	19
In_2O_3	Evaporation	6.75	90	51.6	20
In_2O_3	Evaporation	88	80	1.22	21
In_2O_3:F	CVD	50	85	3.94	22
In_2O_3:F	Ion plating	53	90	6.5	23
ITO	CVD	6.74	80	15.9	24
ITO	CVD	8.37	90	41.66	25
ITO	Spray	26	90	13.4	27
ITO	Spray	9.34	85	21	26
ITO	Spray	10	90	34.9	28
ITO	Spray	3	88	92.8	29
ITO	Spray	4	90	87.2	30
ITO	Spray	4.4	85	44.7	31
ITO	Spray	1.9	95	315	32
ITO	Spray	5.6	88	49.7	15
ITO	Spray	3.1	88	89.84	7
ITO	Sputtering	3.1	83	50	1
ITO	Sputtering	12.5	95	47.9	41
ITO	Sputtering	8.5	95	70.0	38
ITO	Sputtering	19	87	13.1	45
ITO	Sputtering	20	85	9.8	46
ITO	Sputtering	18	80	5.9	37
ITO	Sputtering	10	95	59.8	39

ITO	Sputtering	130	83	1.1	40
ITO	Sputtering	5	90	69.7	43
ITO	Sputtering	23.1	90	15	42
ITO	Sputtering	6.6	85	29.83	35
ITO	Sputtering	40	90	8.72	36
ITO	Reactive evaporation	22	90	15.8	33
ITO	Reactive evaporation	25	98	32.6	16
ITO	e-beam	10	90	34.9	34
ITO	Ion beam sputtering	6	84	29.15	47
ITO	Ion plating	10	80	10.74	48
ITO	Ion plating	18.7	94	28.8	49
ITO	Ion plating	20	85	9.8	50
ITO:F	Sputtering	100	90	3.49	51
Cd_2SnO_4	Spray	400	83	0.39	52
Cd_2SnO_4	Sputtering	2.4	83	64.65	2
Cd_2SnO_4	Sputtering	14	93	34.57	53
Cd_2SnO_4	Sputtering	6.6	90	52.83	54
Cd_2SnO_4	Sputtering	1.35	90	258.3	55
Cd_2SnO_4	Sputtering	1	85	196.8	56
Cd_2SnO_4	Sputtering	100	90	0.35	57
Cd_2SnO_4	Sputtering	4.3	90	81	58
Cd_2SnO_4	Sputtering	10.95	80	9.8	59
Cd_2SnO_4	Sputtering	28.6	93.5	17.9	60
ZnO	Sputtering	85	90	4.1	61
ZnO:Al	CVD	4	85	49.2	62
ZnO:Al	Spray	29.4	85	6.7	63
ZnO:Al	Sputtering	3.5–7.7	90	99.6–45.3	64
ZnO:Ga	Sputtering	33.3	85	5.9	65

5.3 Wavelength Selective Applications

Wavelength-selective surfaces have many important applications, especially in solar energy conversion. There are basically two classes of wavelength-selective surfaces: selective black absorbers and transparent heat mirrors. A selective black absorber has a high absorptivity for solar radiation and a low emissivity for infrared thermal radiation. The transparent heat mirror is a surface that transmits visible light but reflects infrared radiation. In solar collectors, an absorber is used to convert solar radiation into thermal energy. The efficiencies of solar heat collectors are limited by thermal losses from the heated absorber due to conduction, convection and radiation. Conduction and convection losses can be minimized by a vacuum between the absorber and the cover plate. Thermal radiation, which is the dominating loss mechanism at high operating temperatures, can be reduced by using a selective black absorber with high solar absorptivity and high infrared reflectivity (i.e. low infrared emissivity). In principle, a single material could have the combination of high solar absorptivity and low infrared emissivity. However, most selective black absorbers are prepared by coating a metal having infrared reflectivity with a thin film that is transparent in the infrared but highly absorbing in the visible region. The most common surfaces for selective black absorbers include Ni-black, Cu-black and Cr-black. Such surfaces are obtained by electroplating the metal substrate. These electroplated surfaces are not stable at high absorber operating temperatures. Selective black absorbers which are stable at elevated temperatures, have been prepared using cermet films [66–69] such as MgO–Au, Cr_2O_3–Cr and Al_2O_3–Pt. The characteristic combination of visible transmission and infrared reflectivity suggests the potential use of transparent heat mirrors to increase the efficiency of solar heat collectors. The transparent heat mirror is positioned in front of the absorbing surface and reflects the emitted heat energy back to the absorber. The possibility of increasing the efficiency of the solar heat collector is due to the spectral separation between the solar radiation and thermal radiation emitted by absorbers. Figure 5.2 shows normalized spectra for the radiation from the sun, with an effective black body temperature of about 5800 K for air mass zero (AMO), and for the radiation emitted by a black absorber heated to 750 K. The separation of the two spectra is so marked that a heat mirror with the ideal reflectivity spectrum given by the dashed line in figure 5.2 would reflect almost all the thermal radiation emitted by a heated absorber while reflecting hardly any incident solar radiation. Incorporation of a transparent heat mirror, in addition to a selective black absorber, increases the efficiency of a solar energy collector. Such transparent heat mirrors can be realized in two ways: (1) by using a thin metal film sandwiched between dielectric films; and (2) by using a thin film of semiconductor having a sufficiently wide band-gap and high free carrier concentration. In both the cases, high infrared reflectivity is

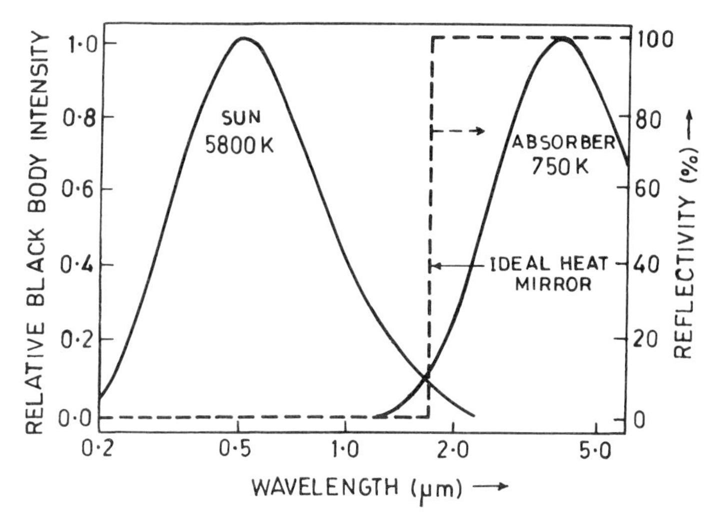

Figure 5.2 Normalized distribution of radiant energy for black body temperatures of 5800 K and 750 K. The dashed line shows the reflectivity of the ideal heat mirror (from [70]).

due to the good electrical conductivity of the materials. The visual transparency in the metal reflector is induced by the dielectrics, whereas in semiconductor heat mirrors, the transparency is a consequence of a wide band-gap. The optical properties of conducting materials are characterized by a plasma frequency for their free carriers. The material shows a reflectance that falls from dielectric behaviour to near-zero and then rises to near unity with increasing wavelength. In the region of dielectric behaviour, the material is transparent and the lower wavelength limit of transparency is governed by interband absorption. Optical properties due to free carriers within a semiconductor can lead, therefore, to regions of transparency between the plasma frequency and the frequency corresponding to the onset of interband absorption. Metals may also be used as transparent heat mirrors if their interband absorption is sufficiently small to allow a film to be thick enough to be coherent while remaining transparent. Metals such as gold, silver and copper, which possess relatively high transmission in the visible range of the spectrum and a relatively high coefficient of reflection in the longer wavelength region, have proved effective. However, the visible transmission of metal films is primarily limited by their reflectivity, therefore their effectiveness as transparent heat mirrors can be improved if they can be coated in a manner that reduces the visible reflectivity without affecting their infrared reflectivity. This is achieved by using additional interference layers of high refractive index materials. The most commonly used coatings [70–72] are ZnS/Au/ZnS, TiO_2/Au or Ag or Cu/TiO_2, Bi_2O_3/Au, or Ag or

Cu/Bi_2O_3, ITO/Au or Ag or Cu/ITO. It should be noted that Ag is a much better metal for transparent heat mirrors than either Cu or Au. Figure 5.3 illustrates absorptance versus wavelength for Ag, Au and Cu films. Both Cu and Au absorb about 50% of the incident energy for wavelengths lower than 0.5 μm, while Ag does not absorb appreciably for wavelengths above 0.3 μm. The low intrinsic absorptance of Ag indicates that a thin film of Ag can have very high solar transmission if the Ag reflectivity is suppressed by an anti-reflection coating such as TiO_2. Figure 5.4 shows typical optical properties of transparent $TiO_2/Ag/TiO_2$ heat mirror film on glass prepared by an RF sputtering technique.

Many workers [68, 73–82] have reported the use of a number of semiconductors which provide the required properties for a thin film transparent heat mirror. Such semiconductors include doped SnO_2, doped or undoped In_2O_3, doped ZnO, and Cd_2SnO_4 or their mixed oxides with suitable carrier concentrations. These metallic oxides and conductors exhibit a very favourable combination of properties, and as such are used for transparent heat mirror applications. The undoped materials have a direct band-gap $\approx$3 eV, thus allowing transmission of solar radiation.

The required infrared reflectance can be achieved by proper doping of these materials without adversely affecting the visible transmission. In order to have low thermal emissivity, the sheet resistivity of the films should be low, usually less than 20 $\Omega\,\square^{-1}$. A typical curve depicting this is shown in figure 5.5 for In_2O_3 films [76], in which the value of emissivity is given for different values of sheet resistivity. The samples were heated to 80 °C and the emissivity was measured under roughly normal incidence with a broadband detector. Dewall and Simonis [83] estimated that for an optimum transparent heat mirror based on SnO_2 coatings, a high electron mobility of at least 40 $cm^2\ V^{-1}\,s^{-1}$ and an electron concentration of about

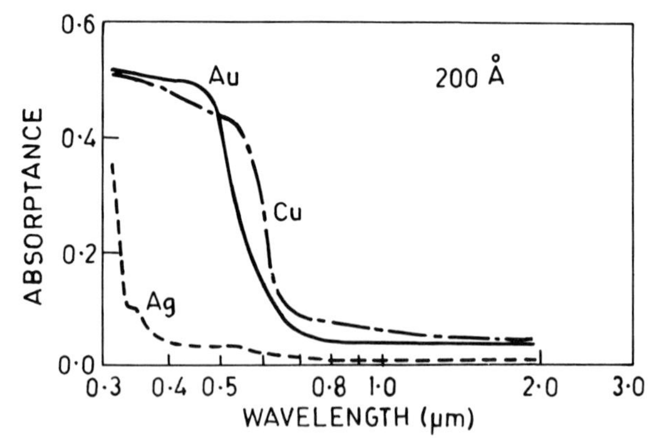

Figure 5.3 Absorptance of Ag, Au and Cu films versus wavelength (from [70]).

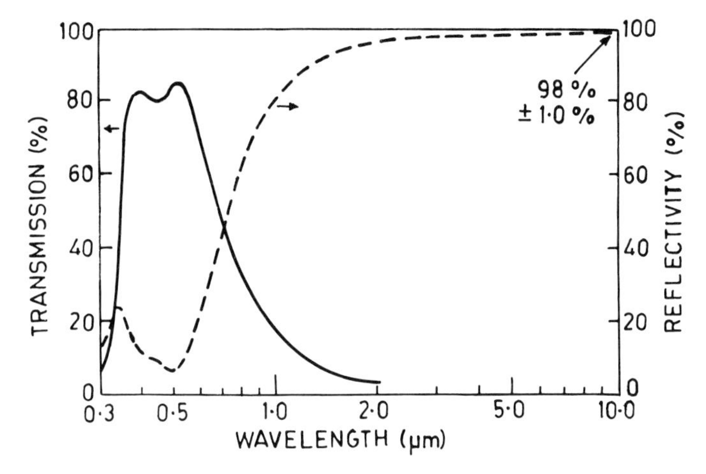

Figure 5.4 Optical transmission and reflectivity of $TiO_2/Ag/TiO_2$ film on glass (from [69]).

3×10^{20} cm^{-3} are required. With the growth techniques described in chapter 2, it seems quite feasible to achieve such properties.

In the last decade, tin-doped indium oxide has become a favoured material for the preparation of transparent heat reflectors. As already discussed in chapter 4 (figure 4.67), ITO films are superior to SnO_2 films because they can be prepared with higher conductivity and infrared reflectivity. The plasma wavelength can be shifted in indium tin oxide to shorter wavelengths than in tin oxide, which is a consequence of the higher free electron densities obtainable in indium tin oxide. In addition, better infrared

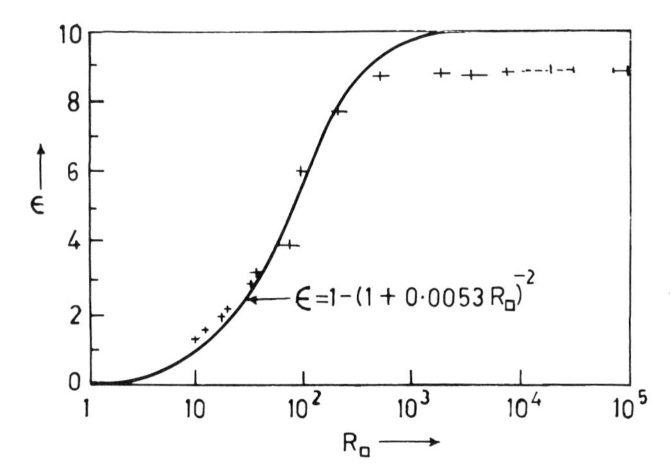

Figure 5.5 The effect of sheet resistivity on the emissivity of In_2O_3 films (from [76]).

reflectance is observed in tin-doped indium oxide because of its higher free electron mobility.

Another type of transparent heat mirror is obtained by fabricating a conducting sheet into micro-grids with openings of appropriate sizes to allow visible and solar radiation to pass through, but not infrared radiation. Fan and Zauracky [66] demonstrated this basic principle of selectivity by etching a micro-grid onto RF sputtered ITO films with square openings of side 2.5 μm and grid width 0.6 μm.

In addition to solar energy conversion applications, the properties of transparent heat mirrors have also been exploited in low pressure gas discharge lamps, incandescent lamps and as heat insulation for windows. In the sodium lamp, the low pressure gas discharge emits yellow light if the discharge tube is hot enough to ensure the required sodium pressure. If the glass enclosure is coated with indium oxide, the heat losses are very small, since the indium oxide heat mirror reflects the heat radiation back to the discharge tube. In a conventional incandescent lamp, the tungsten filament emits nearly 90% heat and 10% visible light and hence is a very inefficient means of producing light. A transparent heat mirror film either In_2O_3 or $TiO_2/Ag/TiO_2$, coated on the inner side of the glass envelope, will reflect the infrared heat radiation back to the filament and allow only light to pass through. Such a lamp may save up to 50% electric power for the same light output.

Heat losses through windows are an important factor in the energy balance of buildings. This is true even when double-glazing is used. In keeping with the present demand for lower energy consumption there is an urgent need for a form of glazing whose overall heat transfer coefficient is considerably lower than that of contemporary double-glazing. In such windows about two-thirds of the heat transport takes place via thermal radiation between the glass surfaces. This can be reduced by coating the inside of window glass surfaces with a heat-reflecting layer, e.g. tin oxide or indium oxide.

5.4 Solar Cell Applications

There has been considerable interest in recent years directed towards the development of conducting transparent oxide-based solar cells [37, 85–94, 99–112, 115–124]. These oxides offer the possibility of fabrication of solar cells with performance characteristics suitable for large-scale terrestrial applications. Transparent semiconducting oxides are particularly effective in solar cell applications because of the following advantages: (a) the conducting transparent film permits transmission of solar radiation directly to the active region with little or no attenuation, so that solar cells based on these materials result in improved sensitivity in the high photon-energy portion of the solar spectrum; (b) ease of fabrication of the junction because of lower junction formation temperatures; and (c) these films can serve simultaneously as a low resistance contact to the junction and an antireflection coating for the active region. Since a solar cell is a low impedance device, even a slight increase in its internal resistance considerably decreases its power output through change in the curve fill factor. The application of a transparent conducting coating can effectively reduce the internal series resistance of the cell. This is possible as a consequence of the shunting of the solar cell diffused layer resistance by the highly conducting coating. Also the refractive index n of the film is related to the refractive index n_s of the substrate by the relation: $n = (n_s)^{1/2}$, for an antireflection coating. The refractive indices of these oxides and silicon are ~2 and 4, respectively, which satisfies this condition well. The theory and performance of photovoltaic devices based on transparent conducting layers have been discussed in detail by many workers [85–88]. These oxide coatings have been used either as an active element in heterojunction solar cells or as a transparent electrode.

Heterojunction solar cells with a conducting transparent film of silicon, indium phosphate, cadmium telluride, gallium arsenide, etc. have been reported by many workers. Solar cells with an efficiency of 12% have been obtained with sputtered [89] ITO/Si(p-type) or sprayed [90] SnO_2/Si (n-type) devices. The results are in the same range as those obtained with the more mature technology of the diffused p–n^+ solar cell. The heterojunction between transparent conducting oxide (TCO) and active semiconducting material such as silicon, InP, etc. provides the bending of the energy bands necessary for the production of photocurrent. Figure 5.6 shows simple energy band diagrams of one such heterojunction, namely SnO_2/Si. The figure also gives the current–voltage characteristics in the dark and under illumination, for such a heterojunction. These heterojunction devices are not only comparable to diffused p–n junction silicon cells but also have the potential for lower cost fabrication.

Moreover, the processing temperature is low and is especially suited for polycrystalline silicon since preferential diffusion along grain boundaries can be avoided. In addition, deposition techniques are consistent with high yield and fast throughout processing, thus avoiding degradation of semiconductor minority carrier properties and suggesting compatibility with film and polycrystalline substrates [92, 93]. The important parameters in a solar cell are short-circuit current density, open-circuit voltage, fill factor and conversion efficiency. The various characteristics of TCOs influence the photovoltaic parameters significantly [94]: (a) the open-circuit voltage changes linearly with the electron affinity of TCOs; (b) an increase in TCO layer thickness decreases the short-circuit current; (c) an increase in TCO donor density decreases short-circuit current; and (d) high conversion

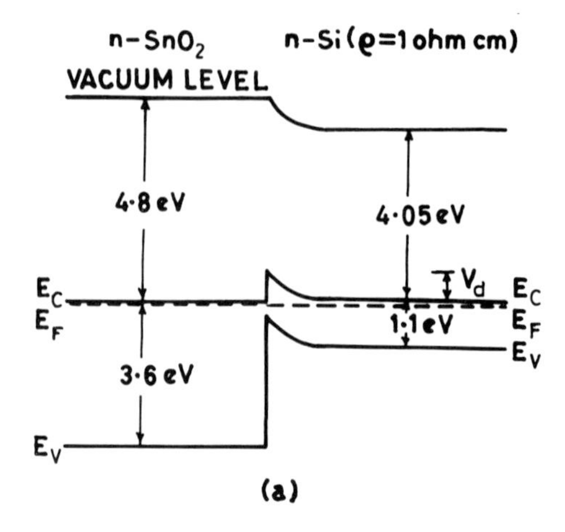

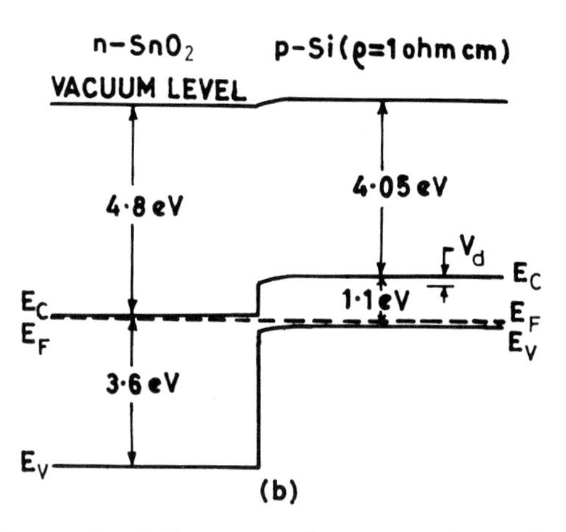

Figure 5.6 Energy band diagrams and current–voltage characteristics for a SnO_2–Si heterojunction solar cell (from [75]).

efficiencies can be achieved by matching the band-gap of TCO with the solar spectrum. The choice of oxide layer depends on its work function and the nature and electron affinity of the active semiconductor. The reported values of work function for transparent conducting coatings lie in the range 4–5 eV [95–98].

There has also been considerable interest recently in solar cells incorporating an amorphous insulating interfacial layer between the TCO and the active semiconductor heterojunction [99, 100]. Structures of this type are known as semiconductor–insulator–semiconductor (SIS) junctions (figure 5.7). The purpose of using the insulating interfacial layer is to increase the

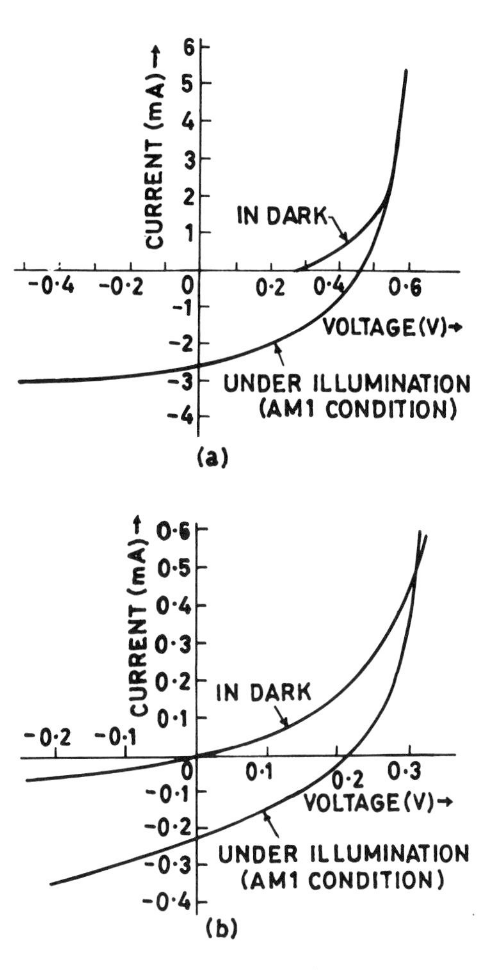

Figure 5.6 *Continued.*

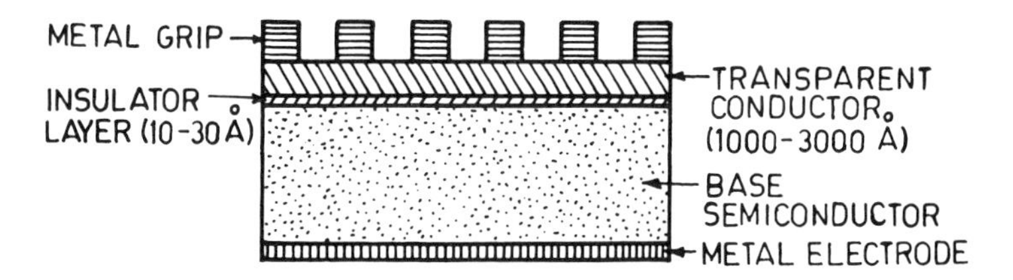

Figure 5.7 Schematic diagram of SIS solar cell.

Table 5.2 Comparison of solar cell properties based on conducting transparent layers produced by different techniques.

Structure	Method of preparation of conducting film	Area (cm^2)	Open-circuit voltage V_{oc} (V)	Short-circuit current I_{sc} (mA cm^{-2})	Fill factor	Efficiency (%)	Reference
SnO_2/Si	Spray	3.8	0.615	29.1	0.685	12.26	90
SnO_2/nSi	Spray	0.153	0.47	18.6	0.42	5.0	101
SnO_2/Si	Spray	—	0.52	21.0	0.53	7.2	102
nSnO_2/p-CdTe	Spray	—	0.50	12.5	0.475	1.8	94
SnO_2/Si	Sputtering	4.0	0.515	31.25	0.55	8.8	88
In_2O_3/InP	Sputtering	0.103	0.69	23.4	0.65	12.4	105
ITO/Si	Spray	—	0.275	48.0	0.31	8.2	37
ITO/Si	Spray	1.0	0.534	41.0	0.46	11.3	106
ITO/Si	Sputtering	0.1	0.51	32.0	0.70	12.0	89
ITO/Si	Ion beam sputtering	—	0.52	32.0	0.71	12.8	98
ITO/InP	Sputtering	—	0.76	19.5	0.7	14.0	107
ITO/GaAs	Sputtering	—	0.42	10.1	0.33	1.9	107
ITO/InP	Sputtering	0.7–0.8	0.768	26.9	0.767	15.8	108
ITO/InP	Ion-beam sputtering	0.2	0.76	21.55	0.65	14.4	109
ITO/SiO_x/Si	Ion-beam sputtering	11.46	0.526	27.39	0.759	11.9	110
ZnO/$CuInSe_2$	Spray	0.1	0.3	23.0	0.29	2.0	111
ZnO/CdTe	CVD	1.0	0.48–0.53	12.5	0.35	3.7	112
SnO_2/CdS/CdTe	Spray	1.0	0.763	20.1	0.72	11.0	104

open-circuit voltage by decreasing the dark saturation current. This SIS structure is potentially more stable and theoretically more efficient than either a Schottky or a metal–insulator–semiconductor (MIS) structure. The origins of this potential superiority are the suppression of majority carrier tunnelling in the SIS structure, the absence of thin metal which absorbs light and is subject to environmental degradation, and the wide choice of conductivity and band-gaps allowed in the top layer. Table 5.2 lists the results of some of the work on heterojunction solar cells in which the conducting transparent layer has been fabricated using different techniques.

Thin films of transparent conducting oxides have been extensively used as a window layer and as one of the electrodes for heterojunction solar cells, e.g. a–SiC:H/a–Si:H, CdS/Cu_2S, CdS/$CuInSe_2$ etc. Figures 5.8(a) and (b)

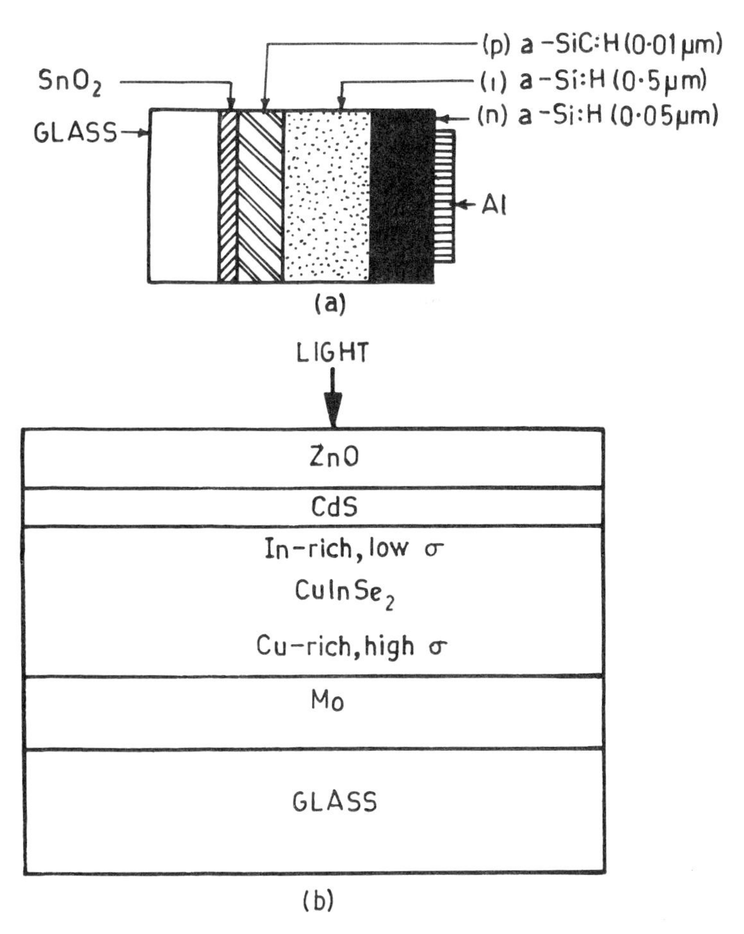

Figure 5.8 (a) Schematic diagram of a–SiC:H/a–Si:H solar cell. (b) Schematic diagram of CdS/$CuInSe_2$ solar cell.

show schematically the structure of a–SiC:H/a–Si:H and CdS/$CuInSe_2$ solar cells. In general, the efficiencies of these cells are of the order of 10–12%.

It should be noted that the TCO layer is degraded during deposition of the p-layer in glass/TCO/p–i–n/metal amorphous silicon solar cells in an RF glow discharge deposition system. Degradation may also occur as a result of contamination of the amorphous layer by diffusion of TCO elements, due to migration of atoms and/or ions via defects in the matrix structure [113, 114]. Degradation lowers the efficiency of solar cells [115–118]. Recently bilayers of SnO_2 and ITO have been shown to be suitable for reducing degradation in solar cells [115].

5.5 Protective Coatings

5.5.1 Abrasion-resistant coatings

Increasing demands are being placed on the serviceability of articles fabricated from glass or plastics. Specifically, high rates of production as well as increased bottle line-filling speeds demand improved methods to maintain the inherently high strength of moulded glass surfaces. It has been reported [125–131] that the application of a metallic oxide coating on glass containers appreciably reduces the coefficient of friction of the glass surfaces, facilitating the movement of containers through high-speed fitting lines. It has now become common practice to apply these metallic oxide coatings to glass containers immediately after forging. The subsequent application of an organic lubricant to a cooled and annealed container gives a highly desirable change to the glass surface, particularly with respect to abrasion resistance and lubricity. The application of tin oxide is made when the containers are hot (400–600 °C), and the process is normally known as hot end treatment. The containers are consequently treated with an insoluble coating such as emulsified polyethylene when they are at much lower temperatures (100–150 °C) and this process is referred to as cold end treatment.

There is always a significant reduction in the strength of glass due to the existence of submicroscopic flaws. These flaws have dimensions of the order of 10^{-6}–10^{-8} cm, compared with flaws of the order 10^{-1}–10^{-3} cm which can be induced during mechanical handling and contact between containers. The coefficient of friction is one factor which decreases the strength of glass containers by affecting the depth of the scratch or flaw produced in them. Figure 5.9 shows the effect of coefficient of friction (F) on the tensile strength of glass surfaces. The experimental data are for rods made of soda-lime silica glass which is normally used for glass bottles and containers. Figure 5.9 shows that as the lubricity of the given surface decreases, shown

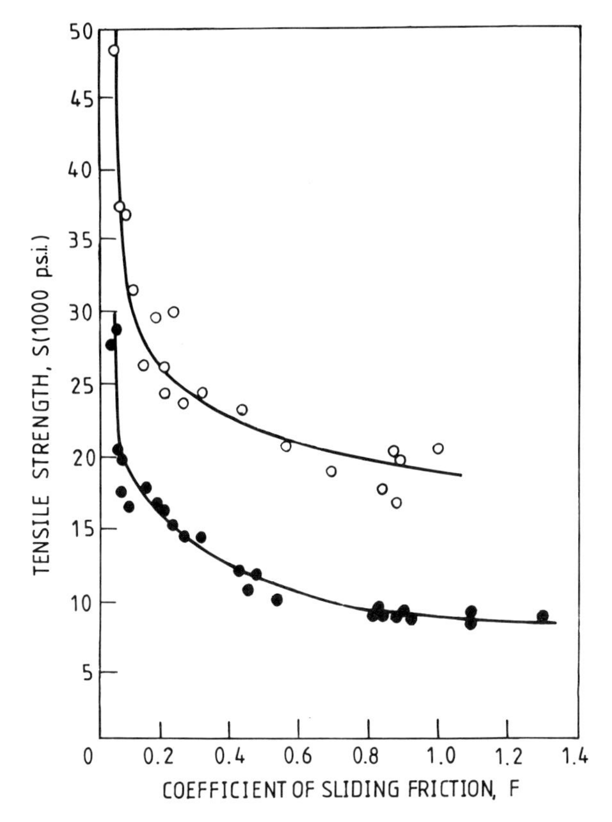

Figure 5.9 Effect of coefficient of friction on tensile strength: ●, in water and ○, in liquid nitrogen (from [126]).

by the increasing value of coefficient of friction *F*, the resultant surface strength also decreases.

Friction forces result in an increase in tensile stress on the glass surface. When this stress exceeds the strength of the glass surface, cracks are produced. The shape of the cracks takes on a crescent form. Figure 5.10 shows the dependence of crack depth on friction coefficient *F*. It can be observed that the crack depth is directly proportional to $F^{1/2}$. It can, however, be observed from this figure that for high values of *F*, the crescent crack depths reach a maximum constant value of 0.000 26 inch. For values of *F* less than 0.1, the crack depth decreases sharply from the linearly predicted value. At low values of *F*, sliding stresses may not be consistently of sufficient magnitude to produce crescent cracks.

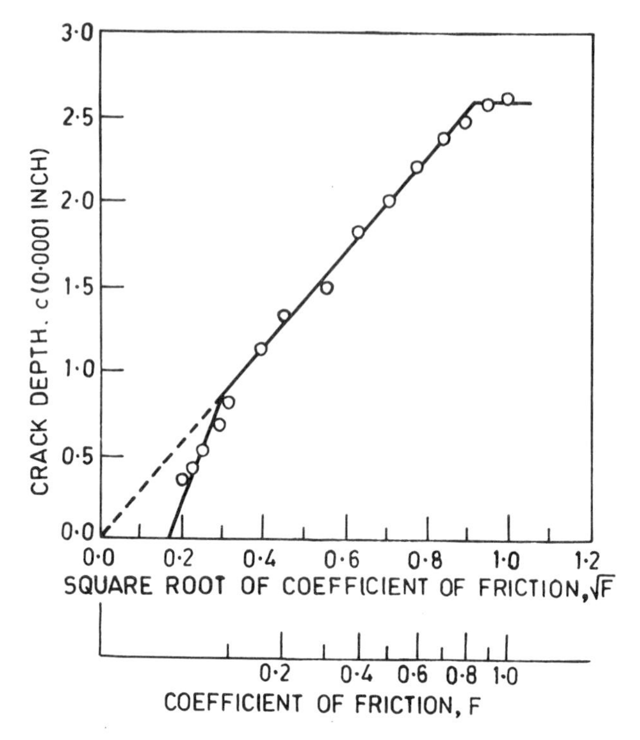

Figure 5.10 Crack depth versus friction coefficient (from [126]).

In order to improve the abrasion resistant properties of glass containers, two different types of treatment have been suggested. The first process (cold end) involves the application of lubricants, namely soaps, waxes, silicones, stearates, polyethylene and octadecylamine. The second process (hot end) involves the application of a tin oxide coating to the glass containers. Figure 5.11 shows the effect of applying these coatings, individually as well as simultaneously, on the coefficient of friction of the containers. In the figure, CE relates to cold end process while HE relates to hot end process. It can be observed that the bottles receiving a commercial lubricant, e.g. polyethylene only, show an initial F value of approximately 0.45, indicating a decrease of about 0.35. On repeated friction testing, this value rises to a constant value of about 0.8 showing that glass-to-glass contact occurs. Any lubricating qualities initially present have either worn away or the organic coating has been cut through by repeated contacts. It can be further observed from figure 5.11 that for bottles coated with tin oxide, the initial value of F is about 0.38, which remains constant for 50 tests. When both tin oxide and organic lubricants are applied, an initial friction coefficient of ≈ 0.3 is obtained and this value remains constant over repeated tests. The experimental results

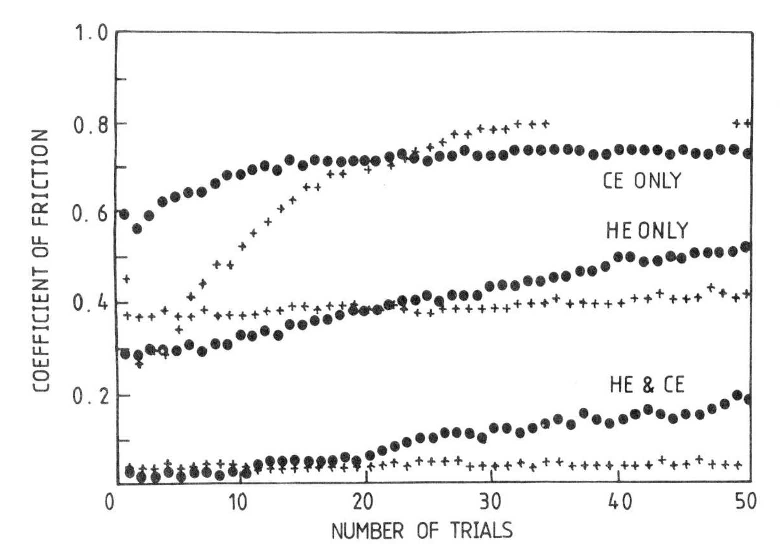

Figure 5.11 Coefficient of friction versus number of trips +, tested dry; ●, tested wet. The interrupted lines indicate the range of F values for uncoated glass tested dry (from [126]).

demonstrated that the tin oxide coating acts to bond the organic coating more firmly to the glass than when no tin oxide is present. The values of the friction coefficient under wet testing conditions were slightly higher. Typically, the values observed for CE, HE and CE+HE treatments were 0.73, 0.5 and 0.2, respectively.

Figure 5.12 shows the effect of different organic coatings (stearic acid and octadecylamine) on the coefficient of friction F of containers. It can be observed that the stearic acid is more wear resistant when it is applied to a glass surface than when applied to a tin-oxide coated surface. However, the opposite behaviour is shown by octadecylamine. It can be further observed that twin coating of tin oxide and octadecylamine is far superior to other coatings in wear resistant properties.

It is worth mentioning that for repeated use of glass containers, they require to be washed with alkaline solution. The effect of such a wash is to remove the coatings from the glass surface, and in some cases an opaque surface is produced. This is usually caused by penetration of alkali through holes or gaps in the tin oxide layer. The alkali subsequently reacts at the interface of the tin oxide and glass, resulting in a third layer which is opaque. In order to overcome this problem, certain inhibitors are generally used to reduce the extent of the attack of detergent on coatings during washing. The use of these inhibitors and the exact procedure for cold end and hot end treatment is patented information. For example, Titan 20 is the name given

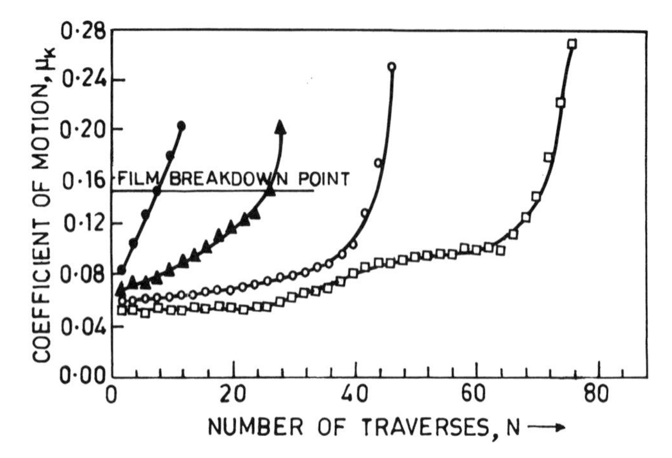

Figure 5.12 Effect of repeated traverses on coefficient of friction for: ▲, stearic acid coatings; ○, octadecylamine coatings; ●, stearic acid + SnO_2 coatings; □, octadecylamine + SnO_2 coatings (from [127]).

to the treatment process [128] used by United Glass Ltd, UK to provide bottles which are reasonably resistant up to about 20 trips. One treatment which is reported to have been commonly applied is an oleic acid vapour treatment. This is carried out by misting oleic acid in hot air, which is then blown onto the bottles while they are passing through a special hood. It has been observed [28] that simultaneous coating with low density polyethylene and oleic acid produces surfaces which have extremely good resistance to NaOH washing. Figure 5.13 shows the results of a comparative study of friction coefficient F versus number of trips for untreated and Titan-20 treated surfaces. Bottles were washed with 1.1% NaOH at 60 °C after every

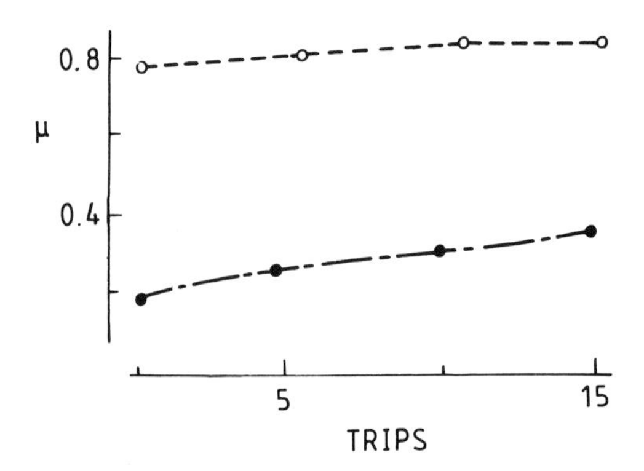

Figure 5.13 Coefficient of friction versus number of trips: ○, untreated and ●, Titan-20 treated (from [128]).

trip. It can be seen that the value of friction coefficient F reaches only 0.4 after 15 trips, indicating that the treatment is quite effective in overcoming the problems of diffusion of NaOH and removal of tin oxide coatings.

In a combined cold and hot end treatment, there is an optimum thickness of tin oxide which is fully effective in increasing the wear resistance of the lubricating film. Figure 5.14 shows the effect of thickness of tin oxide coating on the number of traverses required to break down a lubricant coating of octadecylamine applied on top of the tin oxide coating. It can be observed that the minimum required SnO_2 thickness varies almost linearly with the increase in number of trips. Typically, for 15 trips the required thickness is only 130 Å. On the other hand, the tin oxide layer thickness should not be too great, otherwise problems of opacity become more prominent. Moreover, the organic layer thickness also reduces with the number of trips and washings with NaOH, thus making it essential to give a repeat treatment.

Although SnO_2 can be deposited by any of the methods described in chapter 2, the most commonly used techniques are spray and CVD, because of the ease of using these methods for large and irregular surfaces. Most companies use a special type of hood to deposit SnO_2 on bottles and other containers. Each hood is mounted on a frame which positions the hood over the conveyor carrying bottles from the forming machine. A pneumatic system is provided as a means of raising the hood to permit on-line maintenance of the system. Stannic chloride vapour is generated by bubbling dry air through anhydrous stannic chloride liquid, and is then passed into an internal mixing chamber within a coating hood. There it is diluted with a secondary supply of air under pressure before being blown across the surface of the containers to be coated, which are passing through the coating hood on the machine conveyor.

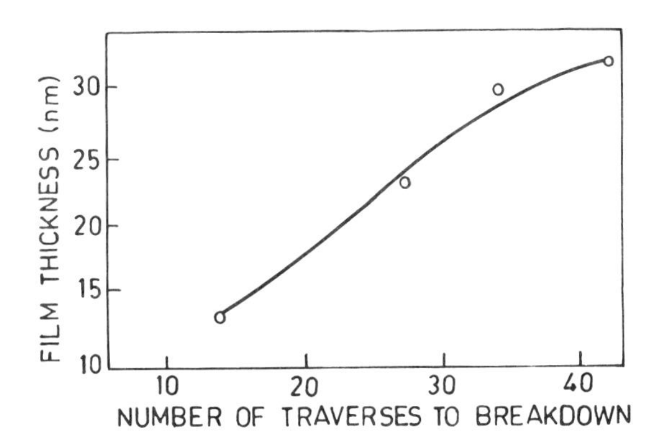

Figure 5.14 Effect of SnO_2 film thickness on the number of traverses it can withstand (from [127]).

5.5.2 Corrosion-resistant coatings

Aluminium has a high and fairly constant reflectivity in the visible region and a reflectivity ≈99% in the thermal infrared region. Because of this and its light weight and low cost, aluminium is being used in a variety of optical applications, e.g. mirrors of high optical quality, and large decorative panels on buildings. However, it corrodes easily in a humid atmosphere and its advantageous optical properties are lost. To overcome this problem, a protective layer is deposited on the aluminium. This is achieved either by an anodization technique, where an aluminium oxide layer is electrolytically grown on the aluminium surface, or by depositing a tin oxide layer on it. However, the anodized surface has some drawbacks: the chemical stability of the anodic layer is insufficient and can be etched by some chemicals, for example sulphuric and acetic acids. Also, this process is not suitable for thin film surfaces.

Tin oxide coatings exhibit protective properties which are superior to those of anodic layers [132–135]. Exposure to sulphuric acid and sodium hydroxide solution has no effect on tin oxide coated aluminium surfaces. Figure 5.15 shows the reflectance spectra for tin oxide coated aluminium before and after corrosion tests in sulphuric acid. The chemical stability test was carried out by immersing the samples in 10% H_2SO_4 for 3.5 h. It can be seen that the reflectivity of the tin oxide coated samples has not changed at all. It should be noted that, under similar conditions, anodic layers are completely etched and the reflectivity decreases significantly [132]. In addition to providing better chemical stability, a tin oxide coating also enhances the mechanical strength of the aluminium surface [134].

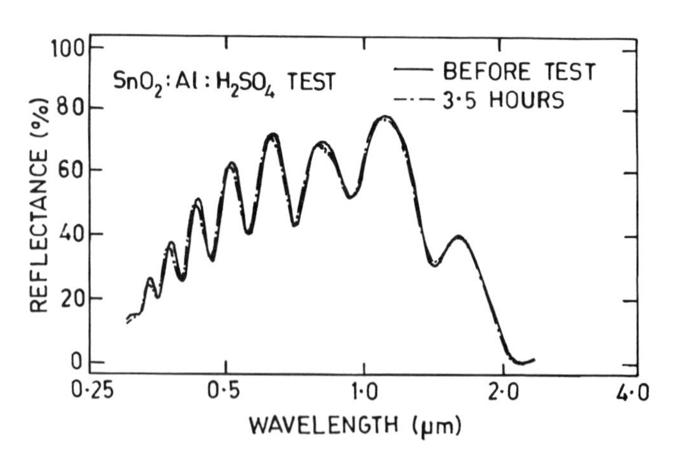

Figure 5.15 Reflectance spectra for tin oxide coated aluminium before and after chemical corrosion test in H_2SO_4 (from [132]).

Metallic films which are used in thin film magnetic recording do not have adequate mechanical durability and corrosion resistance and are normally overcoated with protective films in order to improve their reliability. Tin oxide has been used for overcoating metallic films in magnetic recording media [136]. Such overcoated metallic films have low static friction and good lifetimes in start/stop tests.

5.6 Gas Sensors

Semiconductor materials whose conductance is modulated directly by interaction with an active gas have been studied for many years. There is reversible chemisorption of reactive gases at the surfaces of certain metal oxides and chalcogenides, which is accompanied by reversible changes in conductance. Unlike metal films, where the conductance modulation due to absorption is very small and is caused by changes in mobility due to changes in surface scattering, the conductance changes in semiconductor materials are large and are caused primarily by changes in conduction band electron or valence band hole concentration. The electron concentration in semiconductor sensors varies more or less linearly with pressure over a range of up to eight decades, while the variation in mobility is less than a factor of two over the same pressure range. This suggests that gas chemisorption onto the surface of the semiconductor material influences its electrical conductivity significantly. It is this large and reversible variation in conductance with active gas pressure that has made semiconductor materials attractive for the fabrication of gas sensing electronic devices. For any device to be useful, it should be able to detect the active gas component in the presence of non-interfering background constituents. Moreover, it should have controllable levels of sensitivity, i.e. the change in conductance should be directly proportional to the change in pressure of the air component. Finally it should have reasonably good detectivity. The recent concern over pollution and safety in industrial activities involving poisonous gases has stimulated substantial research in the area of semiconductor gas sensors. Semiconducting transparent coatings such as tin oxide, indium tin oxide and zinc oxide are some of the most promising candidates for these applications. However, tin oxide is currently the most commonly used semiconductor material for gas sensors. Semiconductor-based gas sensors are being presently used for detecting H_2, CH_4 and LPG to prevent leakage, for detecting CO for pollution control, and as alcohol sensors for controlling drunken driving. Catalytic metallic additives such as palladium are often used to increase the selectivity and to enhance the response of these tin oxide based gas sensors. The most important use of SnO_2 sensors in the home is in gas leak alarms. For example, about 5 million gas leak alarms based on tin oxide gas sensors are produced every year in Japan.

5.6.1 Methane and propane gas sensors

The reversible, appreciable change in conductivity of transparent films (SnO_2, ITO, etc.) upon exposure to reducing gases such as propane and methane suggests the potential use of these materials as LPG sensors [137–146]. The gas sensing characteristics of these materials are usually defined in terms of gas sensitivity, which is given by

$$\Delta G/G_{air} = (G_{gas} - G_{air})/G_{air}$$

where G_{gas} is the film conductance in the presence of the test gas and G_{air} is the film conductance in the absence of the test gas. In general, the gas sensitivity of sensors based on thin films of these materials strongly depends on both film thickness and film conductivity. With an increase in film thickness, the sensitivity decreases, whereas it increases with a decrease in film conductivity. Figure 5.16 shows the variation of sensitivity with propane gas concentration as a function of thickness of CVD-grown SnO_2 films. The films were grown on silicon wafers at 500 °C. It is quite evident that the thinner films exhibit higher sensitivity to propane. Figure 5.17 shows the comparison in sensitivity between two SnO_2 films with the same thickness (1000 Å) grown at 500 and 700 °C. The film deposited at 700 °C is of low conductivity and is of much lower carrier concentration than the film deposited at 500 °C. However, the mobility difference between the two samples is small [137]. Figure 5.17 shows that the change in sensitivity with gas concentration is more prominent in films with low carrier concentration and low conductivity. Sberveglieri *et al* [138] used bismuth-doped tin oxide gas sensors for the detection of CH_4. The doped SnO_2 films were grown by evaporating Sn and 3–7 at% bismuth and then oxidizing these in the

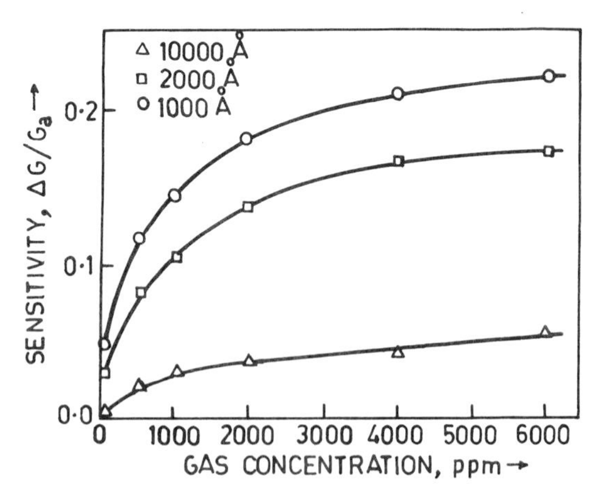

Figure 5.16 Variation of gas sensitivity $\Delta G/G_a$ as a function of propane gas concentration for various SnO_2 film thicknesses (from [137]).

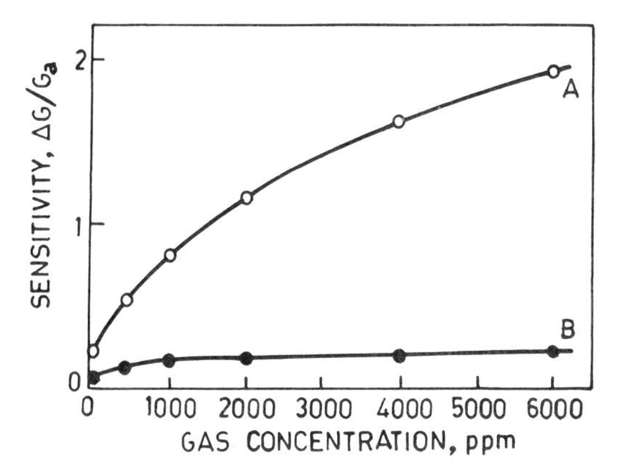

Figure 5.17 Variation of gas sensitivity $\Delta G/G_a$ as a function of propane gas concentration for SnO_2 films deposited at (A) 700 °C and (B) 500 °C (from [137]).

presence of oxygen. They observed a sensitivity of 15% for 1000 ppm CH_4 at an operating temperature of 450 °C.

Recently, Sberveglieri *et al* [139] demonstrated the suitability of sputtered ITO films for gas sensing applications. Figure 5.18 shows the variation of sensitivity with CH_4 gas concentration at an operating temperature of 720 K. It can be observed that for a gas concentration of 1000 ppm, the sensitivity is ≈0.8 (80%), which is significantly higher than has been reported for methane sensors based on SnO_2.

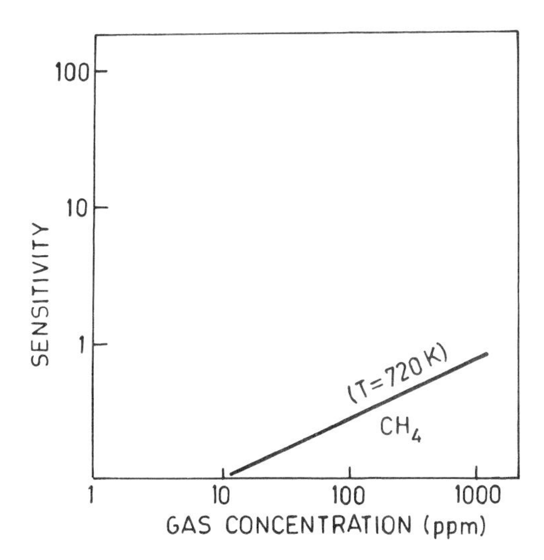

Figure 5.18 Variation of gas sensitivity with methane gas concentration for ITO films (from [139]).

Nitta *et al* [140] used sintered SnO_2 powders doped with Nb, V, Ti or Mo for the detection of propane gas. Stannic oxides mixed with 1 wt% $PdCl_2$ and $Mg(NO_3)_2$, were used as host materials. The powders, to which Nb or other transition metals were added, were presintered in air for 1–24 h at 600–800 °C. These powders were dispersed uniformly in alcohol, painted between the electrodes on the alumina substrate and then again sintered at 800 °C for 1 h. The gas concentration in the chamber was kept constant during the measurements. The sample was heated using a thick film resistor heater at 280 °C. A pick-up resistor was connected in series with the sample. Figure 5.19 shows the change in resistivity as a function of propane concentration measured with 30 V applied to the sample and various Nb contents at 280 °C. It can be observed that the resistance of these samples decreases with increasing gas concentrations. The relations for the sample with Nb 20 wt% or less than 7 wt% are linear. For samples with 9–17 wt% Nb, however, an inflection point appears on these lines. Nevertheless the relation between resistance and gas concentration is still linear up to about 0.3–0.4% of propane gas. For the samples doped with V, Ti or Mo, similar inflection properties were observed. However, the step phenomenon found for 9 wt% Nb-doped samples was not observed in sensors doped with other transition metals.

From the above discussion, it can be inferred that for low concentrations, (<1000 ppm) of propane or methane gas, thin film based gas detectors are

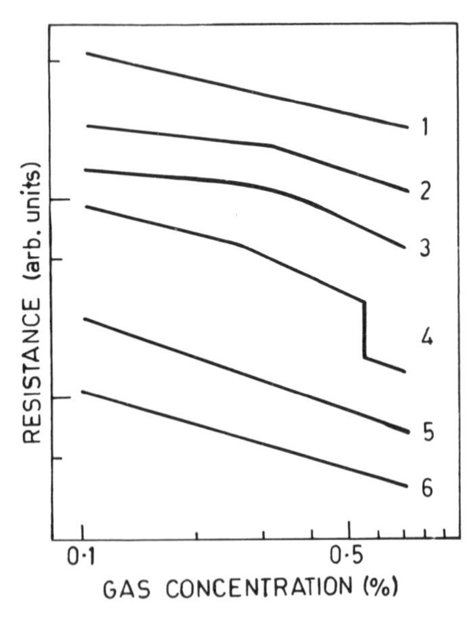

Figure 5.19 Resistivity versus gas concentration for Nb concentration of: (1) 20 wt%; (2) 17 wt%; (3) 15 wt%; (4) 9 wt%; (5) 7 wt%; (6) 5 wt% (from [140]).

more suitable, whereas for higher gas concentrations (1000–4000 ppm), transition metal doped sintered powder SnO_2 detectors are more reliable.

5.6.2 Carbon monoxide detectors

Carbon monoxide sensors, both in bulk form and thin film form, have been extensively studied [139,142,144,147–153]. Although, these sensors are based on a basic sensing mechanism, i.e. surface conduction modulation by adsorbed gas molecules, the sensitivity and selectivity depend mainly on the device operating conditions. Windischmann and Mark [149] developed carbon monoxide sensors based on SnO_2 thin film. These sensors were fabricated onto alumina substrates using an RF sputtering technique. Film thickness was typically 500 Å. The films were thermally biased at 200–500 °C. The sensors were able to detect CO in the presence of background gases such as O_2, SO_2, NO, CO_2 and N_2 at levels in the range of 1–100 ppm with reproducible results. Within the temperature range 200–500 °C, the conductance of the sensor increased with partial pressure of carbon monoxide (P_{CO}) in the ambient gas according to the relation:

$$G = G_o + \beta(P_{CO})^{1/2} \tag{5.9}$$

where G_o is the background conductance in the absence of CO in the ambient gas and β is a constant. Figure 5.20 shows this behaviour for one sensor. The results were analysed on the basis of the following model: (a) there is a surface reaction between associatively adsorbed carbon

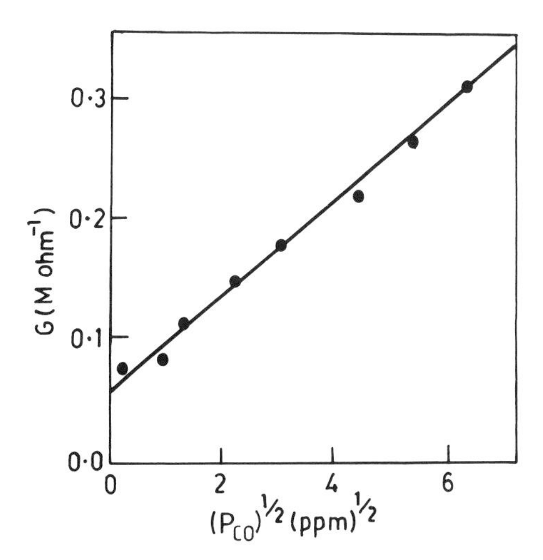

Figure 5.20 Dependence of sensor conductance on pressure of carbon monoxide $(P_{CO})^{1/2}$ (from [149]).

monoxide from the ambient gas with the chemisorbed oxygen to produce CO_2; (b) this is followed by return of the electron from CO_2 to the conduction band, owing to exothermic reaction energy of the co-oxidation; (c) there is subsequent thermal desorption of neutral CO_2. The temperature over which the sensors can operate is restricted. If the temperature is too low (<200 °C), the product species, i.e. CO_2, will not be desorbed, thus poisoning the site for further adsorption of oxygen. On the other hand, if the temperature is too high (>500 °C) oxygen will not be physically adsorbed in the recombination reaction and hence CO will also not be absorbed. Figure 5.21 shows the response of a SnO_2 thin film sensor to a step input of CO of varying concentration. It can be observed that the speed of response increases as the CO concentration increases. This can be explained on the basis that the charge transfer rate between the conduction band and the chemisorbed surface states is exponentially related to the surface potential barrier height V_s. The larger V_s is, the slower the rate of charge transfer and hence speed of response. Since the co-adsorption of CO decreases the surface barrier, it follows that the response speed will increase as the CO concentration increases.

SnO_2 films prepared using a laser ablation technique are reported to be sensitive to a number of gases such as CO, ethanol and oxygen [150]. Ion implantation at a dose of 10^{16} Ar^+ ions/cm^2 is observed to enhance the selectivity of the sensor response to CO gas. The sensitivity of ion-implanted SnO_2 films increases from 10.46 to 46.69 for CO gas, whereas the increase in sensitivity for other gases is negligible. Recently ITO films have also been demonstrated to act as CO gas sensors [150]. It has been reported that ITO films exhibit a significant increase in conductivity when exposed to CO gas.

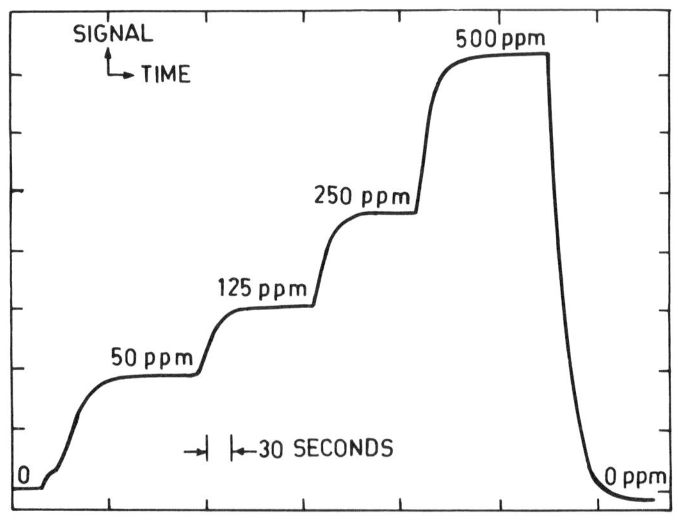

Figure 5.21 Dependence of response speed on CO concentration (from [149]).

ITO films of thickness 2000 Å were deposited using RF sputtering at a substrate temperature of 520 K. Typically, the sensitivity of these ITO films to CO gas increased from 0.5 to 8.0 when the CO gas concentration increased from 10 ppm to 1000 ppm at an operating temperature of 650 K [139].

The addition of a suitable dopant such as bismuth oxide or thorium oxide in SnO_2 further improves the sensitivity and selectivity of CO gas sensors. The addition of bismuth to SnO_2 increases the resistance thereby improving the sensitivity to CO gas. Nomura and co-workers [142, 151] observed that the use of bismuth alters the operating temperature of SnO_2 films for CO gas sensing from 450 to 250 °C. Further, conductivity measurements on Sn–Bi oxide films at 250 °C in various ambients (at 1000 ppm level) such as CH_4, CO, H_2, C_2H_5OH etc., indicated that there was an appreciable change in conductivity in CO. Other gases showed little or no change in conductivity, indicating the selectivity of Sn–Bi oxide films for CO gas. It is worth mentioning that the higher sensitivity and selectivity is only for films with bismuth concentrations up to 20%.

Detectors based on SnO_2 doped with thorium oxide show the presence of an oscillation phenomenon [152] when the devices are exposed to carbon-monoxide gas in air. Sensors are fabricated by mixing 93% SnO_2, 1% $PdCl_2$, 1% $MgNO_3$ and 5% ThO_2 and sintering the mixture at 800 °C for 1 h. The presintered powder is dispersed in silicasol and screen printed on alumina substrate followed by further sintering at 600 °C for a period of 4 h. An AC voltage (200 V_{rms} 50 Hz) is applied between the sample electrodes and the voltage developed across the pick-up resistor, which is connected in series with the sample, is recorded as a function of gas concentration. The sensor temperature is usually in the range 140–250 °C. Figure 5.22 shows the V_{rms}

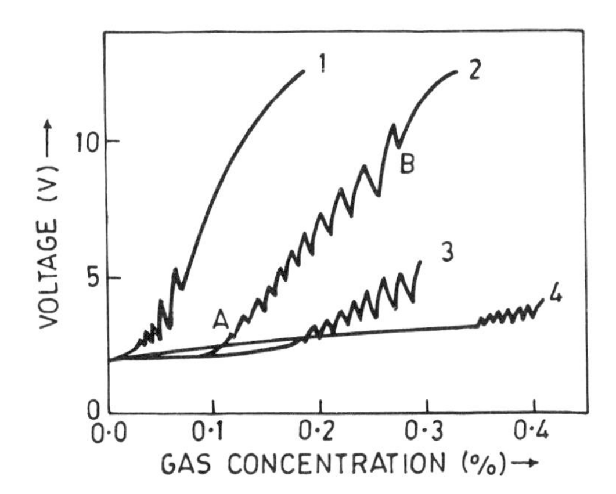

Figure 5.22 RMS voltage appearing across the pickup resistor as a function of CO gas concentration at different substrate temperatures: (1) 146 °C; (2) 183 °C; (3) 200 °C; (4) 236 °C. Applied voltage to the sample is 20 V rms (from [152]).

appearing across the pick-up resistor as a function of CO gas concentration at different sensor temperatures. It can be observed that the oscillations are limited to a certain CO gas concentration region, which depends on the sensor temperature. The CO gas concentration at which the oscillation occurs is fairly low. The oscillation is of sawtooth waveform, with rise time typically 4.5 s and decay time about 13.5 s. The amplitude and frequency of oscillation depend on the environmental gas concentration. This behaviour is shown in figure 5.23. In the gas concentration range of 0.2–0.3%, the frequency of oscillation decreases linearly from 0.68 to 0.54 Hz with increasing CO gas concentration. On the other hand, the amplitude increases from 0.3 to 3.0 V in proportion to the gas concentration. The presence of oscillations is probably due to the periodic repetition of adsorption and desorption of a part of the CO gas on the surface of the sample.

5.6.3 Hydrogen gas sensors

The fabrication of hydrogen gas sensor devices using SnO_2 often requires annealing and/or doping with suitable materials such as bismuth oxide or thorium oxide to increase selectivity and to enhance sensitivity [137, 138, 143, 147, 154, 155–164]. Ultrathin tin oxide films grown using an ion-beam sputtering technique show an exceedingly high sensitivity of $\approx 10^4$ for H_2 gas [156,157]. Such a high sensitivity is due to: (i) development of microstructures during annealing; and (ii) control of the ultrathin film thickness. Figure 5.24 shows the variation of sensitivity (R_o/R_g) with annealing time for SnO_2 films annealed at 500 °C in air. R_o and R_g are the film resistance in air and in hydrogen gas, respectively. It can be seen that the sensitivity increases remarkably with increase in annealing time. Electrical studies [156] reveal that the resistivity of the film in air increases rapidly with annealing time,

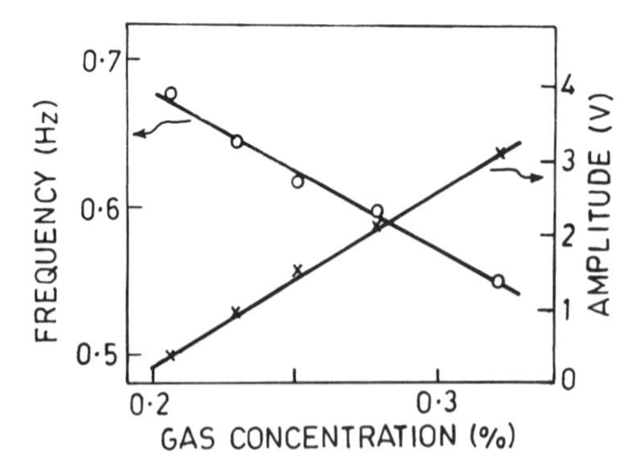

Figure 5.23 Dependence of the frequency and amplitude of oscillation on CO gas concentration (from [152]).

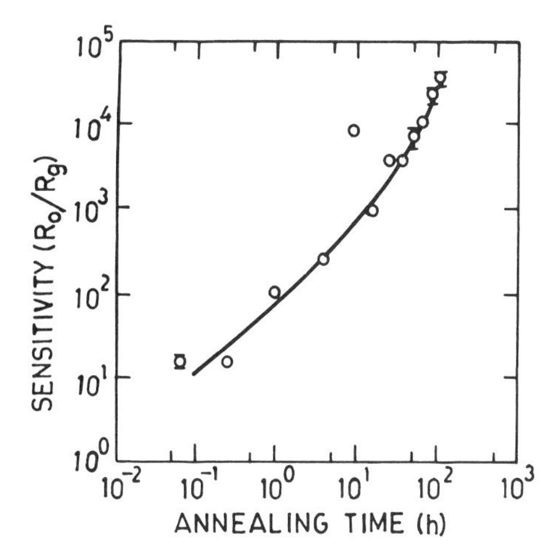

Figure 5.24 Variation of sensitivity with annealing time for SnO_2 films. Annealing temperature, 500 °C; hydrogen gas concentration, 0.5%; sensor operating temperature, 300 °C (from [156]).

whereas there is no appreciable increase in resistivity with annealing time in hydrogen. Therefore the high sensitivity for higher annealing durations is mainly due to the increased resistivity in air after prolonged annealing. The effect of film thickness on the sensitivity of air-annealed SnO_2 films to hydrogen is shown in figure 5.25. As can be observed, the sensitivity is

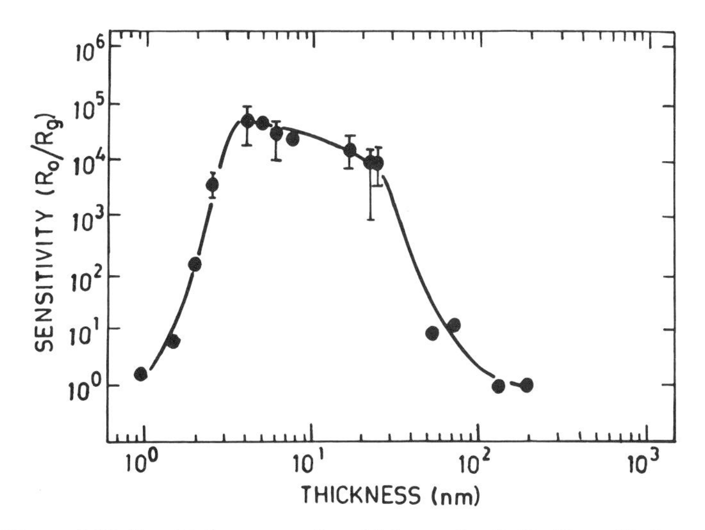

Figure 5.25 Sensitivity versus film thickness for SnO_2 films. Hydrogen gas concentration, 0.5%; sensor temperature, 300 °C; annealing temperature, 500 °C; annealing duration, 50 h (from [157]).

maximum ($\sim 10^4$) over the narrow thickness range (30–200 Å), and thicker and thinner films are relatively insensitive. The possible reason for this may be that the difference in resistivity of films annealed in air and in hydrogen is extremely large only in this narrow thickness range.

In addition to annealing, use of suitable dopants [138, 158, 159, 160] or metallization with a palladium layer as a surface catalyst [147,161] also improves the sensitivity and selectivity of SnO_2 hydrogen gas sensors. Figure 5.26 shows the variation of sensitivity with bismuth concentration in SnO_2 thin films for 100 and 1000 ppm hydrogen concentration. It is evident that the maximum sensitivity is for 5–6 at% bismuth doping. The sensitivity is considerably higher ($\approx$2000%) for 1000 ppm hydrogen concentration. The sensitivity of SnO_2 films strongly depends on operating temperature and the nature of the surface of the film, i.e. whether it is smooth or wrinkled. Sberveglieri *et al* [138, 160] studied this effect in thin film hydrogen gas sensors based on bismuth-doped SnO_2. Tin and bismuth metallic films were grown using a conventional evaporation technique, and a thermal oxidation process was used to transform the metallic films into SnO_2:Bi_2O_3 films. SEM studies revealed that films thermally oxidized for 5–10 min had a wrinkled structure, whereas oxidation for longer duration resulted in almost flat surfaces. Figure 5.27 shows the variation of hydrogen sensitivity with operating temperature for SnO_2 films with flat and wrinkled surfaces. It is evident that the sensitivity is much higher in the case of films with a wrinkled surface. Further, the sensitivity is maximum at an operating temperature of 675 K.

Kanefusa *et al* [158] used ThO_2-doped SnO_2 sensors for the detection of hydrogen gas. The detection sensitivities to hydrogen gas as a function of gas concentration are shown in figure 5.28. Curve 1 shows that at a temperature of 280 °C, there is positive sensitivity, i.e. the resistivity of the sample decreases as H_2 gas concentration increases. On the other hand, the sample

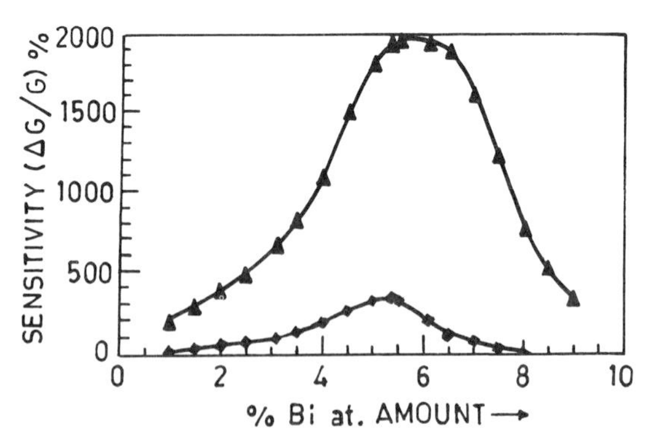

Figure 5.26 Sensitivity versus bismuth concentration: ●, 100 ppm; ▲, 1000 ppm; operating temperature, 450 °C (from [138]).

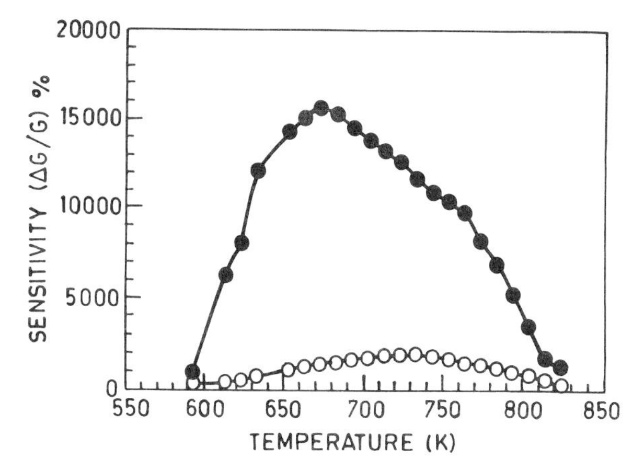

Figure 5.27 Sensitivity versus operating temperature for ○, flat and ●, wrinkled SnO_2 thin film. Hydrogen concentration, 1000 ppm (from [160]).

has negative sensitivity at a sample temperature of 80 °C, as shown in curve 3, i.e. the resistivity of the sample increases with increase of hydrogen gas concentration. At intermediate temperatures 80–250 °C, typically at 200 °C (curve 2), there are two distinct regions; a positive characteristic when the sample is exposed to hydrogen at very low concentration and a negative sensitivity at high concentration. In later work Kanefusa *et al* [162] reported that when sensors based on SnO_2 doped with palladium, rhodium and MgO are tested in air at 92 °C, the oscillation phenomenon exists only when

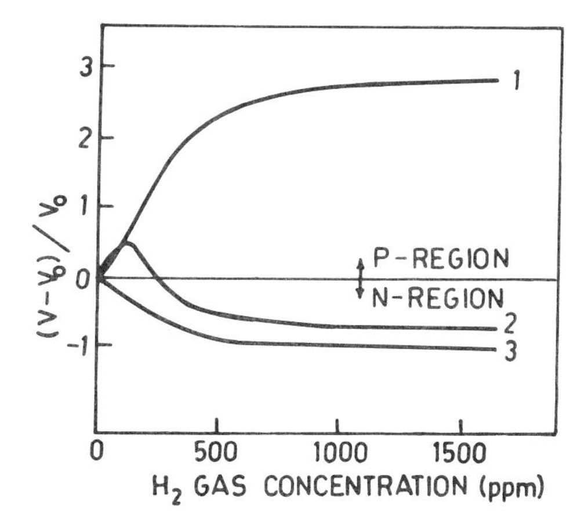

Figure 5.28 Detection sensitivity as a function of hydrogen gas concentration for ThO_2-doped SnO_2: curve (1) 280 °C; (2) 200 °C; (3) 80 °C (from [158]).

hydrogen gas is present. This suggests the possibility of using these devices for sensing hydrogen in atmospheres containing various other gases. Figure 5.29 shows the detection sensitivities of the device as a function of H_2 gas in air at 92 °C. The DC voltage applied across the film was 44 V. V_o and V indicate the voltage across the pickup resistor before and after exposure to H_2 gas, respectively. As shown in the figure, the oscillation appears only when the gas concentration is between 0.23% and 0.94%. A solid line indicates stable non-oscillatory states. The phenomenon has been observed only in samples doped with Pd and Rh. This oscillation phenomenon is not significantly influenced by the concentrations of N_2 and O_2 in air, indicating that oscillations are not due to adsorption or desorption of O_2 on the surface of the device.

5.6.4 Ethanol detectors

In general, sensors suitable for detecting CO, H_2, CH_4 can be used for the detection of ethanol C_2H_5OH. Tin oxide, doped and undoped, has generally been employed for the detection of ethanol [142, 150, 165–167]. Films are normally prepared by a spray pyrolysis technique. Pink *et al* [166] used SnO_2 films for the detection of ethanol, propane and carbon-monoxide gases. Figure 5.30 shows the sensitivity of SnO_2 films to CO, C_3H_8 and C_2H_5OH as a function of deposition temperature. It can be observed from the figure that

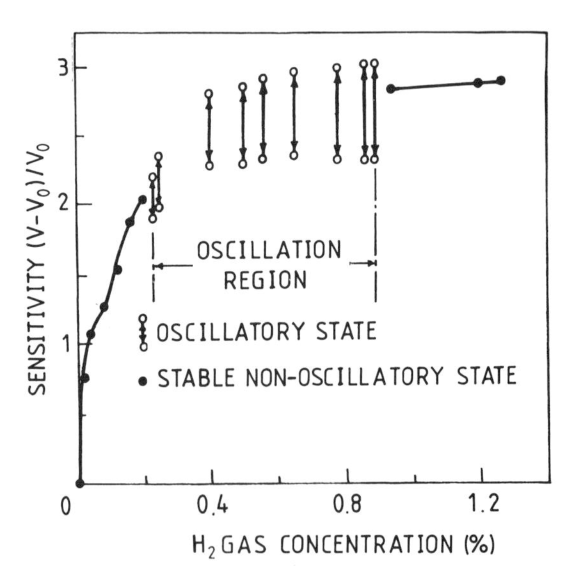

Figure 5.29 Detection sensitivity as a function of hydrogen gas concentration at 92 °C for Rh, Pd and MgO doped SnO_2 (from [162]).

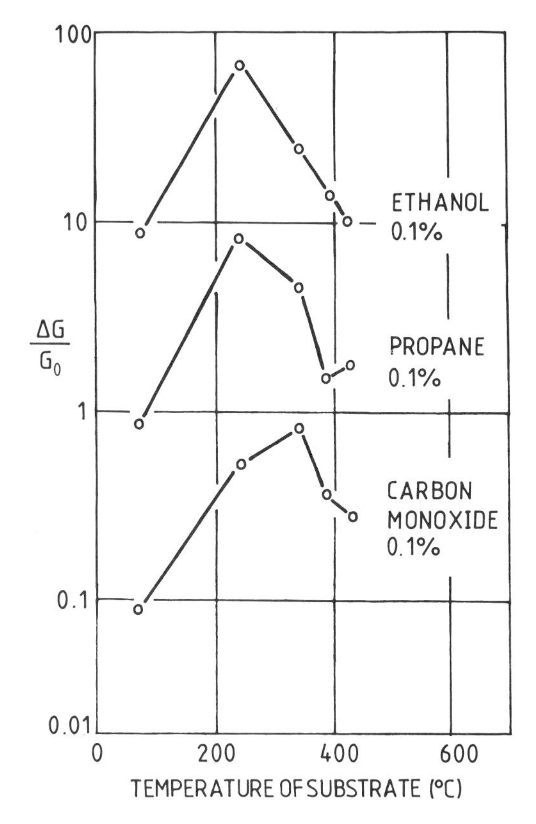

Figure 5.30 Conductance sensitivity of SnO_2 films as a function of substrate temperature for various gases (from [166]).

the highest sensitivities are for films deposited at about 250–300 °C. Sensitivity is also strongly dependent on heat treatment at higher temperatures as can be seen from figure 5.31. It is evident that both conductance and sensitivity decrease with heat treatment beyond 400 °C. Comparison of detectors based on SnO_2 films with detectors based on sintered material reveal that thin film detectors can sense as little as 10^{-1} vpm, whereas bulk material sensitivity is limited to about 5 vpm. These results are shown in figure 5.32. Moreover, the sensitivity of thin film detectors is an order higher than sensors based on bulk materials. The response time of these ethanol detectors is of the order of 30 ms.

It has already been mentioned that Sn–Bi oxide films are highly sensitive and selective to CO (rather than C_2H_5OH) when operated at about 250 °C [142]. However, the sensitivity to C_2H_5OH increases when the operating temperature is increased. These results are shown in figure 5.33. This increase in sensitivity at higher temperatures (>400 °C) is due to the fact that

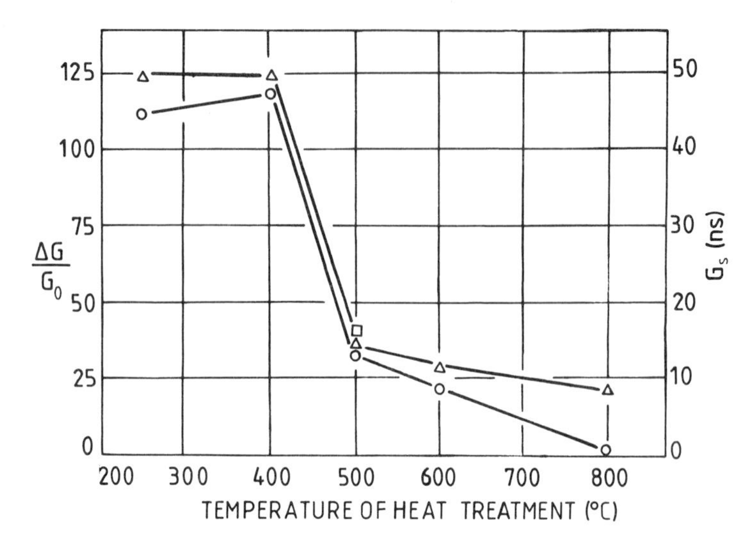

Figure 5.31 Conductance G_S and sensitivity $\Delta G/G_0$ as a function of heat treatment in SnO_2 films for ethanol detection (from [166]).

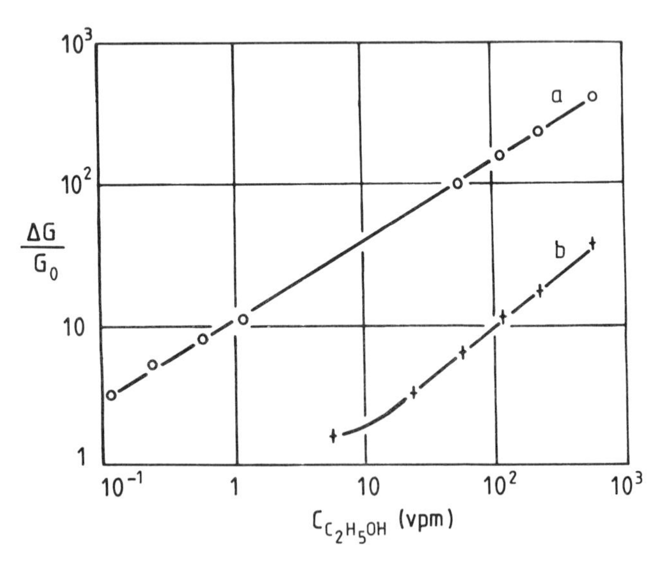

Figure 5.32 Sensitivity of (a) SnO_2 films and (b) sintered SnO_2 for ethanol detection (from [166]).

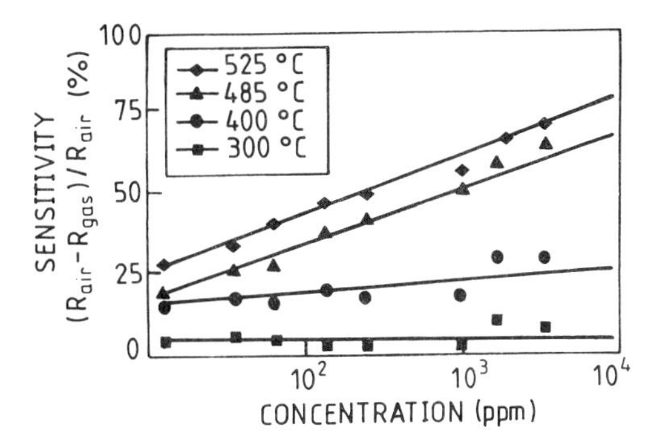

Figure 5.33 Variation of sensitivity with C_2H_5OH concentration for Sn–Bi oxide films at various operating temperatures (from [142]).

C_2H_5OH interacts locally with oxygen ions adsorbed at the surface, thereby forming reaction products such as CO, CO_2, H_2 and H_2O, which are electron donors. At low temperatures, weak chemisorption predominates and as such the surface coverage of donor species is small. This results in negligible change in conductivity. As temperature increases, the fraction of donor species increases and large changes in conductivity occur, resulting in an increase in sensitivity.

5.6.5 Nitrogen oxide sensors

Detection of nitrogen oxides (such as NO and N_2O) resulting from combustion processes is essential with the development of actuators which are able to control and minimize the emission of these exhaust gases. Initial attempts to develop sensors to detect these gases were made by Chang [168], using reactively sputtered SnO_2 thin films operating at 520 K. Though these sensors were highly sensitive, their working stability was very poor. Recently Sberveglieri *et al* [139, 169, 170] demonstrated the possibility of using ITO films for detection of these gases. ITO films were grown by RF sputtering technique. Typical growth parameters for ITO films having optimum response to nitrogen oxides are: bias voltage, 25 V; deposition rate, 1–3 Å s^{-1}; thickness, 2000 Å; and substrate temperature 520 K. Figure 5.34 shows the variation of film sensitivity to NO as a function of operating temperature for several gas concentrations. It is quite evident that the optimum temperature for maximum sensitivity is 580 K. As already discussed, these ITO films can be used for detection of CO and CH_4, however, the sensitivity of these films to NO and NO_2 is far greater. Figure 5.35 shows the sensitivity versus gas concentration curves for NO, NO_2, CO and CH_4

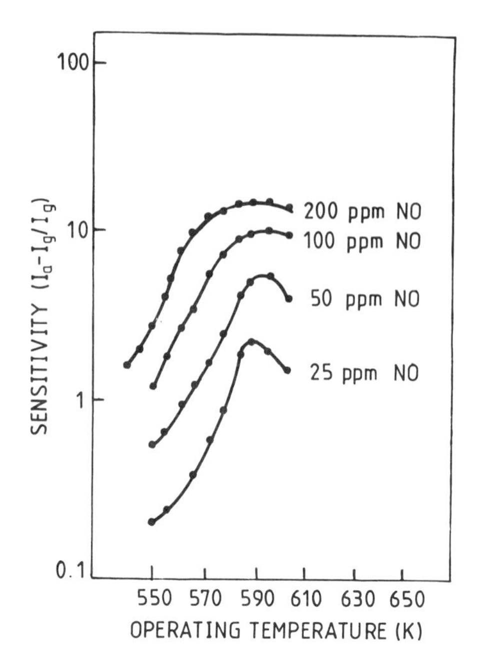

Figure 5.34 Sensitivity versus operating temperature for ITO films for different NO concentrations (from [139]).

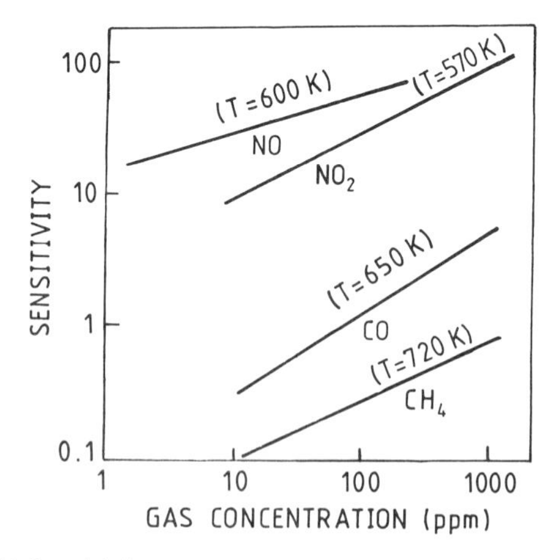

Figure 5.35 Sensitivity versus gas concentration of ITO thin film sensors (from [139]).

gases for operating temperatures at which maximum sensitivity is reached. It is obvious that high selectivity for NO and NO_2 can be achieved using ITO films.

5.6.6 Sensors for other gases

In addition to the above mentioned gas sensing applications, SnO_2 and ZnO are also finding use in the detection and measurement of odours [171, 172]. These sensors, being non-destructive in nature, are basically employed for the detection of freshness of food, which is an important task in the food industry. ZnO:Al and SnO_2:CaO based thin film gas sensors have high sensitivity and selectivity for trimethylamine gas, which is one of the main gases emitted during the deterioration of foods.

Recently, ITO films have been shown to have potential for the detection of chlorine gas [155, 173]. Galdikas *et al* [173] observed that the presence of chlorine in the ambient gas results in an increase in the resistance of ITO films. However, the sensitivity is poor, and more work is required to prove viability for chlorine sensing application.

5.7 Display Devices

Transparent conducting oxide coatings have been extensively used as transparent electrodes in various display devices, e.g. liquid crystal displays (LCD), electroluminiscent (EL) devices, image sensors based on amorphous silicon, light emitting diodes (LED), etc. [45, 174–182]. Figure 5.36 shows the typical structure of an electroluminiscent display device. The electrode material for display devices must be conducting and optically transparent. It is also essential to pattern the transparent conducting layer,

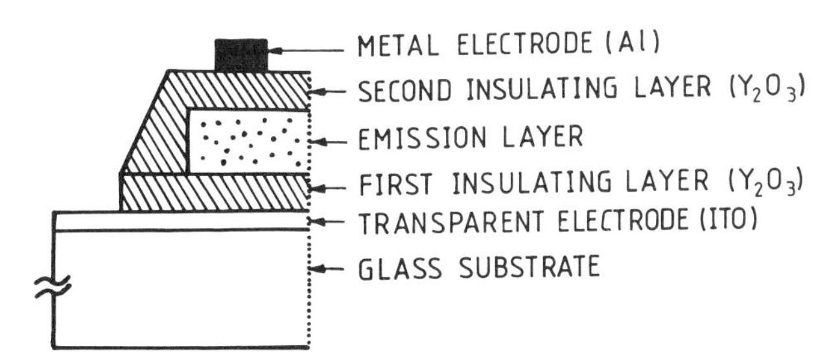

Figure 5.36 Typical structure of an electroluminescent device.

for which etching is required. Different techniques have been employed for etching transparent conducting films [174, 183–188]. ITO is generally preferred for these applications because it can be easily etched. The most commonly used etchants for ITO are 5% H_2SO_4 or 5% HCl in H_2O or a mixture of dilute HCl and HNO_3. Strong acids cannot be employed for etching ITO as they may damage the underlying layers of the devices. In the case of tin oxide, due to its high chemical stability, wet chemical etching is more difficult, and an electrolyte or effervescent action is required. In general, it is difficult to form fine, homogeneous patterns over a large area using wet etching. Recently, various dry etching techniques have been used to produce the required electrode patterns on ITO, tin oxide and zinc oxide. These include plasma etching [183, 185–187] and laser etching [184,188]. In the case of plasma etching, different source gases (electron cyclotron resonance excited hydrogen, chlorine, CH_3OH, CF_2Cl_2, CF_3Cl, CF_4, CH_4/H_2 mixture) have been used to etch ITO and SnO_2 films. Typically, the sizes of patterns on ITO, etched by CH_4/H_2 plasma, are 1.5 μm line, 1.5 μm space and 0.3 μm thickness [185]. Recently, Adesida *et al* [186] produced structures with submicron dimensions in ITO using a CH_4/H_2 plasma etching technique.

Laser etching is emerging as an alternative to plasma etching. The technique has been used to pattern high temperature superconductors and spin-on glass [189,190]. Recently, an excimer laser has been used to etch thin films of tin oxide and indium tin oxide [184]. The laser pulse width and energy density were 23 ns and 1.5–3.0 J cm^{-2}, respectively. Line patterns of the order of 20–50 μm have been demonstrated in these materials. Praschek *et al* [188] demonstrated the possibility of etching ZnO films using a frequency-doubled Nd:YLF laser. The work on laser etching, though limited, has tremendous potential as laser etching can be used for etching any transparent oxide material effectively. As such, further investigations are needed to establish this technique for large-scale commercial applications.

In some display devices, such as LEDs and imaging CCD arrays, a transparent conducting coating on top of active areas is required. For such a requirement, the maximum temperature for growth of the electrode is limited. ITO is commonly used since it can be deposited at room temperature with excellent electrical and optical properties using ion-assisted deposition techniques as discussed in chapter 2.

5.8 Other Applications

Transparent conducting coatings are also used to remove electrostatic charges that accumulate on glass windows. A growing concern about non-uniform electric charge build-up on the exterior surfaces of orbitting satellites has prompted investigations into the preparation and properties of electrically conducting transparent coatings for use as surface layers on temperature-control coatings. Field intensity fluctuations in the earth's magnetosphere, especially at geosynchronous orbital altitudes continuously irradiated with UV and charged particle radiation, may differentially bias spacecraft areas to potentials as high as 10^4 V [191]. Spacecraft elements charged to potential differences of this magnitude may perturb the electromagnetic field and charged particle environment that they were launched to measure, and may also cause catastrophic failure of operating functions due to spontaneous discharges. Electrical conducting transparent coatings can ensure potential uniformity on the exterior of satellites. It has been [192,193] shown that In_2O_3 and ITO coatings are well suited for the purpose, and the coated surfaces have been shown to have excellent stability when exposed to combined UV and charged particle irradiation in both laboratory and actual space flight experiments. Transparent conductors have also been extensively used as transparent heating elements on the glass windscreens of automobiles and aeroplanes to prevent the formation of mist, and as microwave shielding material.

In addition to the above well-established applications, transparent conducting oxides have numerous other potential applications. These applications are still at the laboratory experimental stage. ITO films are shown to be resistant to laser damage and are potential candidates for use in Pockel's cell [194]. Electro-optic shutters based on the longitudinal Pockel's effect are required in high energy solid state systems for fusion research to suppress target-damaging amplified spontaneous emission and to protect the system from target retroreflections. Pockels cells now used consist of cylindrical KDP crystals with a metal ring around the crystal near the parallel faces. To achieve adequate electric field uniformity over the aperture requires a crystal length-to-diameter ratio of $\geqslant 1.1$. Such crystals with a clear aperture of 5 cm, and even somewhat larger, are available, but are difficult to grow and are expensive. A transparent conductive film electrode over the crystal aperture eliminates the length-to-diameter constraint. Disc crystals can therefore be used for the same purpose, provided they are coated with transparent conducting films which can withstand high laser powers. RF sputtered coatings of $In_{1.9}Sn_{0.1}O_3$ on fused silica exhibit damage thresholds of 2.2 and 3.1 J cm^{-2} for 0.15 and 1 ns neodymium-glass laser pulses, respectively [194].

Other potential applications of these materials are as optical waveguides and electrodes in electro-optic modulators [195,196], buffer layers in superconducting thin film [197], gate electrodes in MOS capacitors [198], active elements in transparent thermocouples [199], electrodes in ferroelectric films [200], optical memory materials in optical recording [201] and as alternative materials for varistors [202].

5.9 Conclusions

Table 5.3 lists the transparent conducting materials and growth techniques suitable for various established applications. The important features are:

(i) ITO is the preferred material for optoelectronic device applications.
(ii) SnO_2 is the most reliable material for gas sensing and protective coating applications.
(iii) For wavelength-selective applications, any of the transparent conducting materials can be used. The choice of a particular material and process are governed by such parameters as substrate material, cost, film uniformity, etc. (refer to table 2.6).

It can be concluded that transparent conducting films will continue to play an increasingly important role in the fast developing area of optoelectronics.

Table 5.3 Suitable materials/processes for various applications.

Application	Material	Process
Wavelength-selective	Doped SnO_2	Spray
	In_2O_3	Evaporation/ion-plating
	ITO	CVD/spray/sputtering/ ion-plating/evaporation/
	CTO	Sputtering
	ZnO:Al	CVD/sputtering
Solar cell	SnO_2	Spray/sputtering
	In_2O_3	Sputtering
	ITO	Spray/sputtering
	ZnO	Spray/CVD
Protective coatings	SnO_2	Spray/CVD
Gas sensors	SnO_2/doped SnO_2	CVD/sputtering
	ITO	Sputtering
Display devices	ITO	Sputtering

References

[1] Fraser DB and Cook HD 1972 *J. Electrochem. Soc.* **119** 1368
[2] Haacke G 1976 *J. Appl. Phys.* **47** 4086
[3] Murty NS, Bhagavat GK and Jawalekar SR 1982 *Thin Solid Films* **92** 347
[4] Murty NS and Jawalekar SR 1983 *Thin Solid Films* **108** 277
[5] Afify HH, Momtaz RS, Badawy WA and Naseer SA 1991 *J. Mater. Sci.: Mater. Electron.* **2** 40
[6] Vasu V and Subrahmanyam A 1991 *Thin Solid Films* **202** 283

[7] Manifacier JC, Szepessy L, Bresse JF and Perotin M 1979 *Mater. Res. Bull.* **14** 163
[8] Karlsson T, Ross A and Ribbing CG 1985 *Solar Energy Mater.* **11** 469
[9] Manifacier JC, de Murcia M, Fillard JP and Vicerio E 1977 *Thin Solid Films* **41** 127
[10] Howson RP, Barankov H and Spencer AG 1991 *Thin Solid Films* **196** 315
[11] Stjerna B and Granqvist CG 1990 *Solar Energy Mater.* **20** 225
[12] Kane J and Sweitzer HP 1975 *Thin Solid Films* **29** 155
[13] Kulaszewicz S 1980 *Thin Solid Films* **74** 211
[14] Shanthi E, Dutta V, Banerjee A and Chopra KL 1980 *J. Appl. Phys.* **51** 6243
[15] Siefert W 1984 *Thin Solid Films* **120** 275
[16] Nath P and Bunshah RF 1980 *Thin Solid Films* **69** 63
[17] Agashe C, Takwale MG, Marathe BR and Bhide VG 1988 *Solar Energy Mater.* **17** 99
[18] Mavrodiev G, Gajdardziska H and Novkovski N 1984 *Thin Solid Films* **113** 93
[19] Shanthi E, Banerjee A and Chopra KL 1982 *Thin Solid Films* **88** 93
[20] Pan CA and Ma TP 1980 *Appl. Phys. Lett.* **37** 163
[21] Mizuhashi M 1980 *Thin Solid Films* **70** 91
[22] Maruyama T and Fukui K 1990 *Japan. J. Appl. Phys.* **29** L1705
[23] Avaritsiotis JN and Howson RP 1981 *Thin Solid Films* **77** 351
[24] Maruyama T and Fukui K 1991 *Thin Solid Films* **203** 297
[25] Maruyama T and Fukui K 1991 *J. Appl. Phys.* **70** 3848
[26] Vasu V and Subrahmanyam A 1990 *Thin Solid Films* **193/194** 696
[27] Gouskov L, Saurel JM, Gril C, Boustani M and Oemry A 1983 *Thin Solid Films* **99** 365
[28] Manifacier JC, Fillard JP and Bind JM 1981 *Thin Solid Films* **77** 67
[29] Blandenet G, Court M and Lagarde Y 1981 *Thin Solid Films* **77** 81
[30] Pommier R, Gril C and Marucchi J 1981 *Thin Solid Films* **77** 91
[31] Saxena AK, Singh SP, Thangaraj R and Agnihotri OP 1984 *Thin Solid Films* **117** 95
[32] Ryabova LA, Salun VS and Serbinov IA 1982 *Thin Solid Films* **92** 327
[33] Nath P, Bunshah RF, Basol BM and Staffsud OM 1980 *Thin Solid Films* **72** 463
[34] Krokoszinski HJ and Oesterlein R 1990 *Thin Solid Films* **187** 179
[35] Buchanan M, Weble B Jr and Williams DF 1980 *Appl. Phys. Lett.* **37** 213
[36] Smith JF, Aronson AJ, Chen D and Class WH 1980 *Thin Solid Films* **72** 469
[37] Croitoru N and Bannett E 1981 *Thin Solid Films* **82** 235
[38] Sreenivas K, Sudersena Rao T and Abhai Mansingh 1985 *J. Appl. Phys.* **57** 384
[39] Bawa SS, Sharma SS, Agnihotry SA, Biradar AM and Chandra S 1983 *Proc. SPIE* **428** 22
[40] Jachimowski M, Brudnik A and Czternastek H 1985 *J. Phys. D: Appl. Phys.* **18** L145
[41] Theuwissen AJP and Declerck GJ 1984 *Thin Solid Films* **121** 109
[42] Habermeier HU 1981 *Thin Solid Films* **80** 157
[43] Tueta R and Braguier M 1981 *Thin Solid Films* **80** 143
[44] Ishibashi S, Higuchi Y, Ota Y and Nakamura K 1990 *J. Vac. Sci. Technol.* **A8** 1399
[45] Latz R, Michael K and Scherer M 1991 *Japan. J. Appl. Phys.* **30** L149
[46] Harding GL and Window B 1990 *Solar Energy Mater.* **20** 367

[47] Fan JCC 1979 *Appl. Phys. Lett.* **34** 515
[48] Howson RP, Avaritsiotis JN, Ridge MI and Bishop CA 1979 *Appl. Phys. Lett.* **35** 161
[49] Cui Yuanri and Xu Xinghao 1984 *Thin Solid Films* **115** 195
[50] Machet J, Guille J, Saulnier P and Robert S 1981 *Thin Solid Films* **80** 149
[51] Geoffroy C, Campet G, Portier J, Salardenne J, Couturier G, Bourrel M, Chabagno JM, Ferry D and Quet C 1991 *Thin Solid Films* **202** 77
[52] Ortiz A 1982 *J. Vac. Sci. Technol.* **20** 7
[53] Miyata N, Miyake K, Koga K and Fukushima T 1980 *J. Electrochem. Soc.* **127** 918
[54] Rohatgi A, Viverito TR and Slack LH 1974 *J. Amer. Ceram. Soc.* **57** 278
[55] Haacke G, Mealmaker WE and Siegel LA 1978 *Thin Solid Films* **55** 67
[56] Haacke G 1976 *Appl. Phys. Lett.* **28** 622
[57] Miyata N, Miyake K and Nao S 1979 *Thin Solid Films* **58** 385
[58] Miyata N, Miyake K, Fukushima T and Koga K 1979 *Appl. Phys. Lett.* **35** 542
[59] Leja E, Budzynska K, Pisarkiewicz T and Stapinski T 1983 *Thin Solid Films* **100** 203
[60] Enoki H, Satoh T and Echigoya J 1991 *Phys. Status Solidi* a **126** 163
[61] Webb JB, Williams DF and Buchanan M 1981 *Appl. Phys. Lett.* **39** 640
[62] Hu J and Gordon RG 1992 *J. Appl. Phys.* **71** 880
[63] Aktaruzzaman F, Sharma GL and Malhotra LK 1991 *Thin Solid Films* **198** 67
[64] Igasaki Y and Saito H 1991 *J. Appl. Phys.* **70** 3613
[65] Choi BH, Im HB, Song JS and Yoon KH 1990 *Thin Solid Films* **193/194** 712
[66] Fan JCC and Zauracky PM 1976 *Appl. Phys. Lett.* **29** 478
[67] Fan JCC and Spura SA 1977 *Appl. Phys. Lett.* **30** 511
[68] Fan JCC 1981 *Thin Solid Films* **80** 125
[69] Craighead HG, Bartynski R, Buhrman RA, Wojcik L and Sievers AJ 1979 *Solar Energy Mater.* **2** 105
[70] Fan JCC and Bachner FJ 1976 *Appl. Opt.* **15** 1012
[71] Fan JCC, Bachner FJ, Foley GH and Zarracky PM 1974 *Appl. Phys. Lett.* **25** 693
[72] Kienel G 1981 *Thin Solid Films* **77** 213
[73] Granqvist CG 1990 *Thin Solid Films* **193/194** 730
[74] Howson RP and Ridge MI 1981 *Thin Solid Films* **77** 119
[75] Simonis F, Leij M and Hoogendoorn CJ 1979 *Solar Energy Mater.* **1** 221
[76] Frank G, Kauer E and Kostlin H 1981 *Thin Solid Films* **77** 107
[77] Dislich H and Hussman E 1981 *Thin Solid Films* **77** 129
[78] Jin ZC, Hamberg I and Granqvist CG 1988 *J. Appl. Phys.* **64** 5117
[79] Hamberg I and Granqvist CG 1986 *J. Appl. Phys.* **60** R123
[80] Takaki S, Matsumoto K and Suzuki K 1988 *Appl. Surf. Sci.* **33–34** 919
[81] Haitjema H, Elich JJP and Hoogendoorn CJ 1989 *Solar Energy Mater.* **18** 283
[82] Gitlitz MH, Dirkx R and Russo DA 1992 *Chemtech.* **22** 552
[83] Dewall H and Simonis F 1981 *Thin Solid Films* **77** 253
[84] Fan JCC, Bachner FJ and Murphy RA 1976 *Appl. Phys. Lett.* **28** 440
[85] Singh R, Green MA and Rajkanan K 1981 *Solar Cells* **3** 95
[86] Shewchun J, Dubow J, Myszkowski A and Singh R 1978 *J. Appl. Phys.* **49** 855
[87] Shewchun J, Burk D and Spitzer MB 1980 *IEEE Trans. Electron. Devices* **27** 705

[88] Ghosh AK, Fishman C and Feng T 1978 *J. Appl. Phys.* **49** 3490
[89] Dubow JB and Burk DE 1976 *Appl. Phys. Lett.* **29** 494
[90] Feng T, Ghosh AK and Fishman C 1979 *Appl. Phys. Lett.* **35** 266
[91] Bhagavat GK and Sundaram KB 1979 *Thin Solid Films* **63** 197
[92] Cheek G, Inoue N, Goodnick S, Genis A, Wilmsen C and Dubow JB 1978 *Appl. Phys. Lett.* **33** 643
[93] Mizrah T and Adler D 1977 *IEEE Trans. Electron. Devices* **24** 458
[94] Calderer J, Esta J, Luquet H and Savelli M 1981 *Solar Energy Mater.* **5** 337
[95] Feucht DL 1977 *J. Vac. Sci. Technol.* **14** 57
[96] Singh R and Shewchun J 1978 *Appl. Phys. Lett.* **33** 601
[97] Balasubramanian N and Subrahmanyam A 1991 *J. Electrochem. Soc.* **138** 322
[98] Shewchun J, Dubow J, Wilmsen CW, Singh R, Burk D and Wager JF 1979 *J. Appl. Phys.* **50** 2832
[99] Ashok S, Sharma PP and Fonash SJ 1980 *IEEE Trans. Electron. Devices* **27** 725
[100] Malik AI, Baranyuk VB and Manasson VA 1980 *Appl. Solar Energy* **16** 1
[101] Badawy W, Decker F and Doblhofer K 1983 *Solar Energy Mater.* **8** 363
[102] Nagatomo T, Endo M and Omoto O 1979 *Japan. J. Appl. Phys.* **18** 1103
[103] Seo JW, Ketterson AA, Ballageer DG, Cheng KY, Adesida I, Li X and Gessert T 1992 *IEEE Photonics Technol. Lett.* **4** 888
[104] Dhere NG 1990 *Thin Solid Films* **193/194** 757
[105] Tsai MJ, Fahrenbruch AL and Bube RH 1980 *J. Appl. Phys.* **51** 2696
[106] Schunck JP and Coche A 1979 *Appl. Phys. Lett.* **35** 863
[107] Bachmann KG, Schreiber M Jr, Sinclair WR, Schmidt PH, Thiel FA, Spencer EG, Pasteur G, Feldmann WL and Sreeharsha K 1979 *J. Appl. Phys.* **50** 3441
[108] Coutts TJ and Naseem S 1985 *Appl. Phys. Lett.* **46** 164
[109] Sreeharsha K, Bachmann KJ, Schmidt PH, Spencer EG and Thiel FA 1977 *Appl. Phys. Lett.* **30** 645
[110] Genis AP, Smith PA, Emery K, Singh R and Dubow JB 1980 *Appl. Phys. Lett.* **37** 77
[111] Tomar MS and Garcia FJ 1982 *Thin Solid Films* **90** 419
[112] Tomar MS 1988 *Thin Solid Films* **164** 295
[113] Kitigawa M, Mori K, Ishihara S, Ohno M, Hirao T, Yoshioka Y and Kohiki S 1983 *J. Appl. Phys.* **54** 3269
[114] Kudoyarova VK, Kulikov GS, Terukov EI and Khodzaev KK 1987 *J. Non-Crystalline Solids* **90** 211
[115] Ray S, Dutta J, Barua AK and Deb SK 1991 *Thin Solid Films* **199** 201
[116] Banerjee R, Ray S, Basu N, Batabyal AK and Barua AK 1987 *J. Appl. Phys.* **62** 912
[117] Major S, Kumar S, Bhatnagar M and Chopra KL 1986 *Appl. Phys. Lett.* **49** 394
[118] Kuboi O 1981 *Japan. J. Appl. Phys.* **20** L783
[119] Golan A, Bregman J, Shapira Y and Eizenberg M 1990 *Appl. Phys. Lett.* **57** 2205
[120] Morgan DV, Aliyu Y and Bunce RW 1992 *Phys. Status Solidi* a **133** 77
[121] Subrahmanyam A and Balasubramanian N 1992 *Semicond. Sci. Technol.* **7** 324
[122] Vasu V and Subrahmanyam A 1992 *Semicond. Sci. Technol.* **7** 320
[123] Berger PR, Dutta NK, Zydzik G, O'Bryan HM, Keller U, Smith PR, Lopata J, Sivto D and Cho AY 1992 *Appl. Phys. Lett.* **61** 1673

[124] Das SK and Morris GC 1993 *J. Appl. Phys.* **73** 782
[125] Puyane R (ed) 1981 *Proc. Thin Solid Films* **77** 1–3
[126] Southwick RD, Wasylyk JS, Smay GL, Kepple JB, Smith EC and Augustsson BO 1981 *Thin Solid Films* **77** 41
[127] Jackson JD, Rand B and Rawson H 1981 *Thin Solid Films* **77** 5
[128] Budd SM 1981 *Thin Solid Films* **77** 13
[129] Jackson N and Ford J 1981 *Thin Solid Films* **77** 23
[130] Blocher JM Jr 1981 *Thin Solid Films* **77** 51
[131] Trojer FJ 1981 *Thin Solid Films* **77** 3
[132] Roos A and Hedenqvist P 1991 *Appl. Phys. Lett.* **59** 25
[133] Roos A and Hedenqvist P 1990 *Proc. SPIE* **1275** 148
[134] Hedenqvist P and Roos A 1991 *Surface and Coating Technol.* **48** 41
[135] Roos A 1991 *Thin Solid Films* **203** 41
[136] Novotny VJ and Kao AS 1990 *IEEE Trans. Mag.* **26** 2449
[137] Kim KH and Park CG 1991 *J. Electrochem. Soc.* **138** 2408
[138] Sberveglieri G, Groppelli S, Nelli P and Camanzi A 1991 *Sens. Actuators* B **3** 183
[139] Sberveglieri G, Benussi B, Coccoli G, Groppelli S and Nelli P 1990 *Thin Solid Films* **186** 349
[140] Nitta M, Kanefusa S and Haradome M 1978 *J. Electrochem. Soc.* **125** 1676
[141] Oyaby T, Ohta Y and Kurobe T 1986 *Sens. Actuators* **9** 301
[142] Sharma SS, Nomura K and Ujihira Y 1992 *J. Appl. Phys.* **71** 2000
[143] Ippommatsu M, Ohnishi H, Sasaki H and Matsumoto T 1991 *J. Appl. Phys.* **69** 8368
[144] Oyabu T 1991 *Sens. Actuators* B **5** 227
[145] Martinelli G and Carotta MC 1992 *Sens. Actuators* B **7** 717
[146] Butta N, Cinquergrani L, Mugno E, Tagliente A and Pizzini S 1992 *Sens. Actuators* B **6** 253
[147] Semancik S and Cavicchi RE 1991 *Thin Solid Films* **206** 81
[148] Gutierrez FJ, Ares L, Robla JI, Horrillo MC, Sayago I and Agapito JJ 1992 *Sens. Actuators* B **7** 609
[149] Windischmann H and Mark P 1979 *J. Electrochem. Soc.* **126** 627
[150] Lal R, Grover R, Vispute RD, Viswanathan R, Godbole VP and Ogale SB 1991 *Thin Solid Films* **206** 88
[151] Nomura K, Ujihira Y, Sharma SS, Fueda A and Murakami T 1989 *J. Mater. Sci.* **24** 937
[152] Nitta M, Kanefusa S, Taketa Y and Haradome M 1978 *Appl. Phys. Lett.* **32** 590
[153] Gutierrez FJ, Ares L, Robla JI, Horrillo MC, Sayago I and Agapito JJ 1992 *Sens. Actuators* B **8** 231
[154] Sanjines R, Demarne V and Levy F 1990 *Thin Solid Films* **193/194** 935
[155] Piraud C, Mwarania EK, Yao J, Dwyer KO, Schiffrin DJ and Wilkinson JS 1992 *J. Lightwave Technol.* **10** 693
[156] Suzuki T and Yamazaki T 1990 *J. Mater. Sci. Lett.* **9** 750
[157] Suzuki T, Yamazaki T, Hayashi K and Noma T 1991 *J. Mater. Sci.* **26** 6419
[158] Kanefusa S, Nitta M and Hardome M 1979 *J. Appl. Phys.* **50** 1145
[159] Cooper RB, Advani GN and Jordon AG 1981 *J. Electron. Mater.* **10** 455

[160] Sberveglieri G, Groppelli S, Nelli P and Camanzi A 1991 *J. Mater. Sci. Lett.* **10** 602
[161] Arya SPS, Amico AD and Verona E 1988 *Thin Solid Films* **157** 169
[162] Kanefusa S, Nitta M and Hardome M 1981 *J. Appl. Phys.* **52** 498
[163] Reddy MHM and Chandorkar AN 1992 *Sens Actuators* B **9** 1
[164] Matsushima S, Maekawa T, Tamaki J, Miura N and Yamazoe N 1992 *Sens. Actuators* B **9** 71
[165] Gardner JW, Hines EL and Tang HC 1992 *Sens Actuators* B **9** 9
[166] Pink H, Treitinger L and Vite L 1980 *Japan. J. Appl. Phys.* **19** 513
[167] Fang YK, Chen FY, Hwang JD, Fang BC and Chen JR 1993 *Appl. Phys. Lett.* **62** 490
[168] Chang SC 1979 *IEEE Trans. Electron. Devices* **26** 1875
[169] Sberveglieri G, Groppelli S and Coccoli G 1988 *Sens. Actuators* **15** 235
[170] Sberveglieri G, Faglia G, Groppelli S and Nelli P 1992 *Sens. Actuators* B **8** 79
[171] Nanto H, Sokooshi H, Kawai T and Usuda T 1992 *J. Mater. Sci. Lett.* **11** 235
[172] Fukui K 1991 *Sens. Actuators* B **5** 27
[173] Galdikas A, Martunas Z and Setkus A 1992 *Sens. Actuators* B **7** 633
[174] Hoheisel M, Mitwalsky A and Mrotzek C 1991 *Phys. Status Solidi* a **123** 461
[175] Hoheisel M, Heller S, Mrotzek C and Mitwalsky A 1990 *Solid State Commun.* **76** 1
[176] Kempter K 1986 *Proc. SPIE* **617** 120
[177] Miura N, Ishikawa T, Sasaki T, Oka T, Ohata H, Matsumoto H and Nakano R 1992 *Japan. J. Appl. Phys.* **31** L46
[178] Kuck N, Lieberman K, Lewis A and Vecht A 1992 *Appl. Phys. Lett.* **61** 139
[179] Hagerott M, Jeon H, Nurmikko AV and Gunshor W 1992 *Appl. Phys. Lett.* **60** 2825
[180] Hsu CT, Li JW, Liu CH, Su YK, Wu TS and Yokoyama M 1992 *J. Appl. Phys.* **71** 1509
[181] Chubachi Y and Aoyama K 1991 *Japan. J. Appl. Phys.* **30** 1442
[182] Hamada Y, Adachi C, Tsutsui T and Saito S 1992 *Japan. J. Appl. Phys.* **31** 1812
[183] Kuo Y 1990 *Japan. J. Appl. Phys.* **29** 2243
[184] Lunney JG, Richard R, O'Neill and Schulmeister K 1991 *Appl. Phys. Lett.* **59** 647
[185] Mohri M, Kakinuma H, Sakamoto M and Sawai H 1990 *Japan. J. Appl. Phys.* **29** L1932
[186] Adesida I, Ballegeer DG, Seo JW, Ketterson A, Chang H, Cheng KY and Gessert T 1991 *J. Vac. Sci. Technol.* B **9** 3551
[187] Minami T, Miyata T, Iwamoto A, Takata S and Nanto H 1988 *Japan. J. Appl. Phys.* **27** L1753
[188] Praschek SR, Riedl W, Hoermann H and Goslowsky HG 1991 *22nd IEEE Photovoltaic Specialists Conference* Vol 2, 1285
[189] Inam A, Wu D, Venkatesan T, Ogale SB, Chang CC and Dijkkamp D 1987 *Appl. Phys. Lett.* **51** 1112
[190] Hogan M and Lunney JG 1987 *Appl. Phys. Lett.* **53** 631
[191] Deforest SE 1972 *J. Geophys. Res.* **77** 651
[192] Hass G, Heaney JB and Toft AR 1979 *Appl. Opt.* **18** 488

[193] Ito K and Nakazawa T 1979 *Surf. Sci.* **86** 492
[194] Pawlewicz WT, Mann JB, Lowdermilk WH and Milam 1979 *Appl. Phys. Lett.* **34** 196
[195] Ioannidis ZK, Giles IP and Bowry C 1991 *Appl. Opt.* **30** 328
[196] Chen RT and Robinson D 1992 *Appl. Phys. Lett.* **60** 1541
[197] Kellett BJ, Gauzzi A, James JH, Dwir B, Paveuna D and Reinhart FK 1990 *Appl. Phys. Lett.* **57** 2588
[198] Weijtens CHL 1992 *IEEE Trans. Electron. Devices* **39** 1889
[199] Kreider KG 1992 *Sens. Actuators* A **34** 95
[200] Kwok CW, Vijay DP, Desu SB, Parikh NR and Hill EA 1993 *Integrated Ferroelectrics* **3** 121
[201] de Andrade MC and Moehlecke S 1994 *Appl. Phys.* A **58** 503
[202] Gould RD, Hassan AK and Mahmood FS 1994 *Int. Electron.* **76** 895

Index